# SLICING
## *the*
# SILENCE

Tom Griffiths voyaged to Antarctica as a humanities fellow with the Australian Antarctic Division and writes and teaches history in the Research School of Social Sciences at the Australian National University. His previous books include *Forests of Ash* (2001) and the multi-award-winning *Hunters and Collectors* (1996).

# SLICING
### *the*
# SILENCE

*Voyaging to*
## ANTARCTICA

## TOM GRIFFITHS

**Harvard University Press**
Cambridge, Massachusetts
London, England
2007

Copyright © Tom Griffiths 2007
First published in Australia by UNSW Press, UNSW, Sydney, Australia

Griffiths, Tom, 1957-
  Slicing the silence: voyaging to Antarctica / Tom Griffiths.
      p. cm.
  Simultaneously published in Australia by UNSW Press.
  Includes bibliographical references and index.
  ISBN-13: 978-0-674-02633-9 (alk. paper)
  ISBN-10: 0-674-02633-0 (alk. paper)
  1.  Griffiths, Tom, 1957---Diaries. 2.  Griffiths, Tom, 1957---Travel--Antarctica.
  3.  Antarctica--Description and travel.  I. Title.
  G8502002.G75 G75 2007
  919.8'9--dc22
  [B]
                            2007006549

A Cataloging-in-Publication record for this book is available from the Library of Congress

ISBN-13:  978-0-674-02633-9
ISBN-10:  0-674-02633-0


This book is printed on chlorine-free paper.

*Design*  Di Quick
*Printer*  Everbest, China
*Maps*  Tom Griffiths
*Line drawings*  Bill Wood
*Jacket*  Tom Griffiths sitting on ice reading *Sitting on Penguins, photo* Jenni Mitchell

# CONTENTS

*For Charles Menzies-Wilson*
*and all his family*

# | PROLOGUE

Two moments of voyaging shaped this book. One was the experience of standing at the bow of the first Australian ship to slice the silence of a year at Casey station, Antarctica. We were there to renew the Antarctic generations; to deliver the new team of winterers and to take home the old. The other key moment was bracing myself against the swell of the Southern Ocean as I gave a talk about history to my fellow expeditioners. There on a heaving sea, surrounded by an endless blue horizon and sailing off the edge of the temperate Earth, we desperately needed to know what we were doing down there and why. Both moments helped me feel the vitality and urgency of history as a form of knowledge and understanding, as part of the essential currency of daily life.

This book interlaces diary entries and essays in an attempt to evoke the meditative dimensions of voyaging. Each chapter begins with a diary entry from my trip to Antarctica in the summer of 2002–03 and then offers an essay about an aspect of human experience down south. The diary entries enact the universal encounter with Antarctica, which is one of visiting and returning. Humans always hope to come home from the ice. The essays, however, work against the flow of the travelogue and keep moving towards what Ernest Shackleton called 'The Heart of the Antarctic'. Just as the present is constantly in dialogue with the past, each shaped by the other, so too are the diary and essays similarly in tension and conversation. They aim to help us travel in time as well as to the end of the Earth.

Another tension that weaves through the book is that between Australia and the rest of the world. When I visited Antarctica, I was still unsure whether I wanted to write about Australians on the ice or humanity in general. Being there helped me to understand that one of the inspiring aspects of Antarctic experience is that humanity is so marginal and vulnerable that narrow nationalism appears both comic and inapt. Furthermore, to make nationalism itself a subject of study on the ice, one has to draw back and see it comparatively. So my historical perspective is international and human but with an unashamed Australian bias. There are sensible reasons for this bias other than my own citizenship. Australia is Antarctica's closest geological cousin and they were the two Great South Lands of European voyaging and exploration. They remain the world's two continental deserts. And Australia claims almost half of the continent of ice. Because that claim is not universally recognised, Australia strenuously supports the Antarctic Treaty which sets aside arguments over sovereignty and thus protects Australia's sovereign interest. My window on Antarctica is made in Australia, but I look through it and try to see other figures and societies on the ice.

As the historian Stephen Murray-Smith has argued, history is not a luxury in Antarctica. At the end of his book *Sitting on Penguins*, he made this plea:

> What is needed is the writing of real 'professional' history about Australia's involvement in Antarctica ... We shall lack the essential tool to our understanding of Australian Antarctica until those with the interest and capacity to write its history are found. And not just one history. Preferably several, or at least a history that will provoke a debate.[1]

History down south, as in any society, is a practical and spiritual necessity, but especially so in a place where human generations are renewed every summer and the coordinates of space and time are warped by extremes. And on a continent claimed by various nations but shared by the world, history carries a special international obligation. It is the fundamental fabric of a common humanity.

Antarctica has become, in the words of Barry Lopez, 'a place from which to take the measure of the planet'.[2] It is a global archive, a window on outer space and a scientific laboratory; it is also a political frontier, a social microcosm and a humbling human experiment. It offers us an oblique and revealing perspective on modern history, an icy mirror for the world. To voyage to Antarctica is to go beyond the boundary of one's biology towards a frightening and simplifying purity. It is a land of enveloping silence. How does life sustain itself in the face of such awesome indifference? In Earth's only true wilderness, the fundamentals of existence are exposed. To survive, you need food, you need warmth, and you need stories.

## Wednesday, 18 December

**45 degrees 45 minutes south,
143 degrees 48 minutes east**

Yesterday the *Polar Bird* sailed into the sunset, around the
southern edge of Tasmania and out into the vast Southern
Ocean. This is a routine voyage; Voyage 3 of the Australian
Antarctic Division's 2002–03 summer, the annual resupply of Casey station. But everyone on board – even the
experienced campaigners, even those returning for their
fifteenth summer south – sails with a sense of wonder and
some apprehension. After all, we are journeying to the end
of the Earth! Does a cyclone await us in the furious fifties?
Might the ship again be trapped in the ice, as it was last
summer for 36 days? Why are we ushered down into the
hold and taught how to haul one another out of crevasses?

A few hours before we sailed, all expeditioners
were subjected to a lengthy briefing about the perils of
Antarctica from the man known as 'Doctor Death', the
Antarctic Division's medical officer Peter Gormly. He told
us not only what could happen, he showed us what *has*
happened, offering us a gruesome medical history lecture
illustrated with close-ups of named bodies and corpses. He
aimed to shock us into awe, respect and reasonable fear,
and he succeeded. His humour was black, he made jokes
about injury and death, he left us pale and looking for the
exits. 'There are three things you mustn't forget when you
visit Antarctica', he declared: 'It's cold. It's dangerous. And
please be kind to one another.'

A fourth mantra has been drilled into us: *always* carry
your *ANARE Field Manual* and *ANARE First Aid Manual*.

These were issued to us with our polar clothes, and 'are to be carried on our person at all times'. Whatever fix you are in, they will help you out of it, provided they are in your pocket and you have time to consult them. They are the distilled wisdom of generations of Australian expeditioners. Here you can find how to build a snow cave or tie a double fisherman's knot, how to haul someone from a crevasse, how to deal with snow blindness or frostnip, or when to tie down a helicopter. Over the years, expeditioners have competed to send Dr Gormly photographs of themselves in the most outlandish positions, still grasping the essential *Field Manual*. Our deputy voyage leader on the *Polar Bird* first appeared before us in a slide shown by Dr Gormly. There, hanging from climbing ropes on a precipitous ice cliff in Antarctica was a completely naked Luke Vanzino, his free hand brandishing a strategically-placed *ANARE First Aid Manual*.

We had a moving farewell at Hobart's Macquarie Wharf yesterday. Quite a crowd had gathered to see the ship off; friends and families of expeditioners, and staff of the Australian Antarctic Division including the director, Tony Press. It was a tender, old-fashioned ritual: the ship edged slowly away from the wharf, streamers were thrown and sometimes caught, or they arched helplessly into the dark water. Catching a streamer, you spin it out like a fishing line or a kite, playing it for as long as you can against the pull of the ship and breeze, until the fragile lifeline snaps, the colour coils downwards and the parting is final. Some people on board have farewelled families and the temperate earth for over 12 months.

In keeping a diary, of course, I am working within a great tradition of voyaging. Our libraries and archives are full of diaries kept on ships by people who rarely otherwise wrote about their lives. I bought a moleskine notebook and packed my laptop. Two other diaries are travelling with me. One is an unpublished journal kept by a young doctor, Fred Middleton, who voyaged south with Captain John King Davis and Ernest Shackleton on Douglas Mawson's old ship, the *Aurora*, in the summer of 1916–17. Their task was to rescue the other half of Shackleton's expedition. Two years earlier, Shackleton had sought and failed to make the first crossing of the continent of Antarctica from the Weddell Sea to the Ross Sea. While he and his men were abandoning their doomed ship, the *Endurance*, in the ice of the Weddell Sea and making their way to Elephant Island, the *Aurora* had taken a party to the other side of the continent, the Ross Sea, from where they sledged to over 80 degrees south to lay supply dumps for the anticipated arrival of Shackleton's trans-Antarctic expeditioners. But the *Endurance* men never reached the continent to start their trek – instead they were caught up in a rather different quest for survival – and the supply dumps on the other side awaited them in vain. Shackleton famously rescued all his men from the *Endurance* and returned to New Zealand in late 1916 where he joined the *Aurora* on its return journey south to discover the fate of the ten men waiting for nobody in the Ross Sea. The *Aurora* needed a doctor, and Fred Middleton from Melbourne suddenly found himself playing bridge with Sir Ernest Shackleton in a rolling ship on the Southern Ocean, and writing the wonder of it all into his diary.

The other diary that I am travelling with is a published account by the historian and editor, Stephen Murray-Smith, who was sent south in the summer of 1985–86 by the Commonwealth minister for science, Barry Jones, as an official observer and commentator on the work of the Australian Antarctic Division, then in the throes of a controversial and massive building program. Murray-Smith wrote a reflective, thought-provoking journal of his six weeks on the supply ship, the *Icebird*, which was published as a book entitled *Sitting on Penguins: People and Politics in Australian Antarctica*. I only recently realised, and with delight, that I'm voyaging on the same ship, renamed the *Polar Bird* in 1996 when it passed from German to Norwegian ownership. Murray-Smith sailed on it during one of its first summers in Australian service, and I'm sailing in its final summer.

Stephen took the title of his book, *Sitting on Penguins*, from an entry in the *ANARE Field Manual* on emergency sources of food. 'Penguins may be killed', the manual explained, 'by breaking and cutting the neck or by squashing the air out of their lungs by sitting on them for a fairly long period. Penguin stew is very palatable …' But for Stephen the term 'sitting on penguins' was a powerful metaphor for Australia's giant territorial claim over Antarctica, and it conjured the worrying spectre of a country sitting in a proprietorial, indolent and purely strategic way on the life of Antarctica. It raised the question: Is Australia intellectually and politically investing in Antarctica, or just sitting on it?

I have checked my ever-present 2002 edition of the *ANARE Field Manual* and I find that *sitting on* penguins is no longer recommended as a way of immobilising and killing them, although they are still reported as palatable food in a crisis. This editorial cut may be the most obvious legacy of Stephen's feisty book, but his criticism also brought changes in the management of humans that are less easily mapped.

In Stephen's account, which sits beside my bunk on B Deck, I have a vivid evocation of the floating physical and social environment I've just joined. I recognise my ship in those pages, and even some of the dinner-time conversations. But I nevertheless read his words across a great divide. Stephen's voyage was at the height of negotiations in the 1980s about the future use of the mineral resources of Antarctica. There was to be a Minerals Convention attached to the Antarctic Treaty, but in 1989 Australia decided instead to campaign for a ban on mining. That political about-face dramatically changed Antarctic politics. I am in the same ship as Stephen, in a different era.

In Fred Middleton's diary, I have a private line to the heroic age of Antarctica, even night-time audiences with an anxious Ernest Shackleton as he ponders the fate of his other men. Just as one cannot voyage to Antarctica without negotiating the band of sea ice, so it is impossible to imagine the continent without first navigating a way through the legends of exploration.

# THE FIRE on the SNOW
## Legends of the heroic era

My father-in-law, an impoverished Australian clergyman studying in Cambridge, once accidentally financed a polar expedition. He shared a surname and a middle name (and a distant cousinship) with the director of the Scott Polar Research Institute in Cambridge, and one day, soon after moving there as a mature undergraduate in the 1950s, found his life's modest savings cleaned out of his bank account and expended on snow shoes. The money was soon restored, the institute and its director were horrified, and the bank manager crucified. Two-and-a-half decades later this useful fact resonated sufficiently (the same director being in charge) to enable me, then also impoverished and also a student, to gain prompt and privileged access to the institute's revered collection of documents from Robert Falcon Scott's last expedition. Such are the uses of history. But I did make one significant, historical *faux pas*. I brightly mentioned the name of Roland Huntford.

Two years before I visited Cambridge, Roland Huntford wrote an account of Scott's last expedition which sent shockwaves through the institute and was debated in the letters pages of *The Times*. In his book *Scott and Amundsen*, Huntford exposed Scott – that model of British moral and physical courage, that tragic, frozen hero – as a vain and incompetent fool.[1] We had all learnt as children how Robert Falcon Scott had led a scientific expedition to Antarctica in 1910 (even embarking from Australia and New Zealand), how he had revered knowledge of the icy continent above the vanity of being the first to 90 degrees south, how he had declined to use dogs because it was cruel and had man-hauled the sledges because it was 'sublime', and of how the Norwegian, Roald Amundsen, had – with deceit, dogs, skis and a ruthless efficiency – contrived a 'race for the pole' which he had won, with tragic effects.

It was an immensely powerful moral tale. It made clear that the worth of exploration or any other act of courage or foolhardiness was a matter of means as well as ends, in fact, more means than ends. There was an unwritten code of honour to be observed. The means could ennoble the end even where it was the white tomb of a tent.

If longitude was the fascination of the eighteenth century, then latitude was the geographical obsession of the nineteenth. Whereas longitude was in essence a problem of time, and was solved by the invention of a sea-going watch, latitude was about space, and the challenge of taking empire to the ends of the Earth. Space, not place: for the poles invited exploration at its most abstract. It was all about bearing.

The 'heroic age' of Antarctic exploration describes the two decades from the final years of the nineteenth century to World War 1 when Antarctica became the focus of intense private and patriotic endeavour. At the Sixth International Geographical Congress in London in 1895, scientists resolved that 'the exploration of the Antarctic regions is the greatest piece of geographical exploration still to be undertaken'. Expeditions sailed south from Argentina, Chile, Uruguay, Australia, Britain, Scotland, France, Germany, Norway and Sweden. A simultaneous quest for the northern pole also took place. Funded more from private than government sources, these expeditions were nevertheless inspired by nationalism. The age was 'heroic' not only because it generated tales of extraordinary individual achievement and sacrifice, but also because the explorers aimed not so much to secure territory as to establish national pride and personal honour on the world stage. The great French polar explorer, Jean-Baptiste Charcot, who led two expeditions south in the first decade of the twentieth century (1903–05, 1908–10), gave eloquent and passionate voice to these ideals. 'Great unselfish endeavours', he wrote, 'are necessary for the glory of our country ... in this peaceful struggle the nation which emerges victorious will gain the esteem and respect of its neighbours'.[2] The experienced leader of the Scottish National Antarctic Expedition (1902–04), William Speirs Bruce, hoped that, by going south under his own flag, he had proved

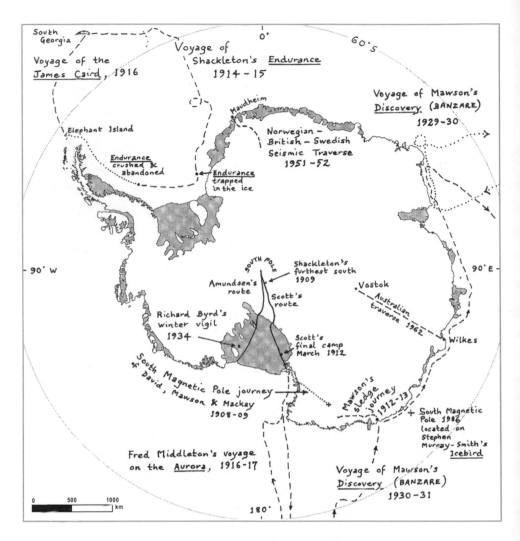

*Voyages and Journeys*

'that the nationality of Scotland is a power that must be reckoned with'. 'It remained for Scotland to show as a nation that her old spirit was still alive', wrote Bruce's scientists proudly.[3] When honour rather than territory was primarily at stake, then 'winning' could take unusual forms.

Every hero invites a debunking, but few have suffered as Scott did at Huntford's hands. Scott, he argued, was a poor leader with little foresight who endangered men in his charge, a reckless and careless planner who trusted to luck, an ambitious naval officer who was competitive and vain. The index to Huntford's book says it all. Under 'Scott, *Characteristics*' comes this litany of accusations: 'absentmindedness, agnosticism, unsuitability for command, refusal to accept criticism, bouts of depression, emotionalism, impatience, belief in improvisation, sense of inadequacy, insecurity, lack of insight, irrationality, isolation, jealousy, defective judgement, failure in leadership, literary gifts, readiness to panic, recklessness, instinct to evade responsibility, sentimentality and vacillation'. The one positive attribute listed was 'literary gifts', a backhander in a portrait of a hollow legend. Scott represented the pretensions and stubborn amateurishness of an empire unwillingly in decline.

Huntford's book reached its climax, of course, in his account of the tragic return journey from the pole. His narrative here was driven by two dangerous ideas. One was that Scott emotionally bullied Captain Oates to his death. The other was that Scott began to realise that it would be better for a bungling leader to die in harness rather than to get home to answer the critics.

Oates' exit is one of the most famous and revered scenes in exploration history, the premier cameo of character vouchsafed to the world from the ice. Weak and frostbitten, he was slowing down his companions in their race for life. He had urged them to leave him in his sleeping bag on the snow. They would not – could not – do so. Waking on the morning of his thirty-second birthday, a morning he had hoped not to see, Oates crawled from the tent into a blizzard with the legendary words: 'I am just going outside. I may be some time.'

Those words are the epitome of British understatement, gentlemanly in their respect for civility above agony, laconic in that they ask for no response. Oates was known and admired for his 'dry smile-less' wit, for his black humour, and this was as black as it could get.[4] Scott wrote in his diary: 'We knew that poor Oates was walking to his death, but ... we knew it was the act of a brave man and an English gentleman.' 'Poor Oates, indeed', exploded Huntford. 'He sat there in the tent, Scott staring at him, with the unspoken expectation of the supreme sacrifice.' Huntford played up the animosity and tension between the two men, quoted Scott's descriptions of the ailing Oates as 'a terrible hindrance' and 'the greatest handicap', and portrayed Oates' accelerating contempt for an incompetent leader. The narrative focus was on Scott's need for a storybook ending and Huntford even doubted that Oates uttered the immortal, final words.

Scott carried his storybook in a secure green wallet on his belt. Although his physical stamina did not outlast that of his companions, his literary stamina did. He had, in Huntford's words, 'been preparing his farewells for some time'. Even as his companions ceased writing and focused their efforts on man-hauling and maintaining morale in a vain effort to survive, their leader was already manipulating the endgame. He poured his remaining emotional energy into his letters and diary, addressing that distant audience of posterity. On his earlier Antarctic expedition of 1901–04, there is evidence that he retired to his cabin to write rather than acknowledge that his ship was being rescued from entrapment in the ice. He wrote 'the rescue' out of history even as it happened around him. Scott was experienced at writing his way out of trouble. Again on the polar journey, he prematurely played the historian. Who is to say how fictive was his non-fiction? As a flawed leader responsible for the death of subordinates, it was in his interests not to get home, argued Huntford; heroism would be purifying. But he needed to take his witnesses with him. In the final week, when Scott himself was ailing and his two remaining companions might have pushed on to the life-saving supplies at One Ton Depot, he constrained them and asked them to lie down beside him in their sleeping bags. He was already creating his metamorphic cocoon. There on the ice, Scott consciously constructed a time capsule.

Huntford's book was denounced by no less a trio than the director of the Scott Polar Research Institute (the one with the accident-prone bank manager), a former director of the British Antarctic Survey (Sir Vivian Fuchs), and a peer of the realm, Lord Kennet (Wayland Young).[5] In some libraries a note was inserted as a frontispiece to Huntford's book entreating the reader to refrain from reading it.[6] Opposition to the book focused on two issues. One was an argument over factual details. Did Scott stare down Oates in the tent, and how can we know anyway? Did Scott's wife, Kathleen, sleep with Amundsen's hero, Fridtjof Nansen, while her husband was facing defeat at the pole? Was Sir Clements Markham, Scott's mentor, a homosexual? Were the 35 pounds (16 kilograms) of geological specimens which were hauled on the sledges to the very end actually worthless?

There is no doubt that Huntford goes too far, and that he fails to alert us to the boundaries of his speculation. It is not so much that he is too careless with facts as that he is too careful about which facts he tells us. His critics were right to be shocked by that in a scholar. In fact, they stripped him of that title. They damned him as a 'journalist'. Just as Huntford depicts Scott as a symbol of a moral weakness of his age – a naval captain puffed up by an imperial past – so do Huntford's critics see him as representative of the ethical decline of *his* age; a paparazzo with a pen.

Huntford was not, however, the first to speculate on Captain Oates' end. Bernard Shaw wondered mischievously about it in 1948, as did people when the news first came through of the expedition's fate. Lord Kennet senior tried to suppress that earlier speculation of Shaw's, just as Lord Kennet junior (Kathleen Scott's son by her second marriage) was outraged 30 years later by Huntford. But Huntford's book undoubtedly revels in a certain excess, a certain reforming zeal, pushing as it does against the weight of more than half a century's scholarship and myth-making. The temerity of the man takes your breath away.

Huntford was unusually selective in his use of evidence, at times over-reading shreds of information in favour of his argument and at other times declining to cite evidence (in the same archives, the same letters, even in the very same envelope) that contradicted his view. In his reading of Kathleen Scott's affair with Nansen, for example, he is overly keen to find her in bed with Amundsen's mentor, sleeping with the enemy, whereas the explicit evidence is contrary, and the general impression at least inconclusive. That they were in love and that they stayed in the same hotel in Berlin in the fateful January of 1912 is not denied. But the debate about this matter degenerates into an unseemly fascination of men for menstruation, as they calculate the monthly arithmetic of Mrs Scott's cryptic diary notations.

More substantial and misleading is Huntford's dismissal of the scientific achievements of Scott's polar expeditions. Whatever his foibles and whatever his motivations, Scott led expeditions that made significant scientific contributions. It's true that he was criticised for taking his second expedition to the same area as the first and that Douglas Mawson declined to join Scott's 1910 expedition probably because he sensed the distraction to science represented by the quest for the pole. But Scott chose scientists for his party, including Australian geologists Griffith Taylor and Frank Debenham, who became leaders in their fields. Huntford disparaged the hauling of the sledge of 35 pounds of geological specimens to the very end as 'a pathetic little gesture', an attempt by the polar party to show themselves martyrs to science and to pretend nonchalance about any race to the pole.[7] But it is a deeply moving act nevertheless.

And Huntford, with a northern hemisphere arrogance, wrongly dismissed the geological specimens as insignificant. As an Australian, I protest! They contained the key to Gondwana. On 8 February 1912, Scott had recorded in his diary the results of some hours geologising at the head of the Beardmore Glacier on their return journey from the pole:

Wilson with his sharp eyes has picked several impressions the last a piece of coal with beautifully traced leaves in layers also some excellently preserved impressions of thick stems showing cellular structure.

The rocks were loaded onto the sledge and hauled by weakening, doomed men hundreds of miles. Eight months later, they were dug out from beside the cairn-like tent by the search party, alerted to their presence by Scott's diary which Edward Atkinson read alone and on the spot for what seemed 'hour after hour' (see *The Changeover*). The 'beautifully traced leaves' were the first *Glossopteris* material to be recorded and recognised from Antarctica, and they became part of the main fossil evidence that gave ancient reality to the popular concept of a 'Great South Land', a Gondwanan supercontinent linking Australia, Antarctica, Africa, India and South America. The specimens became crucial evidence in the early theories of continental drift, which were at first discredited and finally became one of the twentieth century's great scientific revolutions.[8]

The most recent, substantial defence of Scott has come from the pen of another explorer, Sir Ranulph Fiennes. Fiennes has himself accomplished astonishing feats of endurance on the ice, and he makes much of the authority this gives him as a historian of polar achievement. His biography, *Captain Scott* (2003), is dedicated 'To the families of the defamed dead', and he writes of Scott with respect, but as his equal. 'In this book', he writes, 'I have done what my predecessors [the over 50 other biographers of the polar party] could not do; I have put myself in the place of the British explorers and used logic based on personal experience to reconstruct the events. No previous Scott biographer has manhauled a heavy sledgeload through the great crevasse fields of the Beardmore Glacier, explored icefields never seen by man or walked a thousand miles on poisoned feet. To write about Hell, it helps if you have been there.'[9] Regularly throughout the book, Fiennes invokes his own recent polar experience to better understand Scott. 'I have shared Scott's experiences over many years and so can the better comment on his decisions and his behaviour, bringing a sympathy for and practical understanding of the severe problems Scott actually faced.'[10] 'I have always stuck to my guns', writes Fiennes, 'but I can sympathise with explorers who have at least adjusted their memory of the why and the wherefore for public consumption'.[11] He makes a powerful argument for experience as the path to empathy and understanding. Fiennes projects a brotherhood of the ice; he becomes Scott in the telling and so defends him. He disparages Huntford's knowledge of ice and atmosphere in these extreme conditions, and his decision as an author not to go to Antarctica. By contrast, Fiennes claims on his first page that his own exposure to the place and its timeless, physical conditions allows his book to be 'an unbiased account'. This is foolish vanity: his book is riddled with enabling bias. But I am sympathetic with his argument that 'you need to be there'. I'm exploiting that authority myself, by interleaving these historical essays with my own travel diary. I make a fragile claim to be a historian on ice, admittedly thin ice. But Fiennes wants me to accept that, although anyone can be a historian, only experience delivers objectivity. Yet the essence of good history is the balance between empathy and perspective, intimacy and distance.

The other issue raised by Huntford's critics concerned what Sir Vivian Fuchs called 'the tenor of this book'. They resented a hero's 'debunking', and they debunked that very term as a fashion. They allowed no room for a re-reading of evidence, no latitude for speculation. Fuchs felt that these heroes should be 'left to enjoy the position they achieved'; they should be allowed to remain 'legendary examples'. He acknowledged the myth-making and defended its social purpose. Lord Kennet wrote of 'how what is good in a society can be fostered by dwelling on what is good in its history'. Having argued the factual detail, he then set it aside as essentially irrelevant. Every society has this sort of argument about the skeletons in its cupboard and the value of the busybodies who ferret them out: how political is their history and what 'black armbands' are they perceived to wear? What greater good, what beneficial aura might they destroy in their pursuit of the bare truth? They are coldly efficient Amundsens to Scott's idealism. Huntford's black armband was lost in an Antarctic white-out.

What interests me is the latitude for legitimate historical imagination, the speculative space – for instance – in a high latitude tent 18 kilometres (11 miles) from One Ton Depot. As historians, we often cannot know what we most want to know. As Inga Clendinnen has said of the writing of history: 'Were this fiction, I would know that all things said and left unsaid, all disruptions, were intended to signify. But this is not fiction, and I cannot be sure.' The absences in the historical record should not condemn us to silence; they should provoke us to sensitive – and signposted – speculation. If we follow Huntford's interpretation, then Scott knowingly created, enclosed and defended that space, and for him it was a desperate, face-saving site of history-making. We historians should recognise what he was about. I think there is no doubt that he used the space in that way, and there is not necessarily any shame in that – indeed, it reveals an inspiring faith in posterity as well as a far-sighted investment in a great deal of posthumous glory.

Therefore it is fair enough, indeed essential, for us to imagine the stares and gestures of that incredibly public private space. Adrian Caesar felt that he could do so in his book, *The White*, his study of the last days of Scott and Mawson, which he describes as 'an experiment in biography … instead of trying to conceal the fictional nature of biography, I have chosen to emphasise that aspect'. Caesar imagines Scott's final literary outpourings as being powered by nibbles of the opium tablets that Wilson distributed near the end. 'In using my imagination', writes Caesar, 'I have included material that I cannot "know" to be true.' As a result of this transgression, his book was consigned by the Scott Polar Research Institute to the basement 'fiction' collection. What might be a fine line in the use of evidence can, in a large library, become a gulf of several flights of stairs.

Francis Spufford also felt that he could imaginatively recreate life and death in Scott's tent in the final pages of his superb book, *I May Be Some Time: Ice and the English Imagination*. He imagines Scott's wishes, dreams and anguish as he wonders how to die, with his companions cold and silent beside him and his literary powers spent. 'You cannot die in a story; you have to die in your body.' Wilson and Bowers 'died quietly' on the frozen evidence of the search party. Scott had thrown back the flaps of his sleeping bag, inviting the murdering cold within, one arm stretched over Wilson.

The Australian writer and publisher Douglas Stewart also imaginatively entered the tent in his radio play *The Fire on the Snow*. Scott's companions were 'lost in the fog of another man's dream', he wrote. When they saw the Norwegian flag at the pole, the snow on their lips tasted like ash. At the end, Scott talks of the rising moon over Mount Erebus and of the hopes it inspired; he talks to a companion who quietly abandons him:

> Wilson –
> > Wilson!
> > > Agony.
> Two dead men; and a dying man remembering
> The burning snow, the crags towering like flame.

Stewart read Scott's diary 'time and time again', regularly every Australian winter.

Griffith Taylor, a member of Scott's expedition who returned to Australia from Antarctica before the polar party was expected back, wrote to Scott from Canberra in November 1912, the same month that the search party had collapsed the little tent over the bodies and erected above them a cross made of skis. 'It is just over a year since I saw you start for the Pole Journey', he wrote, 'on what I am sure proved to be a successful if terrible experience. I am longing to hear how you got on.' Taylor wrote from his deck chair on the verandah of the Bachelors' Quarters which overlooked the site of Australia's new Federal City. That very morning the last lines were added to the complete 'plan and panorama' of the capital and the precious foundation document was officially delivered to Melbourne, the temporary home of the new

Commonwealth parliament. Taylor, working on the geological survey of the site, set about warming Scott's world with an evocative description of his own:

> The western winds are sighing through the gum trees, the flies have dropped to rest. The moon rises grandly over Mt Ainslie (to the east) and reminds me of her hovering over Erebus – for the mountains have similar outlines, if the heights are somewhat different … The kookaburras laugh lustily in the willows by the river.

On that 'empty plain' before him, he imagines a great city arising: 'It is not often that a city is planned on such a grand scale before a trench is dug.' His memories are of another vast plain, another mountain, another brave attempt to nurture the glow of an imported civilisation.

Historians immerse themselves in context; they give themselves wholly and sensually to the mysterious, alchemical power of archives. As well as gathering and weighing evidence, piece by piece with forensic intensity, they sensitise themselves to nuance and meaning, to the whole tenor of an era, the full character of a person. Their finest insights are intuitive as well as rational, holistic as well as particular; and therefore always invitations to debate. Historians move constantly between intimacy and distance, reading and thinking their way into the lives and minds of the people of the past – giving them back their present with all its future possibilities – yet also seeing them with perspective, from afar. We need history because some things cannot be recognised as they happen. And because stories are privileged carriers of truth. Truth, as Barry Lopez reminds us, cannot easily be stated explicitly. It cannot be reduced to aphorism or formula. It is something alive and unpronounceable. It is not to be found in a chronicle of facts but in the elucidation of the relationships between them. Story creates an atmosphere in which truth becomes discernible as a pattern.[12] Paradoxically, we must strive to make true stories believable.

The outrage provoked by Huntford prevented his critics from acknowledging the book's fundamental service, which was to enable the fascination with Scott to change gear quite fruitfully. Scott's incompetence as a leader is now mostly taken for granted. JW Gregory, who resigned as scientific leader of Scott's first expedition, divined the problem accurately as early as 1901: 'I think he is a poor organiser. His departments are in arrears, & he is so casual in all his plans. He appears to trust to luck things which ought to be a matter of precise calculation.'[13] Even on arrival at the pole, Scott knew it would be 'a desperate struggle' to get home: 'I wonder if we can do it.' During the winter of 1911, the loyal Frank Debenham recorded the power of 'The Owner' (as Scott was called) to paralyse his party with both his temper and his fatalism. There in the cramped quarters, the men could not acknowledge Amundsen's existence, yet they knew that priority at the pole was actually Scott's prime goal. They were not allowed to criticise the surveying work of Scott's earlier expedition, yet they needed better maps. The arithmetic of the proposed sledge journey to the pole did not add up to anything but death, but they were unable to canvass alternatives.[14]

Ursula Le Guin, in a 1986 essay, recognised that she wrote at a time when it had become 'chic to sneer' at Scott. She accepted the evidence of his managerial and technical incompetence but she did not want to lose – and needed to explain – her instinctive admiration for the man. Growing up in the United States, she was not exposed to the British idolisation of Scott, yet he was a hero nevertheless. It was the power of his words, the force of his primary text, that moved her. It was his ability as a story-teller, as a writer, as an artist, that redeemed him. His 'real heroism', fully exposed now that he had been stripped of other versions, was 'what he made of his failure', his literary achievement in the face of defeat and death, his commitment to living and telling a story. It was all about bearing. Once he had lost the race to the pole, he had to find an alternative glorious ending.[15]

It is a shocking and possibly misleading suggestion that Scott may have nurtured a 'death wish'. A literary critic, analysing his carefully preserved words, might say that he sought narrative closure. Le Guin would agree, and she sees his behaviour as, if not deliberate, then certainly as 'sacrificial'. Such an insight is the major intellectual legacy of Huntford's zealous work. It does not condemn Scott, but makes him more human. It does not debunk a hero so much as refine our appreciation of him. Huntford's critics resented his depiction of Scott's literary skills as a 'cover-up'. But, in the elemental purity of the ice, in the white noise of the enshrouding blizzard, the written word assumed extraordinary power.

As Scott's fellow expeditioners collapsed the polar party's tent into a memorial cairn to the three dead men within it, elsewhere in Antarctica another three men sledged towards disaster. Douglas Mawson, Xavier Mertz and Belgrave Ninnis formed one of five exploring parties setting out from the Australasian Antarctic Expedition base at Commonwealth Bay in the summer of 1912–13 to survey unknown territories. The aim of Mawson's small party was to explore the uncharted coastline hundreds of miles to the east, towards the Transantarctic Mountains. They had such a distance to travel that Mawson had reserved the dogs for their expedition, and he had chosen as his two companions the men most experienced as dog-handlers.

Lieutenant Belgrave Ninnis, aged 23 and formerly of the Royal Fusiliers, was nicknamed 'Cherub' on the expedition because of his youthful face and round cheeks. Together with Dr Xavier Mertz, lawyer, engineer and champion skier, he boarded the *Aurora* in Cardiff and they both took on the job of looking after the 50 Greenland dogs on the voyage south to Hobart and beyond. They 'lived in the odour of dogs' and gave them a pedigree of polar names: Scott, Shackleton, Nansen, Ross, Franklin, Bruce, Charcot, Hooker. Thus the names of great explorers lived in the vernacular of the ship: one could swear at Scott or cuddle Shackleton. There was not a soul to see off the *Aurora* in Cardiff, but Ninnis was full of the excitement and significance of what he was doing. 'I can hardly realize that I am absolutely off on a Polar Expedition', he wrote home to his fellow Fusilier, HE Meade. Ninnis declared that he considered Mawson 'a splendid fellow. He is quiet, and a scientist all over; also a gentleman, ditto. My respect for him increases daily.' A week out from Cardiff, it emerged that the ship's cook was suffering from syphilis, and this news was almost as alarming as the force of the roaring forties that engulfed them in the southern latitudes. But young Ninnis was undeterred: 'Whatever I do, however hard or however dirty, however tired, cold or wet I may be, I always have it in my mind that it is Antarctic Exploration I am doing my job for, not lining the pockets of the scum in Whitehall … I am having the time of my life. I feel an object of envy to everyone.' He had unsuccessfully volunteered for Scott's expedition, and now here he was with a place on Mawson's. He felt a rare contentment: 'I cannot realise my luck; at times I lie awake simply hugging myself for joy … Here I am practically in my second childhood, so simple and infantile are our pastimes.' The *Aurora*'s captain, JK Davis (nicknamed 'Gloomy'), was at first unimpressed by Ninnis and Mertz, who he felt were 'idlers' and did not give the dogs enough attention. He confided to his private journal that Ninnis was 'one of those people who go through life always depending on some one else to pull him out of difficulties'.[16] But Ninnis and Mertz, who became close friends, won him over in the course of the voyage. Ninnis, declared Davis, had the 'manner and character … of an enthusiastic public school boy'. When they were guiding their ship through ice towards the Antarctic continent in early January 1912, Ninnis relished his minor role in discovery and declared: 'Hurrah for the Fusilier Columbus.'

A premonition haunted Ninnis's letters home from the ship: 'Should I end my blighted career down a crevasse', he begged Meade, 'always make me out to have been a sweet and sociable companion, popular in the regiment, a good officer, a social star; no one (who does not know me) will put you down as a liar'. Ninnis was a bit of a pessimist. He had a letter for his father already written 'in the event of my being snatched to the skies (or the reverse) during our trip'. 'My "dec." notice is sure to catch your eye', he wrote. 'Welcome me with open arms if I return and weep over my distant and frozen corps[e] if I don't, and cast thoughts after me as I tread my devious and crevasse ridden ways.'

On 10 November 1912, after a winter in the windiest place on the planet, Mawson, Ninnis and Mertz set off from Commonwealth Bay on their far eastern journey. Although they made good progress at 24 kilometres (15 miles) a day, they had to cross two huge glaciers and their journey was perilous. On 21 November, Ninnis fell into a crevasse and was hauled out, and the day after, he fell into another and the day after that, another. On 14 December, he was swallowed without trace by the ice. It was so sudden. One moment, he was there, following on the sledge. The next, when Mertz and Mawson again looked behind, there was nothing but whiteness. He had vanished. Practically all the food and the best dog team had gone down with him. Ninnis's end was just as he predicted and his letters to the future console us that he died happy.

Mawson's and Mertz's fight for survival – stranded with little food and few dogs over 300 miles from safety – is one of the defining stories of Antarctic experience. The set-piece of the 'heroic era' of Antarctic history is always the sledging journey, a raw, elemental fight with nature where humanity is stripped to its essentials. It becomes a fight not just for physical survival, but also for dignity and for history. If you cannot get back yourself, then you must make sure that your frozen body can be found, that some record of your journey finds its way home. In the annals of polar sledging, Mawson's journey was surely the most harrowing, and his survival the most unlikely.

They read the burial service for Ninnis over the yawning, black crevasse and turned immediately for home, eking out the meagre surviving food bag. There were sensuous dreams of the food they could not have, and Mawson also dreamed vividly of his father, not knowing that he had just died at home. Mertz, a Swiss ski-champion, might have skied home quickly, but there was no question of him leaving Mawson, just as there was no question of Mawson abandoning Mertz. Once, a snow petrel visited them, a strange sight so far from the coast. As the dogs collapsed in their traces they were lovingly carried on the sledge to the night's camp and there butchered and boiled. The dogs' livers were precious slippery morsels within these exhausted, tough bags of sinew and bone, and Mawson and Mertz shared them determinedly. Thus, as they succumbed to starvation and cold, they also poisoned themselves systematically. Dr Mawson never knew, right up to his death in 1958, why he and Mertz lost their hair and skin on this journey, why they suffered crippling stomach pains and dizziness, or why Mertz slipped quickly into convulsions, dementia and death. Malnutrition and exposure were part of the story, but also the livers of Greenland dogs contain concentrations of vitamin A that are toxic to humans.[17] Mawson nursed Mertz in his terrible last days and, although he lovingly carried him on the sledge to his final camp, he did not eat him. In desperate circumstances, Mawson, like Scott, was holding on to his humanity. Xavier Mertz was buried solemnly, his grave marked, and a letter attesting to his character and describing Mawson's forthcoming, desperate journey was slipped into the corpse's eternal sleeping bag. Even in death, Mertz might be Mawson's postman. If Mawson did not get home, his messages might. In one of his final diary entries, Mertz unknowingly echoed Scott's words on the ice shelf earlier that year: 'I am sorry, but I don't think I can write any more.' Putting down the pen was death.

There were still more than 100 miles to go. Mawson, like Scott, began to think, if not of personal survival, then of the survival of his diaries and notes. As he buried Mertz, he also mourned that he had not had time to commit his recent geographic observations to paper. The hope of reaching warmth and food and safety drew him on, but so did the secondary aim of getting close enough to be found dead with his records. But, unlike Scott, Mawson wrote a plain, practical account of his troubles, without the spin. And he had not given up, although the temptation was great. In the blunt, cocky words of an Australian science teacher, 'Scott lay back and thought of England; Mawson stood up and walked home.'[18]

Back at the coast, Captain Davis was anxiously waiting in the *Aurora* to take all the men home. 'There is still no sign of Mawson', he wrote in his private journal on 22 January 1913. He went ashore after tea, inspected the wireless mast, and then climbed a snowy slope for a view, in the hope of seeing something. The sight of the hut looking 'like a heap of stones' in the boundless ice was both wonderful and terrible. 'As I stood looking South I tried to tell myself that Mawson and his party were all right but could not help wishing that I could march out myself and make sure.' He felt 'terribly uneasy' at the silence 'one does not know what to think'.[19]

Mawson fell down so many crevasses on his lone trek that he tied himself to his sledge with a rope ladder so that he could constantly rescue himself. The ice and the atmosphere conspired to test his physical capacity and psychological determination in every way. The glacier and the blizzards ambushed him daily. He missed the search party sent out from the hut by just hours, but he miraculously found the cairn they left for him: with its food and precious written words, including the news that Amundsen had reached the pole the December before last, and that Scott had been last seen just 240 kilometres (150 miles) from it in the same month. Over three further days Mawson dragged his sledge – which contained both his provisions and his history – to Aladdin's Cave, the ice shelter named by Ninnis just eight kilometres (five miles) from the hut. There high winds marooned and frustrated him for a further week. On 8 February 1913, he picked his way down the final, steep slippery slope to the hut in time to sight the *Aurora* steaming out of the Bay for another whole year. Six men had stayed behind in case he returned, and one – Frank Bickerton – rushed to greet the lone figure glimpsed on the ice above the hut. Staring into the emaciated Mawson's eyes, Bickerton cried out: 'My God! Which one are you?'[20]

Mawson staggered back to the hut three days before the story of Scott's death broke upon the world. Mawson had pioneered radio transmissions from Antarctica, but the tapping out of his own epic tale from the hut was repeatedly scrambled by an active aurora, and so Scott's story became public first. Even Mawson heard it (by morse code just a fortnight after his return) before he could properly tell his own. There in the lonely hut for a whole year, he shared in the global grief as well as his more private loss. Scott's was an overwhelming tragedy, and the book and film of Mawson's expedition were eventually lost in its wake. And the senseless carnage of the coming war would demand stories of sacrifice rather than survival.[21]

Paul Fussell, in his classic *The Great War and Modern Memory*, movingly reveals how we do not just tell stories, we *live* them. He calls it 'the Curious Literariness of Real Life', 'the simultaneous and reciprocal process by which life feeds materials to literature while literature returns the favour by conferring forms upon life'. Just as Scott's life and end were shaped by the power of narrative, so did his sacred text, retrieved from the preserving ice, prepare his nation for that most grotesque of wars. Polar sledging found a sort of parallel in trench life. Scott's legend could never have gained the power it did if his words had lain undiscovered for a few more years. Such reckless heroism had by then had the shine taken off it by the Western Front.

They died because of bad planning, because the oil evaporated, the stores were unfairly apportioned, they were starved of vitamins, their strongest man (Evans) weakened, the weather was bad, the temperatures lower than expected, the blizzard enclosed them. And they died because they lost the race. And it was not even supposed to be a race. The awful bitterness of defeat, the sight of that Norwegian flag stark against the snow 'like a black crow waiting for the end' in Douglas Stewart's words, a portent they could not deny; these were devastating emotional blows. If they'd been first, would they have got home? This was Amundsen's private burden. Triumph may have made up for the vitamins.

'The causes of the disaster', wrote Scott to the British public from the tent, 'are not due to faulty organisation, but to misfortune'. They died, he explained, because of the weather. The temperatures on their return journey in March were much lower than expected, and the final blizzard blew for more than a week and trapped them. Was this excuse just another failing of character?

How can historians unravel and resolve these mysteries? We have pored over the surviving documents – every precious word – remarkably preserved in the tomb. We have analysed the careers and characters of the expeditioners, all their fears and forebodings, everything that led up to the tragedy and every misgiving of their friends afterwards. We have scrutinised their material culture. Every inflection of social behaviour, every nuance of nationhood, every human motivation has been studied. But a few years ago, North American scientist Susan Solomon also drew upon meteorological data, weather observations that were gathered in Antarctica at the time of the expedition and since. She convincingly discovered that Scott's party was indeed unlucky with the weather in March 1912.[22] They had three weeks of unusually cold temperatures, of 10–20 degrees Fahrenheit lower than normal, the sort of difference that can kill people at the edge of life. And, as Ranulph Fiennes could also testify, such low temperatures also make a critical difference to the behaviour of ice. The loose new snow crystals that form on ice at over −28 degrees Celsius (−20 degrees Fahrenheit) acted as a brake on the sledge. The absence of wind combined with the extreme cold meant that the surface was not swept clean of this hoarfrost; instead, it remained like sandpaper to slow progress. It was so cold that the skis of the sledge could no longer produce their normal liquid lubricating layer. The feet of the men, already vulnerable, froze in their boots. The judgement of Solomon is that it was indeed 'the capriciousness of nature' that delivered 'the stunningly decisive blow' to the survival of the polar party.[23]

But her research also reveals that the final endless blizzard that trapped the men in the tent was almost certainly a figment of exhausted or compassionate imaginations. Scott was immobilised by a frostbitten foot, and perhaps his two loyal companions found an excuse not to leave him. Perhaps they waited – too long – for him to die? Thus, although the meteorological data is definitive, Solomon acknowledges that other aspects of the expedition's fate remain 'a question not for science but for the human heart'.[24]

In his famous book *Walden*, Henry Thoreau urged us to explore our inner selves, to be a Columbus to whole new continents and worlds within: 'Is not our own interior white on the chart?' he wrote. 'Explore your own higher latitudes', he urged. Spiritual and moral barriers needed to be breached as much as geographic ones. Our flawed hero carved a moral universe out of the amoral, inorganic ice. The fire on the snow.

## Thursday, 19 December

**48 degrees 40 minutes south,**
**139 degrees 4 minutes east**

I just went down to the stem of the ship to feel its speed.
From up on the bridge you feel as if you sit atop a factory
and you are enclosed and in command, all the instruments
of navigation smoothing your journey. Down at the water's
edge, at the bow, you can escape the hum and thrum of
the industrial site afloat and hear the sea against steel. You
lurch against the muscle of the ocean and you realise that
you are really moving, in spite of the way the pencil line
on the captain's chart inches imperceptibly away from the
southern Tasmanian coast out into a great landless void.

Fred Middleton retired to the engine room of the
*Aurora* to do his writing, the powerful, warm heartbeat of
the ship. There his hands could be free of his gloves and
his fingers could close around a pen. Stephen Murray-
Smith had his typewriter with him, and battled to return
its carriage against the roll of the ship.

We have an artist on board, Jenni Mitchell, and up
on the bridge she has ensconced herself in a corner of her
own, her hundred or so Derwent pencils spread out before
her, her watercolour box open. Her art is in the thrall
of deserts, and she is drawn to comparing the white salt
crystals of Lake Eyre with the icescapes of Antarctica.
In this endeavour she joins a cluster of Australians (such
as Griffith Taylor, JW Gregory and Cecil Madigan) who
have found fascination in the weird aridity of the two
Great South Lands. Brigid Hains beautifully explores
this theme of Australian cultural and scientific history in

her book, *The Ice and the Inland*, and it includes some of Jenni's photos of the desert. But here, for the moment, our artist is surrounded by water; she is up there on the bridge drowning in its immensity and light.

Ice is already our conversation topic, well before we have any hope of seeing it, for it generates the major uncertainty of our voyage. Kim Pitt is general manager of operations in the Antarctic Division, overseeing the logistics of transport, maintenance and scientific support, and I've just been talking to him on the bridge where he explained to me the looming challenge. The *Polar Bird* is an ice-strengthened ship, but not an ice-breaker. There is a band of concertinaing ice about 80 nautical miles out from Casey. We aim to cross it at its narrowest point. We pray for southerly winds that will push the ice out and allow us to find our way through its reported nine-tenths density. If the wind changes as we move through, we could be caught, as has happened to the *Polar Bird* each of the last two summers. Kim gently touched the wood of the chart table as he spoke of ice and fate.

Fred felt himself to be a young and nervous doctor on board the *Aurora* and always experienced dread whenever summoned by the crew in case he faced a new medical challenge. He did his first ever dentistry job on none other than Sir Ernest Shackleton himself, and this month, 86 years ago, he heard one of Sir Ernest's early renditions of the *Endurance* narrative. 'A most extraordinary trip', wrote Fred in his diary, 'and one which should read well in a book which will sound most impossible'.

We are now deep in the realm of the roaring forties. I saw my first albatross today, black-browed, gliding gracefully and nobly in the wake of the ship, sporting with the quickening wind.

# THE BREATH
# of ANTARCTICA
## *The brave west winds*

After the *Endurance* had been trapped and crushed in the ice of the Weddell Sea, Ernest Shackleton's party made its way across ice floes and open water to Elephant Island, off the Antarctic Peninsula. Leaving their mates sheltering there under upturned boats, Shackleton and five men sailed in their small whaling boat, the *James Caird*, across 1300 kilometres (800 miles) of ocean back to South Georgia. It was not the closest port – that would have been Port Stanley on the Falkland Islands, 870 kilometres (540 miles) away – but it lay in the path of the west winds. While contemplating this desperate journey over 'the most tempestuous storm-swept area of water in the world', Shackleton wandered down to the shoreline to have a good look at the *James Caird*: 'The 20-ft boat had never looked big; she appeared to have shrunk in some mysterious way when I viewed her in the light of our new undertaking.' The carpenter added a deck as a form of shelter, but Shackleton 'had an uneasy feeling that it bore a strong likeness to stage scenery'. The boat had to be baled out before they had even started from Elephant Island because the bottom plug had been left out. They swiftly replaced the cork.

The tale of their famous 16-day voyage was 'one of supreme strife and heaving waters'. 'We fought the seas and the winds and at the same time had a daily struggle to keep ourselves alive', wrote Shackleton. They first navigated north to get clear of the ice and to pick up the westerlies. Sometimes they put out a sea-anchor to keep the *James Caird's* head up as it broached the ocean. They rode the crests of mountainous waves, took turns to huddle under the deck, baled water constantly, chipped leaden ice off the rigging, cooked hot hoosh on the primus stove for a shred of warmth and optimism, and hoped for an occasional glimpse of the sun to take a latitude reading. How sure was such a reading from a pitching boat, and how accurate was their chronometer? Only once a night might a precious match be lit to check the compass. Sea spray froze on the boat, everything tasted of salt and their thirsts raged. Even when they hove to in a storm, the great easterly ocean drift bore the *James Caird* inexorably towards its goal. 'So small was our boat and so great were the seas that often our sail flapped idly in the calm between the crests of two waves', recalled Shackleton. On 5 May 1916, he spied what he thought was a clear line of sky on the horizon. But it was the white crest of a gigantic wave, 'a mighty upheaval of the ocean'. Somehow the *James Caird* rode it out, and was furiously baled back to the surface again. An albatross followed them all the way from Elephant Island – was it always the same one? – and they dared not shoot it. When they joyously sighted the towering black crags of South Georgia – a speck in the Southern Ocean they had somehow found – a hurricane engulfed them, and it took two further agonising days to secure a landing. The whaling stations were on the east coast and they had landed on the west, and so a perilous climb over the icy spine of South Georgia awaited them. But the west winds had saved them.[1]

Moving air is a powerful element. We inhabit a swirling, gaseous soup that is animated by the heat of the sun and the spinning of the Earth. 'Somehow', writes Jan DeBlieu in her meditation on *Wind*, 'out in the elements, the wisdom of science falls a bit short. It is easier to believe that wind is the roaring breath of a serpent who lives just over the horizon.'[2]

The 'roaring forties' were the earliest and most famous southern winds in the European imagination. They shaped the European discovery of Australia, influenced the foundation of British settlement in the Pacific, and brought Australia's nineteenth century immigrants to their new home. Encountering and surviving these wild westerlies on the Southern Ocean was a rite of passage for all newcomers to Australia, a dimension of their new identity.

Europeans who happened upon Australia from the seventeenth century travelled in ships that first voyaged confidently across latitude and then easted anxiously within it. Latitude could be read accurately from the sun, while longitude remained the chief problem in navigation until the late eighteenth century. Sailing south in the Atlantic, there was a known gradation of airs. There were the low-pressure doldrums at the Equator, where ships could be trapped in a sticky calm, then the steady, reliable trade winds to the north and south. Next there was the unruffled air of 'the horse latitudes' around the tropics of Capricorn and Cancer, the latitude of many of the world's deserts. The horse latitudes took their name from the fact that it was here that the crews of becalmed sailing ships threw horses overboard when drinking water became scarce. Or was it that by the time they had reached these latitudes, crews had thrown overboard the 'dead horse' of working off their advance wages?[3] South of the 35th parallel – which was about the Cape of Good Hope, the end of known land – ships began to edge into the swirling atmospheric engine of the Earth. Here crews and passengers found themselves engulfed by an elemental force that was both exhilarating and awesome. The Southern Ocean, observed Jean-René Vanney, a French historian was, above all, 'une realité pour nos oreilles'. Latitude was manifest in the violent music of the winds: in the forties they roared, in the fifties they screamed, and in the sixties they whistled.[4]

The prevailing westerlies of these higher southern latitudes encircle the bottom of the globe and isolate Antarctica. They boom and broil unimpeded by land. The break-up of Gondwana created their vast playground. The Southern Ocean is a pure realm of water and wind, an extensive belt of the Earth's surface where these two fluids encounter and animate one another without interruption. The dance of air and sea is vigorous; each reaches for and agitates the other. Here an ocean swell of greater than three metres (10 feet) is usual, and waves of 10–11 metres (32–36 feet) or more are a regular occurrence, with swells almost as high. A ship might be lost from sight between waves. Clippers running before a gale could travel over 480 kilometres (300 miles) in a day. Mean monthly wind speeds between the latitudes of 40 and 60 degrees south are 20–45 kilometres (12–27 miles) per hour, and gales and storms are frequent.[5] Torrent and current combine to become one inexorable westerly force. A chain of spiralling low pressure systems brings great storms followed by eerie days of calm, a manic climate. Belgrave Ninnis, hugging himself with the delight of his Antarctic adventure with Douglas Mawson in 1911, was sobered by the west winds, which he described as 'Hell, grim Hell, sodden watery Hell. Perchance you have heard of the "Roaring Forties". Well, for eleven days they've been roaring … This is famed as the roughest place on God's earth, and, by Jove, it has lived up to its reputation … Things are coming to a pass when one comes to a meal in oilskins & sea boots.'[6]

In the Southern Ocean, European navigators found that gritty, claimable earth was fictive, wishful, and disappointingly finite.[7] In this region of delusional geography, ice and mist blurred the boundaries between water and air, and sometimes even posed as land. Explorers in the high southern latitudes had to learn to read the sea and the sky for hopeful signs of land: they scrutinised floating kelp, swarms of birds, shallow water, quixotic winds and mountainous clouds. In the screaming fifties, surging and converging ocean currents generate submarine fronts. Here is found perhaps the longest and most significant biological boundary on Earth, the Antarctic convergence, where cold Antarctic waters encounter and slip under the warmer sub-Antarctic waters, and the boundary is manifest in mists and cold and sudden southern marine biodiversity.

The contours and constancy of wind in these latitudes was made visible by a bird. The forties and fifties quickly became known as 'the albatross latitudes' because they were dominated by this magnificent creature that visited and followed ships. Ten of the world's 13 species of albatross live on the open seas of the southern hemisphere. It is particularly the domain of the Wandering Albatross (*Diomedea exulans*), pure white with jet black wing tips, the largest surviving flying animal. The albatross is more than a creature of the wind; it was, in the poet Samuel Taylor Coleridge's words, 'the bird/That made the breeze to blow'.[8] When the North American naturalist, Robert Cushman Murphy, saw his first ever albatross on a voyage south in 1912, he wrote: 'I now belong to a higher cult of mortals … in the morning sunlight, flew the long-anticipated bird, even more majestic, more supreme in its element, than my imagination had pictured … Lying on the invisible currents of the breeze, the bird appeared merely to follow its pinkish bill at random.'[9]

Voyagers to the Australian colonies long anticipated its appearance in the albatross latitudes and recorded the sightings in their diaries and journals, for this bird was the signature of the wind that also filled the ship's sails and took them east. Any vessel was attractive to the bird because it created eddies and up-draughts in air and sea; a ship churned the shrimps and squids to the surface, and there was always galley refuse. Albatrosses are surface feeders and swoop at any floating scrap, particularly if it is white. Dr Edward Wilson on Robert Falcon Scott's ship, *Discovery*, on examining the stomach of an albatross, found in it 'an undigested Roman Catholic tract with a portrait of Cardinal Vaughan'.[10] Shooting albatrosses from ships was a ritual sport, in spite of the warning of Coleridge's 'Rime of the Ancient Mariner' where southern voyagers were punished for killing the bird of good luck. The bird's light, air-filled bones were claimed by sailors for pipe stems. The ease of flight of the albatross was entrancing. It was so big, like a clumsy goose when lured onto the deck with tit-bits from crew and passengers, but in the air it soared, hardly ever beating its very narrow, very long wings. Albatrosses inhabit that thin, lively layer of encounter between ocean and moving air, and they would follow ships for days, carried along by the same medium. In low winds and calms they have to forsake the air and rest on the water. For the Wandering Albatross, land is only important for nesting, and a young bird may fly the open seas for seven years before returning to land to breed.[11]

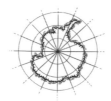

Australian history was shaped by the European exploration of the sea and the winds. In 1519–22, Ferdinand Magellan's fleet – in spite of the loss of three ships and 200 men along the way, including the captain – circumnavigated the world and beheld the astonishing unity of the oceans. The world turned out to be larger than Europeans had supposed, and a third great ocean – the Pacific – was revealed and crossed. In the southern latitudes they found that the sea and the wind flowed with awesome continuity.[12]

If you could master those winds, you could dominate the world. The 'discovery' of the sea introduced an era of maritime empires where expanding trade networks and political influence fell into the hands of those states that could build reliable ships and harness the power of winds. In his book *The Discovery of the Sea*, JH Parry attributes early Portuguese maritime success to its location at 'the street corner of Europe' where it bridged the Mediterranean and North Atlantic worlds, and where, for most of the year, the winds off the coast favoured exploration to the south and southwest.[13] Europeans learned not to sail down the west coast of Africa, where they were repelled by northbound currents and winds, but instead to track across the South Atlantic almost to Brazil where they would catch the roaring forties back east.[14] In the seventeenth and eighteenth centuries, the Dutch, French and English became the maritime superpowers who destroyed the older Arab and Venetian trade between the East and Europe by mastering the alternative sea routes. Dutch seafarers rounded the Cape of Good Hope at 40 degrees south and took the prevailing westerlies as far as they could before curving late and north towards Java. Hence Australia's west coast was an accidental European landfall, its discovery a passive product of a game with the wind. For the same reason, Australia's more fertile east coast was for long protected from the European imperial eye by the west winds, because heading into them from Cape Horn at high latitude was unhappy sailing.[15]

The great historian of the roaring forties is Geoffrey Blainey who wrote these winds into the fabric of Australian history in his book *The Tyranny of Distance*.[16] His abiding interest in climate and culture, and his fascination with the historical physics and geography of the globe, enabled him to offer new and enduring insights into the foundation of British settlement in the Pacific. Australia was colonised by the British less for its land than for its strategic position in an ocean, and it was indeed as an archipelago of islands that it first developed its European character. Safe harbours, timber for masts, flax for sails, proximity to the chief ports and trade routes of Asia: these were the sinews of the eighteenth-century struggle for sea power.

It is worth recalling the sea-consciousness of Australia's early European settlers in the years before their national imaginations turned decisively inland. Australia was invaded by a naval power, its first colonial culture of authority was maritime, whaling and sealing were the colony's earliest productive industries, and it took settlers a quarter of a century to cross the first land barrier, the Blue Mountains.[17] Early Sydney revolved around its port life, its population was 'waterbound' and maintained 'a constant and keen-eyed watch over the harbour', their lives were animated and threatened by the cycles of winds, currents, tides, the dangers, risks and opportunities of the sea.[18]

Experiencing the roaring forties was like an initiation ritual for Australian-bound voyagers. What was it like to be on a ship driven by the 'brave west winds'? Englishman Edward Snell emigrated to Adelaide in 1849 on board the barque *Bolton*. He had always felt a yearning for the sea, and here he is writing in his diary at latitude 36 degrees south, and heading for the forties:

A regular unmistakable thorough hard gale of wind … The ship behaved very well riding over the seas like a duck though they were running as high as our main top. The roaring of the wind was tremendous and the ship rolled and pitched very heavily. After tea I went on deck to get a smoke and was immediately thrown off my legs and struck my right temple against the deck with such a force that I was all but stunned, didn't break the pipe though.[19]

… Came on to blow pretty hard in the evening and before sail could be taken in the fore top gallant sail went to smithereens, the ship flying through the water 11 knots an hour steering ESE … The scene was magnificent for it lightened and rained tremendously but the thunder was drowned by the roaring of the wind, for some time it blew a little hurricane, and I thought of the song

'With lightning above us and darkness below, Through the wild waste of waters right onward we go.'

And what a difference there was between singing it in a drawing room at home and looking at the reality at sea. However without any humbug the thing was beyond a joke, and for sometime the sailors had enough to do to get in the canvass which was threshing and flapping about as if it meant to carry the mast away while the rain came down in buckets full and sailors don't use umbrellas.[20]

Snell returned to England nine years later, in 1858, on the clipper *Norfolk*, a vessel so built for speed that Snell believed that 'it would go ahead even in a calm', although its narrow beam meant that this diarist found it 'rolls like the devil which makes writing rather difficult'.[21] The return journey was by way of Cape Horn, and again the ship dived into the engine of the westerlies, but this time much further south. From Melbourne it dipped immediately south to a latitude of about 60 degrees and maintained this parallel right across the Pacific and beyond Cape Horn. Snell records:

weather alternately calm, and stormy, but always wet, foggy, misty, drizzling, and infernally cold … We are today well round Cape Horn but with the decks covered with snow, sails frozen stiff as boards and every one with a drop on the end of his nose, except the ladies who don't go on deck much.[22]

The latitudinal difference between Snell's two voyages – the first in 1849, the second in 1858 – was not just a function of having to round the more southerly Cape Horn, it was also due to the fact that the 1850s had seen a revolution in the sailing of the Australian route. In 1842 a North American naval officer, Matthew Fontaine Maury began to seek the logbooks of captains so that he could compile an atlas of the world's winds, storm systems and ocean currents, and from 1851 he published a series of wind and current charts to assist sailors. At the same time, an English watchmaker, John Towson, advocated the use of the Great Circle sailing route to Australia, because use of his inexpensive chronometers would enable captains to monitor longitude and thus take a short cut around the globe. As Blainey put it, Towson 'believed that the world was round and that seamen should recognise that truth'.[23] The shortest distance between two points on the Earth was the arc of a great circle, and the largest expanse of ocean, the strongest winds, and the greatest distortions of the old Mercator charts were all on the high latitude Australian route. Captains began arcing down to the furious fifties to save a thousand miles. Maury enthused that the round trip to Australia affords 'the bravest winds, the fairest sweep, and the fastest running' along a route which for more than 300 degrees longitude courses with the westerlies.[24] 'The billows there lift themselves up in long ridges with deep hollows between them', he wrote; 'their march is stately and their roll majestic', 'the scenery among them is grand'. Maury's passion was to understand the planet's 'two oceans of air and water', as he put it, and to show the practical advantages of philosophical research. He noted with pride that his charts and the Great Circle sailing had reduced the voyage from Europe to California by a month and a half, and the round trip between England and Australia by 50 days or more.[25] His famous book, *The Physical Geography of the Sea*, was first published in 1855 and remained in print for 20 years, its public life corresponding to the triumphant years of the clipper.

The American clipper evolved its finest and fastest form in the same years that gold was discovered in the Australian colonies and the science of wind developed such practical dimensions. Gold, tall ships and the mastery of wind together revolutionised Australia's economic growth, and also secured Melbourne's commercial dominance in the nineteenth century.[26] Melbourne was well situated as the major Australian port for the wind travellers for it combined a rich hinterland with proximity to 'one of the great natural propulsion bands of the earth'.[27]

In December 1789, the supply ship, HMS *Guardian*, with 123 people including 25 convicts, and carrying 1003 tons of stores for the infant colony of New South Wales, topped up with 2000 gallons of wine from the Canary Islands, hit a giant iceberg at 40 degrees south. The ship had been painstakingly fitted out in response to Governor Arthur Phillip's written cries for help in the difficult first year of the colony's founding.[28] The commander of the *Guardian* recorded the sudden vision of 'a body of ice full twice as high as our masthead, showing itself through the thickest fog I ever witnessed ... The ship at this time seemed to be entering a cavern which was large enough to receive her entirely.' The bulk of the cargo was got rid of, some of the men abandoned ship and were never seen again, and over eight agonising weeks the *Guardian* eventually limped back to Cape Town, where salvageable stores were forwarded by available vessels to Sydney. The loss of the *Guardian* was a bitter blow to the struggling British settlement, one that Governor Phillip judged to be 'almost fatal to the Colony'.[29] The ice of Antarctica, roaming wild in the roaring forties, had already intervened dramatically in Australian history.[30]

The icebergs were mysterious, far-flung evidence of what lay at the South Pole. What was their source? Where did the roaring forties come from, and go to? Just as the classical cartographers intuited Antarctica's presence from arguments of earthly symmetry and elemental equilibrium, so did early voyagers sense the ice continent before they saw it: they felt its breath. Antarctica has an aura. I don't just mean its mystery, its magical otherworldliness, its implacable grandeur, and its capacity to haunt all who have visited it. I also mean that this land over the South Pole, which is covered by a single mineral, actually emanates ice, water and air well beyond its geographical boundaries. It took people a long time to realise that there was not just land down there but a continent, that it was high and dry and covered thickly in ice, that it was very, very cold, much colder than the Arctic, and that it constantly affects the climate of the rest of the world. Time-lapse photography from satellites now reveals to us that Antarctica is like a giant, breathing organism clamped to the base of the globe. Every winter as the southern hemisphere tilts away from the sun, so much sea ice forms that the size of the white continent appears to double, only to shrink again in the summer, like a billowing creature rhythmically expanding and contracting. When the surface of the sea turns to ice, it releases a dense brine that plunges to the ocean depths, and that thrust of salty water to the sea floor is the piston-stroke that drives the engine of ocean circulation, sending cold Antarctic bottom-water northwards, even infiltrating the northern hemisphere. Meanwhile, continental Antarctica, where ice forms kilometres thick on ancient bedrock, is an ice-making machine of prodigious dimensions. Slowly, inexorably, ice moves from the central heights of Antarctica towards the coast, where great chunks are launched as icebergs, some the size of countries, some big and fast enough to sail to the edge of the tropics. There is so much ice in Antarctica that it skews the Earth into a slight pear shape. Antarctica is 'the world's greatest refrigerator', as Mawson called it in 1935;[31] it is such a cold core of the atmosphere that it 'deflects the meteorological equator of the globe northward nearly ten degrees latitude'.[32]

James Cook circled Antarctica in a dance of awe during his second great expedition. He voyaged distastefully through seas 'pestered with ice' and amongst icebergs that left his 'mind filled with horror'; he made what was perhaps his greatest discovery, the circumpolarity of the Southern Ocean.[33] Cook's three summers probing the Antarctic Circle, in 1772–75 in almost constant bad weather, came near the end of the Little Ice Age, an era of lower global temperatures and more extensive pack ice, and this may explain why this ambitious navigator did not conclusively sight the continent.[34] And when Cook passed close to Heard Island in February 1773 but did not see it, the towering feature may have been disguised not only by cloud, but by a colder climate that clad it in ice to the sea.[35] By the time of Great Circle sailing 80 years later, the Little Ice Age was ending, and high latitude voyaging may have just become, with fortuitous timing, a little less perilous. For much of the nineteenth century, sailors and scientists analysed the sea and the ice and the wind as indirect evidence of the character of Antarctica, and in the early twentieth century, the effect of the southern distribution of ice on Australian weather was one of the key research topics suggested for the officers of the new Commonwealth Bureau of Meteorology.[36]

In the mid-nineteenth century, Matthew Maury had speculated that there was a mild climate about the South Pole, mild by comparison with the Arctic. 'The winds were the first to whisper of this strange state of things', he wrote, and the other clues he drew on were those ships' logs that he knew so well which recorded low barometric pressure at high southern latitudes. He extrapolated their descending oceanic readings into the expectation of a permanent, very low pressure system at the South Pole, whose latent heat and water vapour would suck in moderating winds from the north. He thought that the roaring forties – the winds that drove his sailing ships – must go all the way to the pole.[37]

When the Tasmanian scientist, Louis Bernacchi, sailed south with the *Southern Cross* expedition of Carsten Borchgrevink in 1898, his chief task was to collect meteorological and magnetic records across a whole year from the continent itself. As a Tasmanian, he had grown up with Antarctica breathing down his neck. He sailed on the *Southern Cross* all the way from London, and caught the roaring forties east to Australia. He rode, as he put it, those 'turbulent waves which sweep resistless round the world'.[38] Bernacchi gazed again in wonder at the 'lordly' creature whose flight seemed to define the wind.[39] Meteorology and magnetism lured many expeditions south and these sciences also gave structure, routine and some higher purpose to isolated groups of men wondering what on earth they were doing down there. The tending and reading of the essential instruments was, in Antarctica, heroic and fraught with danger. Both sciences required long, laborious runs of data so that variations over space and time might be mapped. Both were inspired by global Humboldtian science, by the desire to embrace the cosmos with measure and pattern. But they also had quite practical purposes: meteorology was a vital key to survival, and magnetism had long been a tool of empire because understanding terrestrial magnetism and its declination and dips was crucial to better use of compasses and navigational safety.[40]

Bernacchi gained some basic insights into Antarctica through a tough winter; the *Southern Cross* expedition was the first to winter on the land. The South Pole, he discovered, was certainly not warmer than the North, as his year of temperatures showed. And such low mean temperatures suggested that there was extensive land rather than sea further south. And the 'most remarkable feature' was the winds they had experienced. They were relentlessly east-south-east and south-east (not west), which suggested that the Antarctic lands were probably elevated and covered by a great, permanent high pressure system rather than a low pressure system as some expected. Without land at the pole, the great belt of west winds – those famous roaring forties – might well have gone all the way south. Borchgrevink's expedition hardly managed any significant sledging inland, so Bernacchi the scientist had to guess what lay within Antarctica from the wind at the coast. Even on the edge of the continent itself, then, Bernacchi the Tasmanian was still analysing the breath of Antarctica.[41]

Scientists began to realise that understanding the Southern Ocean and the influence of Antarctica would offer clues to Australia's climate. Henry Chamberlain Russell, government astronomer and meteorologist in New South Wales, had already provoked debate with his finding in 1893 that, in the southern hemisphere, anticyclones were not fixed but moving, and that they marched across central Australia in a relentless and predictable progression.[42] He explained this moving chain of pressure systems by the great predominance of ocean over land in these latitudes. A similar progression of cyclones tracked across the Southern Ocean, and while the northern edges of these low pressure systems affected Australia, their southern edges extended over ice, and their centres lay unmapped over the Southern Ocean of the roaring forties and screaming fifties.[43] Writing in 1859, Matthew Maury hoped that one day meteorological stations would be established on Heard Island and other points in the region of the 'brave west winds'. 'I know of no enterprise in the meteorological way that … gives promise of richer rewards than this does, both practically to the mariner and scientifically to the philosopher.'[44] The Australian aviator, Hubert Wilkins, became a strong advocate for such stations in the first half of the twentieth century.

The study of cyclones was revolutionised in the early twentieth century by the Bergen School of Meteorology led by the Norwegian, Vilhelm Bjerknes. He brought a physicist's eye to atmospheric science – seeing air as a fluid – and he pioneered a more sophisticated analysis of air masses and fronts. In the years following World War 1, strategic forecasting became more urgent and air travel common, and improved communications enabled a dense network of weather observations to be amassed quickly so that finer-grained atmospheric structures could be promptly mapped. The frontal concept of the Bergen School introduced three dimensionality and dynamism to the study of air masses, and cyclones were seen as evolving phenomena with life cycles that might be studied. Bjerknes also outlined his idea of a polar front, which he saw as an oscillating battleline between polar and equatorial air, an undulating string upon which the pearls of cyclones developed and moved.[45] Australian scientists did not systematically test these concepts until World War 2 and after, but meteorology was at the forefront of early-twentieth-century Australian investigations of the Southern Ocean and the polar continent. The roaring forties therefore had geopolitical force, for meteorological proximity became one of the justifications for the retention of sovereignty over Australian Antarctic Territory. The links between the Gondwanan partners were sketched out in the atmosphere, as well as in the rocks and fossils.

No continent is more ruled by wind than Antarctica. Mawson famously called it *The Home of the Blizzard*. He established his base at Commonwealth Bay on the Antarctic continent in 1912 on a site that he later found to be probably the windiest place on Earth. The average wind there is an extraordinarily steady gale.

At Commonwealth Bay, the meteorologist Cecil T Madigan left memorable descriptions of his work recording a continental weather pattern ruled by wind: not the west winds, but the violent gravity-driven or katabatic winds that dominate Antarctica itself. They are surface winds formed of cold air pouring down off the high Antarctic plateau; they can become air avalanches. In 1912, complete weather observations were taken every six hours; these consisted of reading the barometer and screen thermometer, estimating the wind velocity and noting its direction on the vane, describing the cloud, recording any phenomena like snowfall, drift, St Elmo's Fire, coronas and halos, sunset colours, state of the sea, and auroral displays. All data was entered in pencil in a meteorological day book, and 15 of these 100-page books were filled. Visiting the anemometer was, admitted Madigan, 'physically a fairly severe undertaking, presenting considerable difficulty in the mere finding of it'. He goes on to explain:

After some practice the members of the expedition were able to abandon crawling, and walked on their feet in these 90 mile torrents of air, 'leaning on the wind'. The picture over this title ... shows a man using a pick in getting ice for cooking purposes, leaning comfortably on the wind as against a wall. For the sceptical one might point out that the body is at rest and in sharp focus, while the pick is moving and somewhat blurred. I found that in a 90-mile wind, if I faced it and kept the body straight, I could touch the ground in front by extending an arm. Of course, one wore more clothing and offered a larger area to the wind than under normal conditions.[46]

Mastery of this 'hurricane-walking', as they called it, became an essential meteorological accessory. Mawson declared that: 'What is certain in my mind is that like difficulties have never yet been experienced in the same degree at any other meteorological station. The bare records in their unfailing regularity are an epic in meteorological achievement.'[47]

Meanwhile at Queen Mary Land, Morton Moyes monitored the weather for the western party led by Frank Wild. On rough days, Moyes had a similar battle to collect the daily observations:

We got back against the wind with difficulty, and I had to stand inside for half an hour with arms up, as hands were frostbitten. Wind so strong, unable to stand up against it, and could not see [my] hand in front because of snow.[48]

On another day, he recorded that 'Many of the gusts must have exceeded one hundred miles per hour, since one of them lifted Harrisson who was standing beside me clean over my head and threw him nearly twenty feet.[49]

For ten weeks Moyes was left completely alone with his weather instruments at base camp, feeling 'like the last leaf of a branch'. Between reading the wind gauge and the snow gauge, he did some other reading: Ernest Shackleton's *The Heart of the Antarctic*, the meteorological notes of the previous British and Scottish expeditions, Thomas Babington Macaulay's *History of England*, Robert Browning's poetry, Baruch Spinoza's *Ethics*, Dante Alighieri's *Inferno*, the *Harmsworth History of the World*, Francis Bacon's *Essays* and Archibald Geikie's *Geology*. He wrote in his diary: 'I'd like a good novel for a change, the shops here don't stock them.'[50]

A priority of the Mawson expedition was to establish a weather station at Macquarie Island, thereby fulfilling Matthew Maury's dream of doing meteorology in the teeth of the westerlies. It was established with the strong support of Henry Hunt, the Commonwealth meteorologist, who loaned one of his officers, George Ainsworth to lead the island party.[51] A wireless mast was erected and daily weather readings were communicated to Australia and New Zealand. An earthquake tremor greeted the party on arrival in December 1911, and the wind 'appeared like an invisible wall'; you had to turn your back on it to take a breath.[52] One of the meteorologists sent down later to Macquarie Island took a football with him but reported that, because of the wind, it 'could not be used'. The original party built their shack on the easterly side of a huge rock that protected them from the relentless westerlies, and at times they had to catch practically all the food they ate. They lived off the fat of the land: the heart, tongue and liver of the locals. Ainsworth confessed to Hunt that 'I shall never willingly look a Sea elephant in the face again.'[53]

The first human did not set foot on mainland Antarctica until the end of the nineteenth century. The continent's meteorology, they were to discover, was distinct, extreme and simple, all of which made its effect on the rest of the world's weather greater. The South Pole harbours a permanent high pressure system and easterly winds blow round the continent's coast. Those outer westerlies of the forties and fifties, acting in concert with the circumpolar ocean current, isolate and amplify Antarctica. They form an atmospheric equivalent to the ocean convergence, and both boundaries – the polar front and the convergence – were not recognised until well into the twentieth century.

All that permanent ice reflects the sun in summer, and the dry clear atmosphere releases what heat there is. Unlike the Arctic, the solar radiation balance of Antarctica is negative all year round. As the

American historian of Antarctica, Stephen Pyne, succinctly puts it, 'The ice amplifies the conditions that generate the ice.'[54] These are the same feedback mechanisms that, on a global scale, can produce ice ages. Over the last century, we have come to realise that Antarctica is not only a powerful influence on Australian and world weather, but it is an essential archive of past climates, and a crucial indicator of global changes today. The roaring forties are both a product and an agent of the ice. And the ice holds secrets of the history of the air.

# Saturday, 21 December

**54 degrees south,
130 degrees east**

The solstice. And we cross the convergence today, somewhere around 54 degrees south and 130 degrees east. The drop in the sea temperature, which we are measuring every four hours, will be the first indicator. At noon today the sea temperature is 7 degrees Celsius (44 degrees Fahrenheit) and the air temperature is now down to 4 degrees Celsius (39 degrees Fahrenheit). We proceed on a straight course from the south-eastern tip of Tasmania, undeviating at 224 degrees, until we reach the band of ice out from Casey. As is traditional in Australian voyaging, a sweep is being conducted on board to guess the timing of the sighting of the first iceberg, still a few days away. The ocean looks benign today, a deep blue, the sky is light on the horizon, visibility is excellent. There is just a soft rumble of the engines heard here on the bridge, and a gentle pitch of the ship. Only a few birds or whales have been seen in recent days. I was lucky enough to be on deck for the first whale sighting, just before 8 am on the first morning of our voyage, a small sperm whale that blew just 30 metres from the ship. But since then, signs of sea life and even birds have been scarce. It is sobering to voyage for days into the Southern Ocean and to realise that humanity has impoverished nature even out here, and to remember that these latitudes were once busy with visible life.

I am talking to as many of the people on board as possible, at least those who are well enough to be on deck or at meals. Many of the scientists are working on cleaning up old tip sites in Antarctica. Since the Madrid Protocol on Environmental Protection to the Antarctic Treaty was signed in 1991, all nations have had to do a better job of returning rubbish and waste matter home and we even need to clean up historic messes. A historian is looked at rather warily by such people. You can see the fear in their eyes that you are about to declare some dump of heavy metal contaminants a heritage site. We are taking down 160 empty containers on this ship so that we can start repatriating some of the contents of the Thala Valley tip site that was the dump of the old Casey station. For two decades from the late 1960s, rubbish was piled up there or just pushed out onto the sea ice where it floated out in the summer melt and sank. Two scientists on our ship are also divers and they are studying the underwater biology near tip sites, and the impoverishment of life is dramatic.

Another combination of science and planning that is represented in the people on this voyage is the project to build an ice runway for aircraft near Casey by the end of next summer. You need glaciology, geology, engineering and courage to bring about such a momentous change in the logistics of Australian Antarctica. Stephen Murray-Smith's journal reminds me that we were talking about an airstrip as an imminent and urgent development in the mid-1980s, and then it was expected to be a rock runway at Davis station. The North American engineer on board who will help us find the right kind of ice has an arresting term for people who fly to Antarctica: 'self-deploying cargo'.

So I am participating in one of the last seasons of Australian voyaging south. Once the runway is built, aircraft will take most people to and from the ice, and the ships will carry large cargo. People like me will fly for several hours and then step out onto glacial ice relatively uninitiated. They will miss attending the University of the Southern Ocean. I got here just in time.

# THE UNIVERSITY of
# the SOUTHERN OCEAN
*Life at sea*

It was the year of Robert Falcon Scott's and Belgrave Ninnis's deaths. Robert Cushman Murphy, a young curator at the Brooklyn Museum of Natural History who desperately wanted to see penguins, had just been offered a berth on a whaling boat sailing for a year to the far South Atlantic, and he had turned it down. Bob, as he was known, was soon to be married and could not contemplate such a long and distant absence. When he wrote to his fiancée, Grace Barstow, of the offer and his reply, she responded by return telegram delivered in the middle of the following night: 'Hold open the seafaring option ...' The next day she caught the first train to Providence so that they could meet and talk. Such a voyage could be the foundation of a naturalist's career, Grace advised Bob; you could be like Charles Darwin sailing in the *Beagle*! His degree from Brown University would be nothing compared to this learning experience. She urged him to go, and declared: 'I'll marry you now.' They were wed in June 1912 and travelled together to Barbados, where Grace found that their separation loomed like a 'rain-wall creeping toward sunshine from across the sea'. Bob departed from Dominica on the *Daisy* in July. He stowed on board a large bag carefully prepared for him by Grace, containing a year's worth of letters and mementos from herself, family and friends. He promised to leave a note in a bottle for her at any uninhabited island they visited and to drop others overboard at the Equator and the Tropic of Capricorn.[1]

The ship on which Bob was suddenly heading south to the stormiest seas in the world was the brig *Daisy*, with four dogs, a pig, a steward with a peg-leg, a crusty captain named Cleveland and 31 other men hunting sperm whale and sea elephants. The American Museum of Natural History was sponsoring the voyage in exchange for Murphy's passage to the sub-Antarctic island of South Georgia and his return home with natural history specimens. Amid the boiling blubber on deck, Murphy read William Shakespeare. He had taken a library of 30 books, including: Herman Melville's *Moby Dick*, James Weddell's *A Voyage Towards the South Pole*, Joseph Banks' journal on board the *Endeavour*, Henry Moseley's *Notes by a Naturalist on HMS Challenger*, Dante Alighieri's *Divina Commedia*, John Bunyan's *Pilgrim's Progress* and, of course, Charles Darwin's *Voyage of the Beagle*. 'How I long to see with the eyes of that matchless man of science, and to write with his pen!' he wrote to Grace. 'When I come home, I must study more geology. I want to be able to grasp something of the whole scope of nature in the lands and seas I visit; to be broad, not narrow; to be both a naturalist and a humanist, not a mere specialist ... Nature is a chain, a million-knotted web or fishnet of life. Nothing exists of or for itself, but only in relation to other organisms, as Darwin seemed to know more thoroughly than anyone else.'

At 21 degrees south, Bob slept under a blanket for the first time and wore a sweater, and at 23 degrees south he changed to a flannel shirt and soon after dropped his bottle for Grace overboard at the Tropic. Bob thought he was sailing into a wilderness. Explorers and voyagers were still mapping the bare outlines of the southern continent, and Roald Amundsen had reached its symbolic heart only the year before. But the circumpolar seas had been raided by humans for a century and Murphy was to be astonished by the detritus of industry and history that he would find in this southern hemisphere of water. He smelled his destination before he saw it, for the whalers' cove was brown with blood, gut and grease. The beaches of his sub-Antarctic isle were stacked high with bleaching bones. At South Georgia, spinal columns, loose vertebrae, ribs and jaws of whales tangled grotesquely on the shoreline. Bob found 100 skulls within a stone's throw. Dead sea elephants, slain, skinned and stripped, their blubber removed, lay red and frozen on the beach. As well as these bones and bodies left on shore by marauding men, Bob found that inland were vast graveyards of the clean-picked bones of terns and petrels, left by the brown rats introduced by sealers a century before. He reflected that 'alas, there is a canker in the Garden of Eden'. Ernest Shackleton, when he arrived exhausted in the *James Caird* on a remote end of the island a few years later was bemused to find rats raiding his food scraps, a strangely welcome signature of his own civilisation. He doubted the evidence of his eyes: 'One would not expect to find rats at such a spot.'

The history of sealing and whaling in the Southern Ocean offers a stunning, repeated pattern of discovery, over-exploitation and rapid decline, as new islands or oceans were invaded and swiftly exhausted. Sealing took off in the far south in the wake of James Cook's reports of island colonies of glistening creatures, and it was sealers who became the incidental (and sometimes secretive) discoverers and explorers of the Antarctic and sub-Antarctic coastlines. Remarkably quickly and with ruthless efficiency, British and American sealing ships stripped the rocks and beaches bare of the animals, slaughtering fur seals for their pelts and elephant seals for their blubber oil. The thick, dark fur of the fur seals was in much demand for clothing, and as a result, many of the northern fur seal colonies had been hunted to extinction by the early nineteenth century. The elephant seals' pelts were considered worthless for clothing, but their blubber was as good as a whale's for the quality of oil it produced. The sealers' arithmetic was that a barrel of blubber made a barrel of oil, and it was this oil that lubricated the machinery and burned the lamps of civilisation.

The fur and elephant seals were easy prey to humans. They hauled themselves annually onto the rocks just in time to greet the hunting ships of summer, and their gregariousness and renowned loyalty to breeding sites ensured that they could be harvested *en masse*. Fifty or 60 seals might be clubbed and skinned by an expert sealer in an hour, and it was usual for the whole of a colony to be exterminated in a raid. The dominant and possessive bull seal – known as 'the beachmaster'

– would often prevent his harem from escaping into the sea, thereby conspiring to keep them within reach of the murderers. Gangs of sealers lived ashore in shelters built against a cliff-face, with whale-rib rooves covered by sailcloth and walls of boulders caulked with sealskin. At one sealers' cave on Livingston Island, whale vertebrae had served as seats, and moccasins of canvas and sealskin remained to be recorded in 1959. A single ship in the South Shetlands took 9000 seals in 15 days in January 1820.[2] Historians lose count beyond 10 million when they try to quantify the total slaughter of southern fur seals.[3] The killing of sea elephants was particularly bloody; they were clubbed or shot, then lanced, and their blubber was cut away and boiled in 'try-pots' or cauldrons. Reducing the thick layers of fat to oil turned a sealer into 'a living greasespot', as one sailor at Kerguelen Island put it: 'If anyone had put a match to me, I should have burned like a tallow candle.'[4] The killing of the ubiquitous crabeater seals was especially dramatic because their domain was the ice-floes. WG Burn Murdoch, an artist on board one of four Scottish whaling vessels sent south in 1892–93, lamented the *'indiscriminate* slaughter' of crabeaters and wrote of his shipmates 'painting the snow vermillion'. Louis Bernacchi, the Australian scientist accompanying Borchgrevink's Antarctic expedition of 1898–1900, described 'dyeing the dazzling immaculate white of the ice-floe with glaring crimson pools of blood'.[5]

The killing of seals and whales was both a cause and product of discovery, and the alliance was breathtakingly brutal. From the late eighteenth century until the early 1820s there was a period of sustained slaughter first of southern fur seals, and increasingly of elephant seals. On Kerguelen Island, more than a million fur seal pelts were taken in the final decades of the eighteenth century, following its discovery in 1772. On Macquarie Island, for example, there was intensive fur sealing from discovery in 1810 to 1813 (perhaps killing 50 000 a year), and by 1815 it was no longer profitable and by 1822 there were no fur seals left.[6] The archipelago of the South Shetlands was discovered in 1819 and sealed out in two summers. In 1821 there were no fewer than 30 sealing vessels anchored off these extraordinarily remote and hitherto unknown islands.[7] Heard Island was discovered quite late in 1853, and intense killing of sea elephants took place in 1857–59. By 1874, when the British oceanographic ship *Challenger* visited Heard, there were still 40 sealers on the island and the expedition naturalist, Henry Moseley, found himself walking amongst thousands of elephant seal skeletons and tusks washed into curved lines by the tide.[8]

The extension of commercial whaling into far southern seas in the late nineteenth century opened the heroic era of Antarctic exploration. By 1904, the famed Norwegian whaler and explorer, Carl Larsen, had established Grytviken, the first coastal whaling station on the island of South Georgia. Its name meant 'Pot Cove' and commemorated the fact that the sealers had been there almost a hundred years earlier and that their rusting try pots remained. Within ten years of Grytviken's foundation, the humpback whale population was commercially exhausted. In the 1920s, Larsen pioneered the use of factory ships big enough to process a whole whale on deck, and this invention (which freed a ship from the need for a shore station) opened up hunting in the unregulated and remote seas close to the Antarctic coast. The 1930–31 summer became the season of the greatest number of vessels ever operating in the Southern Ocean: 41 factory ships and 232 whale catchers manned by 11 000 men as well as uncounted transport vessels. In 1937–38, this phase of pelagic or deep-sea whaling resulted in the killing of 46 039 whales in the Southern Ocean, almost 90 per cent of the worldwide total taken that year.[9] As an older man, Robert Cushman Murphy was shocked by that season. In an article entitled 'Slaughter threatens the end of whales', he argued that 'the number of whales being slain is at least fourfold what the oceans can endure on a long-term basis'.[10] He was right. Pursuing a familiar pattern, the peak was followed by a collapse in the industry due to massive over-exploitation.[11]

So Bob Murphy arrived at Grytviken in 1912 with humpback whaling in full swing and on the brink of decline. In the summer before last – the season of 1910–11 – 6175 humpback whales were killed off South

Georgia, and most of the skeletons that Murphy saw on the beaches were those of humpbacks slaughtered during the previous six years. In just a few hours, Bob watched at least 15 humpback whales being drawn up the slip and flensed. Even Captain Cleveland was 'nothing short of goggle-eyed over the big-scale butchery of modern whaling'. New regulations had been introduced, and Bob recorded his own captain's disgust that 'law and regulation have reached an island that was formerly the Old Man's personal property', and that there were now rules about what size and sex of sea elephants he could kill. Whales which had previously been gutted for their blubber oil and dumped now had to be completely utilised for all the secondary products: soap, perfumes and whalebone. Elephant seals helped keep Grytviken going (more than half of the world's southern elephant seals lived in South Georgia), and once the humpbacks were exhausted, other whales such as the blue, sei, and fin were more intensively hunted. Bob himself found humpback whale meat surprisingly palatable and much superior to sperm whale.

On the day he watched and counted those 15 whales being processed, Bob was invited to lunch with Captain Larsen. The symbiotic relationship between whaling, exploration and science was perhaps best exemplified in the career of this famed Norwegian. Carl Anton Larsen was a whaler with a scientific bent who had also officially been an explorer. He had captained two Norwegian expeditions on the *Jason* in 1892–94 to open up southern whaling grounds, and he had also been a member of Otto Nordenskjöld's Swedish Antarctic Expedition of 1901–04, as captain of their ship, the *Antarctic*. On each of these trips south, Larsen had conducted important science. In the early 1890s, he had brought back fossil vertebrates and a silicified tree from Seymour Island, some of the earliest evidence of Antarctica's warmer ancient history. In the early 1900s he participated in the Swedish party's scientific program, especially during an enforced winter on Paulet Island. Larsen also used the Swedish scientific expedition to make some commercial catches (elephant seals and leopard seals), to explore South Georgia, and to launch his plan for a whaling station at Grytviken.

Captain Larsen had once been a whaler hitching a ride on a scientific expedition, and Bob Murphy was a scientist shoehorned onto a whaling brig. They had things to talk about.

At Larsen's residence on remote South Georgia that day in November 1912, Bob was served an eight-course meal by a butler, and at the end was offered such an excellent brand of Havana cigars that he almost regretted he didn't smoke. 'It calls for a good constitution to endure this Antarctic fare – otherwise a man could die of gout', he confessed to Grace. Larsen's residence featured luxuriant palms and blossoming plants, a billiard room, a salon with a piano, a conservatory, singing canaries, several portraits of the King of Norway and 'the only woman in all South Georgia' in the garden hanging up clothes. 'You must be careful not to repeat all I tell you about the luxury of Cumberland Bay', he begged Grace, 'because, after all, a voyage to the Antarctic is supposed to be a hardship rather than a garden party!' Murphy's 'illusions of the rude Antarctic were shattered'. He had glimpsed the seat of an empire.

Grytviken, in spite of its acrid, whaley stench and graveyard of bones, was an astonishing outpost of civilisation in the Southern Ocean. This township of wooden houses, dwarfed by mountain, glacier and fjord, was the largest of several whaling stations that drew a summer population of 2000 to South Georgia.[12] Bob found that there was quite a round of visiting and dining required at Grytviken, 'a polite necessity' even on this treeless, windy island half-covered with permanent ice and snow, with glacier snouts at sea level. This was a place where the British magistrate would put on a dinner jacket in honour of a royal birthday. As well as being an international community and a centre of industry, Grytviken became a crucial lifeline to Antarctic expeditions. In 1912 the settlement boasted a whaleslip (for winching the whale out of the water), an oil factory, docks, a marine railway, dwellings, dormitories for 200 men, carpentry and cooperage shops, metal workers' forges and machine shops, cattle and poultry shelters, a powerhouse for electric lights and telephone system, a library, a chapel, an infirmary, and – the sight that so gladdened Bob's heart – a fully-fledged Empire Post Office! He could send letters to Grace, and could have received them if only he had known that he would be travelling to such a busy ocean! He returned to his mailbag stowed on board and consoled himself with another letter written before he left. One letter from Grace he never opened was the one he always wore around his neck in an oilskin folder. It contained the message he was to read only if it seemed certain that he would never come back to her.

But one day, when Bob was skiing in the mountains with the doctor and company secretary, they heard a distant, immobilising whistle: it was the mail steamer from Buenos Aires! It brought 16 bags of mail. When back at the ship Bob heard there was indeed a letter for him and he 'jumped ashore among the stinking whale skeletons and ran all the way to the magistrate's home'. It was the only letter he received from Grace during his whole time away (of the scores she wrote to every port on both sides of the Atlantic). There was also news from the rest of the world: Woodrow Wilson had been elected president (Bob, a Democrat, whooped with triumph, while his captain lamented). Ships were expected to pass through the Panama Canal the following year, a Zeppelin had just flown over the Baltic Sea to Sweden, and 106 children had been killed in roller skating accidents in New York City alone in the past year. But it was the 'precarious conditions in the Old World generally' that stirred most discussion. There was conflict in the Balkan States and some commentators forecast a general European war. In early March 1913, just as Murphy was preparing to leave South Georgia, British whaling captains brought news of Scott's death at the pole. Murphy particularly mourned the loss of Edward Wilson, 'the best naturalist who ever worked in the Antarctic'.

The day after his dinner with the captain, Bob took up Larsen's invitation to join a whale-chaser on the hunt. The *Fortuna* steamed for four hours to the best whaling grounds north-east of the island. In these longitudes, the convergence arches north and embraces South Georgia in Antarctic waters, and where the cold currents eddy in the lee of the island is a profusion of krill (*Euphausia superba*), the principal food of the whales. Krill is the foundation of the Antarctic food web and it masses in such profusion that it can turn the ocean pink. The largest congregations are to be found near South Georgia, the South Shetland Islands, and west of the South Sandwich Islands. In the 1980s, the North American biologist, David Campbell, worked for three summers in the South Shetland archipelago just off the coast of the Antarctic Peninsula, and he especially studied krill. The one short link in the food chain between a diatom (a one-celled alga, 'the pasture of the ocean') and a hundred-ton whale, explains Campbell, is the Antarctic krill, and a whale eats three to four tons of it a day.[13] There is a greater biomass of krill than any other species of animal on Earth: a single 'super swarm' observed in 1981 had a surface area of about 150 square kilometres (58 square miles) and was at least 200 metres (256 feet) deep. This profligacy is typical of the Southern Ocean whose biological richness is expressed more in the quantity of individuals than in diversity of species. The krill migrates through vertical movement in the layered ocean, hitching a ride on a southerly current by sinking, or a northerly one by rising.[14]

The *Fortuna* found the krill, and there, too, were the humpback and blue whales feasting, and Murphy watched wide-eyed as three men loaded the whale gun and launched harpoons weighing 100 pounds (45 kilograms) into these gentle giants of the ocean, creatures of 'sheer beauty, symmetry, utter perfection and form of movement', as he described them to Grace. The whales were winched to the surface and pumped full of air so that they floated like balloons beside the *Fortuna*, dwarfing the hull of the boat. 'The process looked more than ever like murder with ease and no trace of uncertainty', he reflected. At least on his own brig, there was the semblance of an old-fashioned fight between human and animal muscle. He now knew he had sailed south in an 'anachronism'. So steeped was the *Daisy* in the traditions of Yankee whaling that, on the voyage home, they cleansed the decks with a barrel of their collective urine, plus some blubber ash and elbow grease. The historian of South Georgia, Robert Headland, declared this voyage of the *Daisy* to be 'the last voyage of the first epoch of South Georgia sealing', an era that had begun shortly after Captain Cook's visit in 1775.[15]

Bob Murphy joined a New England whaling brig in the final days of that industry, saw South Georgia shore whaling at its height and was welcomed into the 'family' of the 'king of whaling', Captain Larsen. What his companions called an 'industry' he insisted on calling 'exploitation'. His little ship sailed in an acre of blood every time a newly caught whale was flensed. Like many naturalists, he found a necessary alliance with the chief exploiters of his wildlife. He took home a herbarium of South Georgian vegetation, a collection of invertebrates and fishes, rock samples, seal skulls and whale embryos, and more than a hundred bird skins. What he observed was in some senses remote and 'pristine', yet he was made keenly aware that he was actually very late on the scene. And, true to Werner Heisenberg's uncertainty principle which was enunciated in the following decade, he found that the observer could not examine his subject without changing it. Murphy was hard pressed to study the elephant seals before his shipmates slaughtered them. His own captain disregarded provisions made to conserve seals on South Georgia and doggedly avoided the payment of dues on seal oil.[16] Their tally of elephant seals taken during the South Georgia visit was 1641, a total that made Bob exclaim: 'I hope with all my heart that no sealer from the United States will ever trouble these shores again.' Murphy went on to combine the fields of oceanography and ornithology and to write *Oceanic Birds of South America*. His most prominent conservation campaign was against the spread of aerial spraying of chemicals in the United States in the 1950s (to control populations of the gypsy moth). In *Silent Spring*, Rachel Carson described how a group of Long Island citizens 'led by the world-famous ornithologist Robert Cushman Murphy' sought a court injunction to prevent the United States Agriculture Department from spraying DDT over 3 million acres in 1957. The suit was unsuccessful but did focus public attention on the deleterious effect of DDT on birds, animals and people, at the same time as noting that it had no long-term impact on the gypsy moth population.[17] Murphy had become a scientist with public clout. Grace had been right: his year on the *Daisy* did make his career.

The voyage also made his marriage, or so they persuaded themselves. When Bob returned to Grace after a year of yearning, they both were dazed by the immediacy and ubiquity of one another. After waiting so long to speak, 'our tongues were tied', recalled Grace. 'Something of the mystery of that strange year has held us two together', she wrote later in her account of their lives. 'It greatly added to our problems, yet forever made our union.' Writing in 1952, when they were both past their sixtieth birthdays, Grace Barstow Murphy reflected on her role in propelling Bob south. It was her first investment in his career, which would become her absorption also. He reaped the consequences, as did she, in a life with a man never forced to choose between his science and his family. She made herself share his interests and sometimes – later – even his fieldtrips. Grace refused to die before she saw an albatross in flight. She was eventually to accompany Bob on a tough expedition to another sub-Antarctic island and finally, in 1970, for a day on South Georgia on a tourist cruise. But she was complicit in his single-mindedness of intellectual purpose and, although she shared his pride in his work, she admitted that she had lost herself once or twice along the way through the effort of 'somehow or other keeping up a companionship when there is often almost nothing to keep it up on'. Once she stared into 'the black pit of a nervous breakdown' because of their extreme difference in temperament. Her feminism was one of universal gender roles and separate spheres: women were uniquely committed to humanity, she argued, and men of science needed to be free of people; so men needed their women as bridges to reality. Women were different but equal, Grace insisted, and she looked forward to the world's release from 'the historic domination of the male'. Yet she argued for female subordination, too, in the service of a greater whole that only women could perceive. She wrote, 'No matter how I bruised myself against the wall that his science built, so very cold to live with when one is warm and hungry, I knew my role in growing clarity.' His was 'the genius's temperament' that it was her duty – no, honour – to cultivate and protect. His science was the salve to her soul. Her love would be distilled into the public benefit of his work. Subversively, she saw the most important aspect of Bob's work to be not his esoteric zoology, but his contribution to understanding the world's food supplies. It was almost as if she was the master of the whaling brig on which Bob had been conceded a place.

Three years after Bob Murphy left South Georgia on the *Daisy*, Sir Ernest Shackleton, Frank Worsley and Tom Crean staggered down from the icy spine of the island into Stromness whaling station. As they proceeded into the settlement, two boys bolted and an old man hurried away from the frightening sight of them. When they were finally led to the manager of the station, Mr Sørlle, Shackleton challenged him with the words: 'Don't you know me?'

'I know your voice', replied the manager. 'You're the mate of the *Daisy*.' He had mistaken Shackleton's Irish lilt for that of Robert Cushman Murphy.

Shackleton and his men had left England in August 1914 on the eve of war, and war had haunted them throughout their trials in the Weddell Sea. In quest of adventure at the bottom of the world, they were embarrassed to find that, suddenly, the prime site of heroics was in their own backyard. Shackleton had offered his ship and men to the Admiralty for immediate war service, but was ordered to proceed south. Nevertheless, the expedition was criticised for doing so, and Shackleton was conscious of the dismay felt in Britain that such a fine body of men should leave their country in its hour of need. It is hard to resist the conclusion that the war and the timing of his departure pushed Shackleton to his own heroics on the ice. He certainly could not afford to be unadventurous, he could not be seen to lack daring or hardship, he could not play safe with the ice as the Norwegian whalers had urged him to do. When Shackleton stumbled into Stromness on that day in May 1916, he begged Mr Sørlle: 'Tell me, when was the war over?' And the crushing answer was: 'The war is not over. Millions are being killed. Europe is mad. The world is mad.' Shackleton knew he was henceforth condemned to write and speak, in the same breath, of the red warfare of Europe and the 'White Warfare of the South'. He celebrated the 'generous self-sacrifice' of his men in Antarctica, and he boasted of how, on their return, they had rushed to enlist and to serve on the western front and that, as a consequence, 'the percentage of casualties amongst members of this Expedition is high'.[18] Shackleton's photographer, the Australian Frank Hurley, reflected: 'Emerged from a war with Nature, we were destined to take our places in a war of nations.'

As the three men came over the mountain into Stromness to learn of endless war, the first human sound they heard was a factory whistle. They had struggled back from wilderness and privation and now suddenly found themselves amidst modernity. Standing on the edge of an industrial plant, they had nothing but an adze, a saucepan, Worsley's log of their journey from Elephant Island and the wet clothes they stood up in. They paused 'to straighten themselves out a bit', not just out of embarrassment at their dirtiness but also sensing the gulf between their desperation and the workaday world they were about to enter. The heroic era was about to tangle again with the commercial imperialism that was Antarctica's backyard. Worsley produced some safety pins and fumbled some repairs to the rips in his clothing. They were, he said, 'ragged, filthy and evil-smelling' and 'a terrible looking trio of scarecrows'. They felt like 'three Rip Van Winkles'. They had been outside time and had awoken in a different world. Their own endeavour seemed archaic, foolhardy, even anti-modernist in its reliance on individual character, honour and sacrifice. But the Norwegian whalers at South Georgia in 1916 knew a legend when they heard it. When they recovered the *James Caird* and listened to the story of its voyage, the whalers solemnly hauled the little boat up the wharf at Stromness and each came forward in turn to shake the hands of Shackleton and his men.[19]

## Monday, 23 December

**58 degrees south,**
**119 degrees east**

Earlier today I stood on the deck in light snow and
snuggled in my black jacket against the cold. I watched as
a white-chinned petrel sailed the quickening wind at the
stern of the ship. Albatrosses have been following us for
days, dipping in and out of the waves. Our progress has
slowed a bit because we've been heading into a 4–5 metre
(13–16 foot) swell and, last night, the wind upgraded to a
'Severe Gale' (that's 45 knots). I spoke to the chief officer
this morning and he's clearly a bit worried about how we'll
get through the pack ice. He told me that the *Polar Bird*
was used for the filming of the Ernest Shackleton movie,
so I can tell you that Kenneth Branagh smokes but is
careful of what he does with his cigarette butts, and he also
willingly helps with the washing up. The English insisted
on having their own cook on board! So they brought an
English chef with them to replace the excellent Norwegian
and Philippine cooking we are enjoying.

On board the *Aurora* with Ernest Shackleton in
1916, Fred Middleton took a little while to find his sea
legs. The ocean swell frequently sent him running to the
ship's rails where he paid his 'homage to Father Neptune'.
'The wonder is that I've been eating any meals at all', the
young doctor wrote in his diary. 'I am rather proud of my
seamanship, considering our poor old wireless operator
is still in his bunk and has hardly eaten anything for the
week, while since Saturday morning, I've been eating my
two eggs for breakfast, boiled mutton and vegetables for
dinner, and scrambled eggs or steak and onions for tea,

as well as afternoon tea'. Soon he settled in and remained ravenous. In the evenings Fred played cards, tenderly tutored by Sir Ernest: 'I'm becoming quite expert under his tuition, as he takes me for his partner and I make some awful blunders, but he smiles through it all and explains patiently my misdeeds.' In the morning a steward would awaken Fred with a cup of tea and buttered biscuits in bed at 7.30 am.

I find curious comfort in *his* comfort because we are not exactly roughing it either. In spite of the swell, I find myself lining up at the servery each day at 8, 12 and 6 with my plate held out. The challenge of modern Antarctic exploration is to remember, amidst all our technology, how fine is the envelope of safety. Hence our training session in the hold of the ship today on how to cope with falling down a crevasse is not only practical, it's also psychological. It's designed to tune our sense of danger. I try to imagine the isolation of our tiny ship in this heaving sea. I certainly have a new respect for the size of our globe. We have already been voyaging long enough to wonder if we are sailing off the end of the Earth and have seen the last of land.

# GREAT SOUTH LANDS
*Reading the rocks*

Was there a Great South Land? That geographical question stimulated exploration of high southern latitudes for 500 years, from the 1400s to the 1900s. From the earliest classical cartographies, belief in the existence of a 'Terra Australis' was tenacious. The very words, wrote one historian, are 'aromatic with old romance'.[1] Every fragment of land encountered by navigators in southern seas was hopefully sketched in as the northern promontory of the long-awaited continent, the tip of the iceberg, as it were. Classical cosmography combined with explorers' expectations to make a virtual southern land. There had to be such a continent down there to balance the geography of the world. The theory of a spherical Earth, championed by Pomponius Mela and Claudius Ptolemy in the first centuries AD, required a southern landmass to match Europe, Asia and Africa in the north. Cartographers elaborated this expectation during the Renaissance and projected tremendous expanses of southern land on the basis of Marco Polo's confusing reports of distant Asian gold, elephants and porcelain shells. Captain James Cook's eloquent biographer, John Cawte Beaglehole, observed that 'geographers no less than nature seem to have abhorred a vacuum'.[2]

Historian Stephen Pyne has mused that exploration merely transformed Antarctica from an unknown white space on the map to a known white space. Curiously, now that Antarctica's coastlines have been drawn and circumscribed, it often falls off the map; although certainly not in Chile or Argentina where it is confidently regarded as domestic space. Lexicographer and historian, Bernadette Hince, has tried in vain to get Qantas airlines to show Antarctica on its inflight-magazine maps, even though its jumbos take tourists on Antarctic flyovers. The Arctic islands are shown in detail, but there is nothing but sea in the south. Greenland is presented in all its glory, magnified by the Mercator projection, but the icy southern continent is completely absent. The North American Antarctic scientist Bill Green owns a globe of the world which has a plate at its base that obliterates everything south of 75 degrees latitude. The plate reads: 'Replace with 15-watt, 120-volt lamp only.'[3]

In spite of these absences, Australia and Antarctica – once linked in deep geological time – were generally united in the geographical imagination. They were both sources of cartographic disappointment, for neither satisfied the expectation of a vast, rich continent spanning the southern hemisphere. Inexorable plate tectonics may have separated Australia and Antarctica 45 million years ago but it was only within the last few hundred years that Abel Tasman and James Cook isolated and defined them as separate lands for European science, politics and strategy.

Cook was renowned as 'the negative navigator', for he 'erased whole continents of a hypothetical geography'. As historian Glyn Williams memorably puts it, 'A study of English enterprise in the South Sea is, to some extent, a study in credulity.'[4] Cook's achievement, allegedly, was to disprove the existence of a Great South Land.[5] That is the way history often records it. But that's a peculiarly northern hemisphere perspective, one that dismisses with a sentence the significance of the two vast southern lands he circumscribed, the two continents whose boundaries he helped to define. Cook had voyaged distastefully through seas 'pestered with ice' and amongst icebergs that left his 'mind filled with horror'. He had circumnavigated the south polar regions and had shown that there was no room for the *Terra Australis* of the classical and medieval cartographers. Yet from his vantage point as the first human to cross the Antarctic Circle, Cook gazed further south with both foreboding and intuition. He was convinced by the character of the ice that there was a nucleus of land at the pole: 'the excessive cold, the many islands, and the vast floats of ice, all tend to prove that there must be land to the south', he wrote. 'If any one go further south than I have been', declared Cook, 'I shall not envy him the honour of the discovery'.[6]

It was not the climax to the expedition that Joseph Banks had imagined. Flushed from the triumph of his earlier voyage around the world in the *Endeavour*, and now a lion of London society, Banks had, for a while, expected to accompany Cook on his second voyage. 'O how Glorious would it be to set my heel upon the Pole!' he wrote to his French friend the Comte de Lauraguais. 'And turn myself round 360 degrees in a second'.[7] Thus encompassing latitude and longitude in a single sweep.

The early voyagers hungered for rock, and for a continent to claim. In January 1840, the ships of the French expedition of Dumont d'Urville and the United States exploring expedition of Charles Wilkes were nosing around the same stretch of Antarctic coastline, each trying to distinguish land from clouds, and land from ice, and land from islands. The visual phenomenon of 'looming' brought distant sights deceptively close, hopes were lifted by the appearance of kelp or albatrosses or particularly large bergs or a change in the colour of the water, big black clouds 'looked like raised land', discoloured icebergs were closely scrutinised, the sailors were 'dying to see' the blackish colour of earth, and having found earth, they wanted to land on it and collect it and claim it and souvenir it. With picks and hammers they secured fragments of 'the mineral realm', loading their pockets with rocks – real earth from the world of ice – and they carried off penguins as trophies, and even emptied the birds' crops to get the pebbles within. Rocks were more than of scientific interest; they were crucial geopolitical footholds. And here at the end of the Earth in 1840, national rivalry remained keen. The French and American expeditions encountered one another but suffered a ridiculous misunderstanding. As the ships manouevred together, both in the hope of communication, sails were rapidly hoisted aloft. Wilkes misinterpreted the French intentions, thought he was being avoided and snubbed, and suddenly sailed off in a huff. The two captains never met again, but traded justifications and frustrations in their journals.[8]

In 1901, the two Great South Lands again shared world attention. The Australian colonies had just federated and formed a nation, the Commonwealth of Australia. And the International Geographical Congress had proclaimed 1901 as 'Antarctica' year. The heroic age of Antarctic exploration was underway, and British, German, Swedish and Scottish expeditions started for the ice. As early as the 1880s, Australian colonists had turned an imperial gaze southwards. GS Griffiths, one of the organisers of an Antarctic Exploration Committee formed in Melbourne in 1886, argued that 'the exploration of these regions is a task which, by our geographical position and our wealth, is thrown on Australia as a duty which we cannot evade if we have any adequate conception of our great position in the southern seas'.[9] Exploring Antarctica was Australia's duty, Australia's 'preserve', Australia's destiny.

In that year, 1901, the geologist JW Gregory (then living in Australia) considered the primary geographical problem of Antarctica still to be 'whether the hypothetical "Terra Australis" has any existence at the present day'.[10] Opinions were divided on this question, he reported: Is there a great southern polar continent, or only a number of comparatively small and widely scattered islands? In 1893, John Murray, on the basis of the evidence of the great Southern Ocean voyages of the *Challenger* (1872–76), argued for the existence of a southern continent.[11] But the great Norwegian Arctic explorer, Fridtjof Nansen, doubted the existence of a continuous continent in the far south. Sir Clements Markham, president of the Royal Geographical Society in London, told Douglas Mawson in 1911 that he believed that he would find not continuous rocky land, but only islands cemented by ice accumulations. It was Mawson's pride that perhaps the most important outcome of his 1929–31 expedition to Antarctica was to confirm 'the limit of a continental shelf throughout the sector' meaning that 'the presence of a real continent within the ice has been finally established'. Thus, reflected Mawson in 1935, 'the continent within the ice, conjectured by Cook, was not confirmed until recent times'.[12]

But was it one continent or two? Was the Ross Sea connected to the Weddell Sea, thus dividing Antarctica into two land masses? This was one of the geographical questions driving the North American expeditions led by Richard Byrd in the interwar period. But Byrd's geologist, Laurence McKinley Gould, found himself stranded from the rock he so much wanted to study because Byrd's base, called 'Little America', was established on the coastal edge of the Ross Ice Shelf. Gould had to undertake the longest scientific dog-sledging expedition in Antarctic history just to hold sedimentary rock in his hand. Their party had sailed to the Antarctic ice shelf in late 1928 and the following spring they learned of the Wall Street stockmarket crash on their weekly radio news. When Gould began his geological sledge journey in the summer of 1929–30, he had already lived and worked on floating ice for a year, and his journey to the Queen Maud Mountains (Roald Amundsen's route to the pole) was across floating ice the whole way. Now, finally, Gould was to have earth under his feet. He declared: 'It was a moment for which we had been living. We were camped in the very shadow of our long sought mountains and *we had reached land!*'[13] The young huskies had never before seen rock and were wary of it. And the first rock Gould picked up there was sandstone. His long journey was justified; he could have turned back then a satisfied man. He stood transfixed by the fragment of rock held reverently in the palm of his hand. 'No symphony I have ever heard, no work of art before which I have stood in awe ever gave me quite the thrill that I had when I reached out after that strenuous climb and picked up a piece of rock to find it sandstone. It was just the rock I had come all the way to Antarctica to find.'[14] He had been fearful that the rock would be volcanic, formed hot and whole in a moment and with little of its history embedded within it. But the fragment of yellowish-gray sandstone he held in his hand was sedimentary, historical and potentially fossil-bearing. It had been laid down over aeons from the time of the great armoured fishes to the age of the dinosaurs. The rock was similar in kind to those found by Robert Falcon Scott's *Discovery* party in 1901–04 on the western edge of the Ross Ice Shelf. Amundsen's mountains of triumph and Scott's mountains of trial were geologically one. Here was a rock that proved that the Transantarctics were great upthrusts of the Earth's crust, 'the most stupendous fault block mountain system in all the world', and closely related to mountain systems in India, South Africa, Australia and South America.[15]

Gould concluded that the Ross and Weddell seas were almost certainly not connected. The North American expedition of 1946–47, led by Finn Ronne, also wished to settle the question of whether or not Antarctica was two land masses. However it was not until the International Geophysical Year of 1957–58 that the unity of land under the ice of East Antarctica was mapped, and the Great South Land truly found.

The Great South Land was to be even greater than the eighteenth-century voyagers imagined. For under the ice lay the key to Gondwana. Those rocks hauled to the end by Scott's doomed polar party, with their fossilised leaves of southern beech trees, were clues to that mystery. The jigsaw fit of continental coastlines had been noted in the sixteenth century, and the idea of a southern supercontinent, with Antarctica at its centre, had existed in sustained scientific form since the late nineteenth century. But for most of the twentieth century, scientists rejected the geological theory that would bring Antarctica into the centre of Earth history. Could separate lands have once been fused? Did the Earth move for them?

When Alexander von Humboldt first experienced an earthquake in South America at the end of the eighteenth century, he noted that it

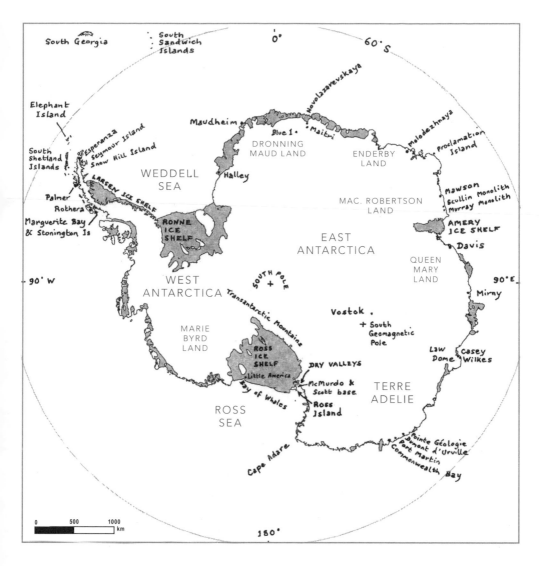

*Places and Features*

was not an undulation, but a positive movement upward and downward, and it was the novelty rather than the fear of danger that made it so unforgettable. 'It is like waking up', he wrote. 'We feel that we have been deceived by the apparent stability of nature; we mistrust for the first time the soil we have so long trod with confidence.'[16] Several decades later, in February 1835 while on the voyage of the *Beagle*, Charles Darwin had a very similar South American experience, and he, too, experienced an intellectual aftershock. He considered that 'A bad earthquake at once destroys the oldest associations; the world, the very emblem of all that is solid, had moved beneath our feet like a crust over a fluid; one second of time has created in the mind a strange idea of insecurity, which hours of reflection would not have produced.'[17] Was it possible that the movement of the earth had shaped the geography of life, that geology could determine biology?

The closest kin to Australian flora and fauna, alive or fossilised, were found to be in South America, South Africa or Antarctica. How could they be related and yet so dispersed? In the late nineteenth century, scientists had begun to speculate about the ancient existence of a southern supercontinent, and therefore about Australia's Antarctic associations. In the 1890s, the naturalist William Saville-Kent, drawing on the writings of Darwin, Thomas Huxley, Alfred Russel Wallace and others, suggested that 'this insular territory' of Australia actually seemed to be a fragment of a pre-existing Antarctic continent.[18] Similarly, the discovery of plant and vertebrate fossils near the Antarctic Peninsula by Carl Larsen in 1892–94 and Otto Nordenskjöld in 1902 suggested not only that Antarctica had once been warmer, but also that it had connections with other landmasses. In December 1902 Nordenskjöld speculated in his diary on the common geological and biological history of the southern continents and the possible existence of Gondwanaland. Meanwhile, another stranded fragment of his expedition, wintering at Hope Bay on the Antarctic Peninsula, gained some consolation from the fossiliferous slabs of shale they had discovered. Throughout the grim winter they would comfort themselves by examining these exciting finds by the light of the blubber burner.[19]

These surprising biogeographic affiliations of far-flung lands were explained by the supposed rise and fall of ocean beds and the existence of land bridges that once existed between continents. Charles Laseron, who had accompanied Mawson to Antarctica in 1911–14, looked across the Southern Ocean from Australia in 1953 and reflected that 'somewhere beneath this expanse of water lies the lost continent of Gondwanaland'. Two hundred million years before, in Permian times, he imagined a vista of endless land: 'Australia then stretched far to the south … [and] high mountains towered over what is now sea.'[20] To many scientists, continental fluctuations in altitude seemed much more likely than their lateral movement across latitude.

In the early 1930s GE Nicholls, the professor of biology at the University of Western Australia, reflected on the 'land bridges' theory in his presidential address to the Royal Zoological Society of New South Wales.[21] To restore the land connections between the continental cousins of Australia and Antarctica, Nicholls argued, would require 'the upheaval of a vast southern continent, (much of it requiring an elevation of at least two thousand fathoms [3.7 kilometres]) and would raise the south polar highlands to heights comparable with the higher peaks of Himalayas or Andes'. And what of Australian fauna and flora with strongly marked South African affinities? 'Another land bridge, and a very extensive one, is wanted here, for [it] … must occupy approximately the area now of the Indian Ocean.' This imagined land was the long lost continent of Lemuria, so named because it might explain the puzzling distribution of lemurs. The earth scientists could not find a mechanism to match the evidence and imagination of the biologists, who were demanding more land bridges than the geophysicists could supply. Nicholls found that Alfred Wegener's Theory of Drifting Continents continued to offer 'a hopeful solution'. Wegener (1880–1930), who was also an Arctic explorer, imagined continents as vast icefloes, 'shedding fragments which fall behind, exactly as a drifting iceberg might ground and calve and drift on again'.[22]

The mutability of the globe that we now accept, and that underpins the new narratives of ancient Australia and Antarctica, was scientific heresy less than a human lifetime ago. American geochemist Bill Green reflects that the new geological story is 'so recent in its writing that it can almost be called contemporary fiction, were it not surely science'.[23] Even those few people who had championed the idea of 'continental drift' from early in the twentieth century had not been able to suggest the means by which landmasses moved around the globe. Wegener had been rebuffed and ridiculed when he suggested that such awesome mobility might be credited to the forces unleashed by the rotation of the Earth or by gravitational forces from the sun and moon. He was himself dissatisfied with these explanations and in 1929 admitted that 'The Newton of drift theory has not yet appeared.'[24] Like Darwin and his theory of the origin of species, the explanatory power of Wegener's idea could not yet be supported by a satisfactory mechanism. Therefore, by the 1950s even sympathetic scientists acknowledged that there was a significant decline in support for the continental drift theory.[25] These were the years when the North American geologist and writer John McPhee was a graduate student and he recalled that 'nearly all the faculty at Princeton thought continental drift was sheer baloney'.[26] But persuasive field evidence continued to overwhelm a sceptical physics.

Plate tectonics, a theory which was developed in the late 1960s and early 1970s, was a revelation arising from palaeomagnetics and deep ocean exploration.[27] World War 2 refined techniques of deep ocean mapping, and the Cold War intensified the collection of seismological data in order to monitor the tremors of nuclear testing. Mountain ranges, it was found, ran like seams through every ocean, and a high percentage of earthquakes were occurring along their submerged spines. New geological dating techniques enabled the discovery that the seafloor is younger than the continental rocks and is being continually transformed. Harry Hess, a North American mineralogist and war-time naval officer generated another geological by-product of war.

He had mapped the ocean floors he voyaged in battle, and by 1960 had decided they were geologically 'ephemeral'. He published his account of the 'History of ocean basins' as 'an essay in geopoetry'.[28]

Palaeomagnetism, an early twentieth-century discovery, dramatically supported the new synthesis. The magnetic poles of the Earth wander and periodically reverse, and rocks of different ages record these changes at the time they were formed. And because compass needles lie flat at the Equator and want to stand straight up on end at the poles, rocks can reveal not only the whereabouts of the magnetic pole but also the latitude of the rock when it formed.[29] Magnetometers dragged across the seafloor in the early 1960s revealed alternating bands of magnetism. But the most striking discovery was that these stripes emanated in exactly symmetrical patterns in opposite directions from ocean ridges, as if this were an ancient scroll unfurling from the centre of the Earth.[30] Vulcanism along the submerged ocean ridges was creating new rock and pushing the seafloor out in both directions. Where continental plates are moving apart, new seafloor comes into being at the mid-ocean ridges, and where they converge, one plate plunges down into the mantle beneath and forms an ocean trench. Inexorable seafloor spreading drives continents apart. The hard rigid masses of the continental plates therefore move on slow but powerful currents of molten rock, surfing the underlying mantle. This startling insight reveals that the Earth's crust is constantly mobile and that modern, separated lands have intimate histories of the most surprising physical relationships.

Just as fossils of the fernlike plant, *Glossopteris*, hauled on sledges, had provided vital clues to Gondwana in the early 1900s, so would Antarctica provide a stunning confirmation of plate tectonics. Encouraged by fossil vertebrate finds in the Transantarctic Mountains, the North American palaeontologist Edwin Colbert found in the same area in December 1969 the fossil jawbone of a small freshwater reptile called Lystrosaurus. It lived 200 million years ago and its fossil remains had been found in abundance in Africa and other southern landmasses.

It could not have negotiated ocean crossings. As Laurence McKinley Gould wrote in 1971, 'If one had gone to Antarctica to find a specific fossil, this would have been it … No further proof was needed for the existence of Gondwanaland'.[31] Southern hemisphere geologists, and especially those who had worked in Antarctica, were early and enthusiastic converts to the new synthesis.[32]

The break-up of Antarctica and Australia 45 million years ago was, as one scientist put it, 'the switch that turned on the modern world'.[33] The calving of the continents created the circumpolar ocean current that isolated and refrigerated Antarctica, which had global climatic implications. Australia journeyed to lower latitudes and became a different sort of desert.

For millions of years Australia had a finely balanced relationship with latitude. It rafted north into warmer climes at a time in planetary history when the Earth grew cooler, thus moderating climatic change on the continent and nurturing great biodiversity.[34]

When Australia broke away from Antarctica, it was covered in rainforest dominated by the ancestors of hoop pine (*Araucaria*) and Antarctic beech (*Nothofagus*). Moving north at seven centimetres (2.8 inches) a year, the Australian continental plate began its defining journey. It has been the fastest moving continent over the last 1 per cent of time, moving at the rate of fingernail growth. The continent began to leach, dry and burn. Australia's ancient soils became degraded and impoverished and were hardly renewed or disturbed by glaciers or volcanoes. The land became more arid and the inland seas began to dry up. The sub-tropical high pressure systems of the thirsty thirties controlled the new weather. Fire became more frequent and dominant. Under the combined assault of soil degradation, aridity and fire, the greenery of Gondwana burnished into sclerophylly. The hard-leaved vegetation emerged from within the rainforest to dominate and diversify, eucalypts dramatically extended their range, casuarinas succeeded auracarias, and grasses replaced ferns, moss and fungi.[35] Just the last tenth of the 400 million year evolutionary history of the Australian flora produced 'the bush'. During Greater Australia's lonely, latitudinal drift (crossing 27 degrees), the continent became embraced by fire just as its abandoned partner, Antarctica, loitering at the pole, became overwhelmed by ice.

Although Australia has been geologically ancient and stable, it has now begun to stir. Over the last 5–10 million years there have appeared signs of a subtle but genuine tectonic restlessness. As the Australian plate rams into the Indonesian one, the continent is shivering, as it did in the coastal city of Newcastle in 1989, building stress and even tilting as it drives under the northern plate. The southern end of the seesaw, the Nullarbor Plain, once under the sea, now ends in southern cliffs 280 metres (918 feet) high. And as Australia ploughed northwards, opening up the Southern Ocean and then squeezing northern seaways, it restricted the flow of warm tropical waters into the Indian Ocean, cooling it, making the African continent drier, and perhaps having a decisive effect on the evolution of early humans.[36]

The science and politics of 'rainforests' provide an interesting example of these changing concepts of southern time and space. In Australia, rainforest was formerly seen to be an exotic element, an un-Australian feature, alien to the image of the wide brown land dominated by eucalypts. So rainforest was characterised as recent and invasive, as outside, as imperial, as northern, as equatorial. It was described as a 'Malayan element', as 'oriental' in character, as 'Indian in its density and massiveness', and as infiltrating the continent via Torres Strait and Cape York. But in the 1970s, in time to empower political activism in defence of rainforests, there was a major scientific paradigm shift, partly due to plate tectonics. It was recognised that rainforest had once been the dominant vegetation type across the continent, that it had an ancient Gondwanan lineage. What was left were precious indigenous remnants, not recent Indo-Malayan 'invaders'. Rainforests were acclaimed as 'living fossils', as our 'green dinosaurs', and conservationists celebrated the glory of 'Antarctic beech'.

Discovery of this 'geopoetry' introduced a revolutionary southern narrative and reunited the Great South Lands. The great Australian ecologist John Calaby put it this way in 1984, with a wry smile playing around his words:

> Once upon a time, about 30 years ago, when nearly all wisdom resided in the Northern Hemisphere, there were certain revealed truths about Australia. For example, in the beginning the continent was devoid of land vertebrates and their ancestors arrived in waves from elsewhere, notably the Northern Hemisphere. It was dogma also that Australia's more distinctive vertebrates such as the marsupials, were physiologically and ecologically inferior to their counterparts in the Northern Hemisphere ... 'Down under' was a good place to banish ... second-class animals.

> Nevertheless there were some simple souls in Australia and elsewhere who studied certain arthropods or marsupials in the Southern Hemisphere and had the temerity to believe in southern origins for the 'old Antarctic' elements in the fauna ...[37]

Plate tectonics literally undermined Australia's history of original isolation. It revealed that the island continent only became a separate entity in the recent geological past and that most of the country's fossil history is cosmopolitan.[38] It focused the attention of Australians on a geological genesis in the southern hemisphere, followed by a relatively brief, formative journey north.

Thus Australians came to realise quite late that their land came from higher latitudes and was not an appendix to the north but a broken heart of the south. Antarctica was not only Australia's destiny, but also its ancient past.

## Tuesday, 24 December

**62 degrees south,**
**113 degrees east**

We are sailing through the ice! It's Christmas Eve, we are in the 60s south, and there are ice floes constantly on our starboard side. These are like the spilt, crumbling contents of an ice chest, humble emissaries, the highest a metre or so above the water, they live on its surface and undulate with it. They emerge from the mist, outliers of a giant floe, and we navigate courteously around them. Our long straight course across the Southern Ocean has begun to weave. Edward Wilson of Robert Falcon Scott's expedition described the movement of a swell in pack ice as constant and gentle, 'like a breathing in sleep'.

And then you see them. The icebergs. They loom into sight, at first a craggy eminence in the mist and then a faceted jewel revealing its edges and cliffs and threatening fragility. These are great fortresses of ice; one might look like the spine of a prehistoric creature, another the crumbling ramparts of a castle. Some are like servings of meringue, vast delectable concoctions of beaten egg-white. Bernadette Hince, whose *Antarctic Dictionary* should be as compulsory as the *Field Manuals*, urges Australians to call this continent 'The Big Pav'. I understand why. It is a wondrous dessert. But these great icebergs make me shiver with more than cold. In their stateliness and marbled grandeur they instil a frisson of fear because of what they come from, because of what they are fragments of. We are heading towards a giant ice-making machine and these are the most trivial of its industrial residue. They make one tremble for what may be around the corner, they are heralds of something awesome.

Fred Middleton leapt out of his bunk to see his first iceberg.

It is more usual to see the icebergs before the floes and the pack because the bergs cannot be stopped by the west wind. It is a surprise to many that in summer there is much open water within the guardian ring of ice around Antarctica. As the days lengthen, the sea ice is loosened from its moorings to the continent by the tide and then drifts out on northerly currents until it reaches the realm of the westerly winds which muster and contain it. The icebergs, however, forge on, bulldozing the wind barrier. There is a beautiful Russian word for these belts of open water within the ice: *polynyas*.

I've just checked with the captain (Sigvald, aged 43), and yes, he considers 60 degrees to be the southerly limit of the westerlies in this longitude. I asked him about the ring of ice and how he negotiates it. The satellite ice maps are useful, he says, when you can get them. But mostly, he says, 'you have to listen to your stomach'. He speaks English well, but with that strange phrasing of a second language. He means gut. A combination of a slightly out-of-date satellite photo and his stomach led Sigvald to take us east around that first tongue of ice floes we encountered today about 60 degrees. He is pleased with his decision, proud of his stomach. He likes the Australian work, even more than working for his native Norwegians. The Australians, he says, are well-organised and friendly. He is sad that the Antarctic Division's leasing of the *Polar Bird* will finish this season.

# HEAVENLY BODIES
*Space weather*

In 1605, Francis Bacon reflected that three inventions had made the modern world: printing, gunpowder and the magnetic compass. For centuries sailors had relied on compasses without knowing quite why they worked. When, from the late fifteenth century, European ships ventured into deep water sailing and began to run down the latitude in daring ocean crossings, they found that their compasses began to bear north-west instead of the usual north-east.[1] They found themselves wrestling with the problem of magnetic variation – the difference between true north and magnetic north – a variation that became most apparent if you sailed along a parallel of latitude, and which also increased with latitude because proximity to the poles sharpened the angle of difference. A careful mapping of magnetic variation, it was hoped, might even provide a way of determining longitude at sea.

In 1600 the English doctor, William Gilbert, argued that compasses worked by aligning themselves with the magnetic force of the planet. In other words, Earth itself was not only a steadily rotating clock on an axis, and not only a heat engine seething with vulcanism; it was also a magnet, with distinct magnetic poles. It was alive with magnetic force. Charged particles energised by the solar wind and caught by Earth's magnetic field animate the polar skies as auroras. This is 'space weather', crackling at the earthly poles. The brilliance and magic of the aurora australis enthralled Antarctic expeditioners: luminous curtains of colour and shafts and waves of light would leave 'a glow in the sky like the dying embers of a great fire'. 'It is impossible for one who has not seen it to even feebly understand its great beauty', exclaimed Louis Bernacchi in 1899.[2]

Earth, then, was a wilful creature, headstrong and full of spirit, and herself a source of magnetism! Gilbert's vitalist terrestrial cosmology of the seventeenth century was an attack on conventional Aristotelian natural philosophy, particularly the idea of a passive Earth in a realm separate from the heavens.[3] Earth, wrote Gilbert, is not to be 'condemned and driven into exile and cast out of all the fair order of the glorious universe, as being brute and soulless'.[4] Gilbert rejected a celestial source of magnetism. He argued for an animate Earth: it had a magnetic core, it had an internal life, a kind of soul. And he argued against it being utterly separate from the perfect heavens. He scorned the ancient Aristotelian division between the terrestrial and the heavenly. Earth herself had virtue; she was a heavenly body.

The drive to understand and map terrestrial magnetism led to an earlier 'race to the pole'. Following the location of the North Magnetic Pole by the British naval officer James Clark Ross in 1831, there was a competitive quest to locate the South Magnetic Pole in the years 1837–43. Three expeditions voyaged south, taking magnetic readings around the edge of the ice: they were led by Dumont d'Urville of France, Charles Wilkes of the American South Seas Exploring Expedition, and James Clark Ross in his British ships, the *Erebus* and *Terror*. Ross happened upon the huge embayment that came to bear his name and was thus able to voyage furthest south. He discovered a mountain of ice and fire – the volcanic Mount Erebus – but he also encountered an awesome shelf of ice towering above the masthead which he called the 'Barrier', for its ramparts stood between him and his goal. He was reluctant to acknowledge that he was butting up against the floating edge of a vast continent for it would mean giving up on his dream to sail all the way to the South Magnetic Pole. In his fine history of 'the magnetic crusade' in the south, Granville Allen Mawer explains that 'many men from many nations went in pursuit of a scientific phenomenon, only to find something rather more substantial in their way'.[5] It seemed that explorers couldn't win: when they sailed towards the South Magnetic Pole they were frustrated by a continent, but when they looked for land they found ice. The unveiling of Antarctica's vast and shifting coastline was a by-product of the desire to understand Earth's magnetism.

In 1908, the Australian scientists, Edgeworth David and Douglas Mawson, together with the English naval surgeon Alistair Mackay, sledged inland from Ernest Shackleton's Antarctic base at Cape Royds in quest of the South Magnetic Pole. It was a moving target. The South Magnetic Pole is highly mobile, typically shifting about 5–10 kilometres per year.[6] It moves in response to fluid motions in Earth's liquid outer core. Between 1841 when James Clark Ross longingly located the South Magnetic Pole at a distance and 1902 when Scott's expedition again calculated its position, the pole had moved 322 kilometres (200 geographical miles) to the east. And at the magnetic poles the whole of the magnetic force is vertical, so the dip circle – the instrument Mawson used for measuring the vertical component – was vital. By 13 January 1909, at 73 degrees south, the horizontally moving needle almost ceased to work and the dip circle was just 15 minutes off the vertical. Edgeworth David noted that the magnetic pole executed 'a daily round of wanderings' and on 15 January Mawson considered that if they were to wait where they were for 24 hours, the pole would come to them. But they marched on – determined to reach it before it reached them – found a mean position for the pole, bared their heads to the Antarctic air, hoisted the Union Jack and briefly pinned their wandering destination for British possession.[7] Later calculations suggest that Mawson's instrument readings were imprecise and that, after all their hardship, they did not actually reach the Magnetic Pole Area of oscillation.[8] Their 2030 kilometre (1260 mile) trek was the longest unsupported manhaul sledging journey of the heroic era.

Edgeworth David was a beloved geologist and teacher and so famously courteous that his students could never conspire to get him to go through a door first. One day, during his journey towards the South Magnetic Pole with Mawson and Mackay, David gently called out to Mawson, who was changing photographic plates in the tent. Mawson's brusque reply made it clear he was busy. There was silence for some time until the 50-year-old professor again called out: 'Mawson! ... Oh, still changing plates, are you?' 'Yes', Mawson answered curtly. Again there was a period of silence. Finally, in slightly more urgent tones but still with infinite politeness, Professor David declared: 'I am so sorry to disturb you Mawson, but I am down a crevasse and I really don't think that I can hold on much longer.'[9]

The South Magnetic Pole has now wandered into the ocean where James Clark Ross would finally be able to fulfill his dream. The first time that the position of the South Magnetic Pole was determined at sea was on Stephen Murray-Smith's voyage on my ship on 6 January 1986, where it was momentarily located at 65 degrees 18 minutes south 140 degrees 1 minute east. It oscillated as they observed it and moved 50 kilometres (31 miles) in a west/north-west direction in one hour.

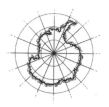

In Antarctica you are intensely conscious of the celestial Earth. The three great twentieth-century frontiers of exploration were the ocean depths, outer space and Antarctica. Superpower rivalry and the Cold War motivated the exploration of all three, and their mutual threshold seemed to be the ice. Along the shimmering edge of Antarctica, the annual formation of sea ice drives the globe's deep ocean circulation, while high on the polar plateau there is the coldest, driest and most stable atmosphere on the globe. There is virtually no daytime heating cycle to generate turbulence, and so the clarity of the air and the darkness of the months-long nights make it an astronomer's paradise. In the deep black Antarctic sky, the stars simply do not twinkle as much.[10] From the mid-twentieth century, Antarctica was the site of the transformation of Earth science into planetary science.[11] The continent of ice was no longer just the end of the Earth; it became a place from which to intellectually encompass the planet and a privileged human window on the universe. The blue ice areas of Antarctica have proven to be a unique preserving and collecting ground for meteorites, as effective as the Apollo program in supplying rocks from space. People on the ice often feel they have left Earth: one American described the residents of South Pole station as 'like men who had been fired off in rockets to take up life on another planet'. Antarctica's dry valleys are used as a useful analogue for the Moon and Mars, for it's hard to find signs of life there too. 'It was strange to watch the moon describing a complete circle in the sky, not setting for days at a time, but just coasting along the summits of the mountain ranges', wrote Louis Bernacchi in 1901:

> One actually had before one's eyes the diurnal revolution of the earth in space. At such times one is mentally struck with the similarity between the moon and that part of the globe; a dead silent world above you, and a dead silent world at your feet; above, the vast mountains of the moon, beneath the cold barren peaks of the southern extremity of our planet; but the lunar mountains are possibly more genial, for there, there are some mountains upon which the sun never sets − mountains of eternal light − whilst on the Antarctic mountains, added to the horrors of cold, and barrenness, and solitude, there is that of darkness.[12]

In Antarctica, the return of the Sun − the beginning of its journey back south, the shaft of its first rays across the ice, the full bloom of its brief, erotic reign − shapes the polar year and its celebrations.

Latitude is an earthly measure of a cosmic fact. It describes not only our place on the globe but also maps our relationship to the Sun. It is a dynamic, rhythmical relationship, this dance of the seasons, because our globe tilts on its axis as it circles the Sun, presently at an angle of 23.5 degrees. For half the year, each hemisphere basks under the most direct sunshine until planetary progress bears it off into the twilight again. The Sun seems directly overhead at the Equator, and the tropics of Capricorn and Cancer (at 23.5 degrees latitude, of course) mark the limits of its apparent seasonal wanderings. Latitude was calculated simply, by measuring with a sextant, theodolite or astrolabe the angle between the horizon and the Sun at midday. When Roald Amundsen beat Robert Falcon Scott to the South Pole in 1911, he left his sextant there, possibly because it was the most potent symbol of his triumph over latitude.[13] That the two sledging parties located the pole in the same featureless acre of ice may be the most unexamined and astonishing fact about the race to the pole. Scott's gift to Amundsen was to make his triumph indisputable. The Geographical South Pole, unlike the magnetic pole, does not move, but the kilometres of ice that cover it do, at glacial pace.

The European encounter with the east coast of Australia in 1770 was a spin-off of the international scientific effort to determine the dimensions of the universe, or more precisely, to measure the distance of Earth from the Sun. The solar system was in some ways better known than the southern hemisphere. James Cook turned his telescope to the Sun and found Australia. He was chosen as commander of the *Endeavour* at least partly because he was a skilled astronomical observer. By observing a solar eclipse in 1766, he had been able to determine his longitude in Newfoundland to within two nautical miles. During the eighteenth century, there was international scientific interest in the size of the solar system and in the possibility of calculating it accurately by measuring what seemed to be one of the natural constants of physical astronomy: the Sun's mean distance from the Earth. This was 'a kind of celestial meter stick', and the great scientist Edmond Halley in 1716 had exhorted members of the Royal Society to watch the skies after his death in the knowledge that close observation of the transit of Venus across the face of the Sun from widely different parts of Earth would provide a way of triangulating this critical distance.[14] Transits of Venus occur rarely, every century or two, but then in pairs eight years apart. The years 1761 and 1769 provided a timely opportunity for the burgeoning scientific spirit of the Enlightenment. For purposes of celestial mathematics and well spread observations, even the southern hemisphere had to be recruited. When in 1768 Cook had to explain himself and his ship, the *Endeavour*, to a suspicious viceroy of Brazil when getting supplies at Rio de Janeiro, his explanation about observing the transit of Venus struck the viceroy as a cock-and-bull story behind which lurked a likely smuggling enterprise. In any case, Cook reported that the viceroy 'could form no other Idea of that Phenomenon (after I had explained it to him) than the North Star passing thro the South Pole'. And what were natural philosophers doing in an alleged naval vessel anyway?[15]

After years of anticipation and preparation, capricious cloud could obscure the crucial view of the planet crossing the Sun's disk. And even if the sky were clear, how exactly could one determine the beginning and end of the transit, especially as 'the black-drop effect' meant that Venus, surrounded by an atmosphere, seemed to kiss the Sun rather than just crisply detach from him.[16] At Tahiti on 3 June 1769, the day was hot and the 'the Air was perfectly clear', but 'an Atmosphere or dusky shade' surrounded the body of Venus such that Cook's various observers all differed on the period of transit 'much more than could be expected'.[17] Then there was the ever-present problem of determining longitude: the mathematics would be awry if you could not precisely confirm where on Earth you and your telescope were. In 1769 there were 151 observers from 77 stations around the globe but the calculations failed to be definitive because of these problems of perception and location. In the nineteenth century, it emerged that the mean distance between Earth and the Sun was not an independent constant.[18] The burning star and its blue planet were involved in a complex dance, and their subtle relationship was the key to understanding ice.

'At 3 am I discovered the cause of the Ice Age.'[19]

So wrote Griffith Taylor in his *Field Journal* on 31 July 1919. The following morning he cabled his teacher and mentor, Professor Sir Edgeworth David, with the news – 'Believe found key past climates and Continents. Theory correlates Archaen complex, Ice ages, Gondwana, ages of planation, rapid evolution, mountain building and races of man. Letter follows, please comment' – and then he penned a letter of seven pages explaining his breakthrough. He speculated that the cycle of ice ages that marked Earth's history was due to the gravitational pull of a returning comet slowing the globe's rotation about its axis. Slower rotation, according to Taylor's theory, would lead to two coincident phenomena: mountain-building due to the flattening out of Earth's equatorial bulge, and less warm air flowing to the poles. Taylor was famous for courageous leaps of scientific faith and was characteristically seeking a theory of everything, one that would correlate climate, geology and evolution. And, like all scientists in the era before the plate tectonics revolution, he was hard-pressed to explain the presence of geologically ancient sediments at low latitude and low altitude. His theory proved to be wrong, but Taylor was looking in the right place for the solution to the ice ages: in the heavens. And he personally knew an ice age in action, for he had visited Antarctica with Scott.

We live in a relatively warm part of an ice age, an interglacial. Calling our own geological age 'the Holocene' and thereby separating it off from the Pleistocene – the 'Ice Age' – is misleading and unduly hopeful. As humans, we are inhabitants and creations of an extended, continuing ice age, the Quaternary. The river in the ocean that is the Gulf Stream, bringing warm tropical currents north, long protected European thinkers from the daily reality that they inhabited an ice age. Ice sheets four kilometres deep sit upon Antarctica, and deposits almost as thick envelop Greenland. Having polar ice caps has been rare in Earth history and having two at once may be unique.[20] But the southern ice cap is by far the most significant in terms of size and influence. Ninety per cent of the world's land ice and 70 per cent of Earth's fresh water are locked up in that Great South Land. That great natural refrigerator (and the lack of a Gulf Stream) denies Tasmania – which is at the same latitude as Spain and southern Italy – a Mediterranean climate. Cook's naturalist on his second voyage, Johann Forster, 'constantly found, that in Southern latitudes the cold is much more intense than in the corresponding degrees of the Northern hemisphere'.[21]

The temporary retreat of the ice which introduced our interglacial just 10 000 years ago left a geological legacy that remained unrecognised until the mid-nineteenth century. The emergence of the theory of continental glaciation shocked the scientific world as deeply as did plate tectonics a century later. This theory claimed not just that glaciers had moved rocks around and later retreated – hence explaining the puzzling presence of isolated boulders in Swiss valleys – but that whole continents had been covered under miles of ice. The great, mad exponent of this theory was a Swiss-born professor of natural history, Louis Agassiz. In 1836 he became convinced by the evidence of his native countryside and the advocacy of the geologist Jean de Charpentier that glaciers actually moved and that ice sheets had once encased the middle latitudes of the globe. The idea of a former ice age took several decades to be fully accepted by the scientific community. It overturned the dominant scientific assumption that Earth was gradually cooling down through history: one had to learn to look at glaciers not only as dynamic but as 'the fossils of a lost age'. Furthermore, the theory challenged the biblical account of a Great Deluge with an alternative apocalyptic cosmology.[22] Frozen water had wiped Earth clean. And there was indeed something obsessive about Agassiz's catastrophist crusade. He was prepared to be lowered into the bowels of glaciers to understand their dynamics. During his later career in North America, Agassiz became a major opponent of Darwin's theory of evolution. He believed that his own account of cyclical large-scale glaciation explained the extinction of species without invoking evolution. 'So here is the end of the Darwin theory', Agassiz was reported to have declared in 1866 in public celebration of his ice age theory. Harvard botanist Asa Gray told Darwin that Agassiz was 'bent upon covering the whole continent [of America] with ice'.[23] Agassiz's friend and mentor, Alexander von Humboldt, warned him against his ambition: 'Your ice frightens me.'[24]

A hundred years later, during the second half of the twentieth century, understandings of the ice ages were still evolving rapidly. The idea of one ice age had quickly been replaced by four ice ages called Gunz, Mindel, Riss and Wurm, but now scientists were mapping climatic 'oscillations' rather than 'stages', and the Quaternary emerged as a distinctive period of Earth history marked by a repeated cycle of rapid global warming followed by more gradual cooling, with the warm peaks about 100 000 years apart. As Australian archaeologist Mike Smith puts it, this was a change from a primordial to a processual view, from a stately, almost Biblical account of ages to charting the oscillations of an unstable system. Furthermore, in the late 1990s ice cores from Antarctica and Greenland revealed that the global climate often switched rapidly, swinging as much as 5–10 degrees Celsius (10–18 degrees Fahrenheit) in 30 years. Graphs of the Quaternary climate began to look like the palpitations of a nervous, sensitive system.[25]

The essential rhythm of the ice ages was found in the heavens. The Serbian climatologist, Milutin Milankovitch (building on earlier work by James Croll), recognised that the key factor was Earth's relationship to the Sun, and he subjected this celestial marriage to the scrutiny of mathematics. Earth's orbital geometry held the key. Milankovitch identified three crucial astronomical rhythms: the variation in Earth's orbit between a circle and an ellipse, the changing tilt of Earth (it wobbles a few degrees on its axis), and the rotation of the equinoxes around the orbit. All of these factors affect the amount of solar energy available to different latitudes of the globe, and each of these cycles has a different periodicity. Together they produce a complex cosmic rhythm of temperature variation on Earth. But the heavenly triggers become exacerbated by earthly feedback mechanisms, by the historical configurations of land and sea, ice and atmosphere, such that small changes in temperature can precipitate large-scale changes in ice sheets and ocean currents. Astronomical cycles interact with plate tectonics to generate climate change. Thus heaven and Earth combine to create the ice ages.

Humanity may be considered a species of the ice ages. About 3.6 million years ago, the first northern hemisphere glaciation of the current ice epoch reduced rainfall in East Africa, and forced the least specialised and most versatile apes to the edges of the contracting forests. There they faced severe pressure to innovate or perish. Interglacials enabled the stronger survivors to consolidate, while the next ice age tested and extended their descendants even further. Deep in the forest, the more specialised and successful tree-climbing apes faced fewer selection pressures. The drier cycles brought by the ice ages – ten or a dozen every million years – combined with the reduction in African rainfall resulting from Australia's drift north into equatorial seaways, pushed early humans out onto the plains, encouraged them to walk upright, and demanded they live on their wits. By the time of the last ice age, fully modern humans (*Homo sapiens sapiens*) had spread across the globe, and the last retreat of the ice heralded a spring of astonishing cultural creativity.[26]

Australia has had an intriguing relationship to ice. Australia's separation from Antarctica 45 million years ago triggered the oceanic and atmospheric isolation of land at the pole that made the growth of a formidable southern ice cap possible. But its swift drift into lower latitudes meant that by the time of the Quaternary ice ages (the last 1.6 million years), Australia mostly escaped the scrape of large ice-sheets. Parts of Tasmania, the highlands of New Guinea and the Kosciuszko Plateau were the only landscapes scoured by the last ice age. Nineteenth-century geologists, their imaginations caught up in vigorous debates about glaciation, searched for signs of ice ages on the face of Australia, especially any that might correspond with the

'Great Ice Age' of the northern hemisphere that was championed so vociferously by Agassiz. 'With its low altitude and low latitude', writes historian Kirsty Douglas, 'Australia was not necessarily a very fertile garden for the budding glaciologist', and even promising evidence was rejected by naturalists. And where glacial landscapes were identified and accepted, such as at Hallett Cove in South Australia, it took time to accept that it was actually evidence of a much earlier glaciation in Permian times (290 million years ago). Douglas suggests that Pleistocene Australia could instead be said to have experienced 'dirt ages' because growth of the ice caps dried the continent, extended the interior deserts, and filled the air with dust. Pleistocene ice age Australia was cold, windy and arid.[27]

George Seddon recommends that a good question to ask of all continents is: Did you have a good Pleistocene? Australia didn't, in the sense that there was little soil-building going on. A non-glacial glaciation meant that you lost dirt rather than gained it. Glaciers grind new soil from rock while windy aridity disperses it. Nor was there much mountain-building going on. Geologists call the swelling up of mountains an 'orogeny'. John McPhee has celebrated the sensual quality of the geological lexicon: 'There was almost enough resonance in some terms to stir the adolescent groin', he reflected.[28] Australia hadn't had an orogeny for quite a while.

But a lot of other things were happening. Glaciation in Australia was manifest in a cycle of drier and wetter conditions and in sea level changes. Aboriginal people had to cope with the expansion and contraction of the deserts and substantial fluctuations in sea levels. By around 18 000 years ago, sea levels had dropped to their lowest level since the arrival of humans, the continent became much drier, colder and windier, rainfall was half what it is today, linear dunefields formed across central and southern Australia, lakes such as Willandra dried up, as did other surface water, and people abandoned the arid lands except for refuges in the ranges for perhaps 10 000 years.[29] Until the 1980s, it seemed possible that the arid centre of Australia had remained uninhabited until the ending of the last ice age, but there is now enough archaeological evidence to establish that major desert uplands and river systems in the arid interior were occupied during the late Pleistocene, before the last glacial maximum; at least as long as modern humans have occupied western Europe.[30] What, then, was the nature of that occupation? Archaeologist Mike Smith argues that the ebb and flow of population in the great sandridge deserts of Australia – the Great Sandy, Great Victoria and Simpson deserts – was determined less by technological, economic or social developments than by fluctuations in climate, especially the availability of drinkable surface water, and that these great deserts would have been recolonised as part of a post-glacial expansion of population in the terminal Pleistocene or early Holocene.[31] One glimpses a great, unfolding human drama in ice age Australia, as people occupied new land left by the retreating coastal seas and themselves retreated from the arid centre as cold droughts held sway. Thousands of years later, the sea regained land as one seventh of greater Australia was inundated, and rains made the central deserts accessible once more. Tasmania and Papua New Guinea were cut off from the continent by the rising seas, and isolated social groups in central desert refuges again encountered strangers.

It was an earlier ice age that first interested Douglas Mawson. For Mawson the geologist, the arid outback of South Australia and the arid ice sheet of Antarctica were linked. It was not just the rocks. For adventurous and scientific Australians of the early twentieth century, two frontiers beckoned: the white ice and the red heart, the south and the outback. Mawson, Cecil T Madigan, Charles Laseron, Edgeworth David and Griffith Taylor ventured in both directions. Baldwin Spencer and JW Gregory went inland after almost going south. Adelaide, where Mawson gained a lecturing post in his twenties, was sand-wiched between these frontiers; it was an Australian capital exposed to both desert winds. The tradition continues: Bloo Campbell, the Casey winterer from my ship who hailed from Humpty Doo in Australia's north, compared blizzards to duststorms, and wrote for *BushMag:*

*Journal of the Outback* about the similarities of his two frontiers, 'the white and red deserts'.[32]

Mawson was one of the first generation of geologists trained in Australia, inspired by the teaching of Professor Edgeworth David at the University of Sydney. He studied the glacial sediments and landscapes of South Australia and recognised the residue of an earlier, much more ancient ice age, and soon he 'desired to see an ice age in being'.[33] He wanted to experience modern glacial processes in action. When he sighted Adélie Land in Antarctica in 1911, with its massive ice cliffs along the coast, he felt elated, and overwhelmed. 'Here was an ice age in all earnestness', he wrote, 'a picture of northern Europe during the Great Ice Age'.[34] The North American Antarctic explorer, Richard Byrd, was also excited: 'For here was the ice age in its chill flood tide. Here was a continent throttled and overwhelmed.'[35]

In these decades following the acceptance of the concept of ice ages, the exploration of Antarctica gathered intensity. The continent seemed to allow one to travel through time to the Pleistocene Earth. Antarctica was a luminous relic, a clue to lost ages. Antarctic explorers, as we have seen, originally wanted rock. The ice was a testing ground for physical endeavour, a source of beauty and fear, and an obstruction. But during the twentieth century, the ice itself became a primary scientific focus and was no longer regarded just as an obscuring and frustrating 'barrier' between visitors and the much-desired land. Ice was a mineral, glaciers were geological and Antarctica was a vestigial landscape, a giant white fossil. The ice had a history, and one that revealed the vicissitudes of the whole Earth and even offered planetary parallels with other bodies of rock and ice.[36]

'It appears out of the fog and low clouds, like a white comet in the twilight.' With this first sighting of a great berg, spinning slowly in the Southern Ocean, North American historian Stephen Pyne begins his 'journey to Antarctica' in *The Ice*. 'To enter Greater Antarctica', he continues, 'is to be drawn into a slow maelstrom of ice. Ice is the beginning of Antarctica and ice is its end.' No other book on Antarctica so relentlessly analyses the continent's essence. Pyne revels in the paradoxes of ice: that it is simple yet baffling, reflective yet dangerously enveloping. He adopts a narrative structure dictated by the major forms of ice, beginning with the berg, then the pack, shelf, glacier, sheet and finally, the source, the interior domes of ice. Pyne's journey is towards 'the heart' of Antarctica, against the flow of the ice. He is determined to take a view that will propel him inland, against the gravity of settlement. He is the only historian of Antarctica who does not come home. He finishes his book in the 'self-referential, autistic' zone of the high polar plateau, a source of ice but an energy and information sink. Only a civilisation rich in energy and intellectual activity, writes Pyne, can enter the region and escape with a residue of information. Historians of Antarctica must find a way to engage with the history of the ice.

In the first half of the twentieth century, the great symbol of heroic endeavour in Antarctica had been the sledging journey. Its aim was primarily geographical: to achieve furthest south, to reach a pole, to extend territorial boundaries or to map features or coastlines. Explorers travelled over the ice and fought its surface, and – like Laurence McKinley Gould in 1929–30 – they hungered for any rock that might defy or penetrate it. They craved nunataks, an Arctic term for the peaks of submerged mountains protruding above the overwhelming ice. But from mid-century, a different type of journey became the centrepiece of expeditions. The traverse by tractor was the dominant heroic feat of the 1950s and '60s and its chief purpose was the study of the ice itself.

The first official international expedition to Antarctica, which headed south in 1949, subjected Antarctic ice to comprehensive investigation. Scientists in the 1940s had observed the recession of glaciers in the northern hemisphere and wondered if those in the south were also receding. They speculated about the influence of such a vast ice sheet on the Earth's climate, and needed to know more about its size, history and stability. The Norwegian-British-Swedish Antarctic Expedition of 1949–52, which also included Australians, met with a Norwegian whaling factory ship on the ice edge to collect dogs, men and equipment, and established a base called Maudheim in Dronning Maud Land where the thickest ice in Antarctica was thought to be found. Maudheim was established on 10 February 1950 on a floating ice shelf that imperceptibly rose and fell with the tide. For two years scientists studied the ice in myriad ways: by placing a grid of stakes over an area of 25 square kilometres (9.7 square miles) in order to measure ice accumulation and any differential movement of the ice shelf; by drilling down 100 metres (328 feet) to obtain undisturbed cores of ice so that they could study density, porosity, grain size and orientation, as well as the pressure of air bubbles trapped in the ice; by systematically photographing the ice sheet from the air; and by mapping the sea edge of the ice barrier from the ship. Glaciology was emerging as its own sophisticated discipline. All accessible mountain ranges and major nunataks were visited and chipped into specimen bags, and mosses, lichens and mites were collected. The British geologist, Alan Reece, chipped some rare rock from a nunatak into his eye and lost the sight in it.

The scientific centrepiece of the expedition was a seismic traverse to measure ice thickness. It was long believed that Antarctic ice was no more than a few hundred metres thick, and so this traverse towards the continental interior, stopping regularly to make soundings, made such surprising discoveries that the scientists doubted their arithmetic. From 18 October 1951 to 6 January 1952, Gordon de Quetteville Robin (an Australian physicist), Charles Swithinbank (a British glaciologist), Peter Melleby (a Norwegian radio operator) and Ove Wilson (a Swedish medical officer), journeyed 600 kilometres (373 miles) towards the continental interior and conducted some 80 seismic soundings. They travelled in tracked vehicles called weasels that also pulled sledges, and they had an emergency dog team in case of mechanical breakdown. Gordon Robin often drove the dog team ahead of the weasels to check for crevasses, controlling the huskies with a 12-metre whip made in Sydney from bullock hide. While preparing weasels for this trip, four men at Maudheim, momentarily disoriented by low sea fog, drove off an ice cliff and plunged into the sea. Only one man was able to swim to safety, where he awaited rescue on an ice floe for 13 hours.

For the seismic traverse, the weather was fine and on most days visibility was limited only by the curvature of the Earth. The men were sustained by porridge and pemmican. To test ice thickness, they set off a small 'earthquake' with explosives and then measured the time the vibrations took to travel through the ice to the underlying rock and back to the surface. Sometimes it took a full day to achieve a reliable reading. On substantial sections of their traverse, they found the ice to be more than 2000 metres (6562 feet) thick and the land beneath them to be below sea level. Robin double-checked his calculations. The men travelled in two dimensions: their weasels chugged up and down the gentle contours of the ice cap while their imaginations (with help from the plotted soundings) followed the sharper topography of the land so far below them. Kilometres of ice smoothed out the roller-coaster ride they would have had on bedrock. The driest of all continents was actually a vast elevated plateau of frozen water. 'We never did get used to the prodigious scale of the landscape', wrote Charles Swithinbank.[37]

This pioneering seismic journey divined another land under the ice; an understorey of surprisingly rugged relief. The greatest ice thickness they plumbed on the inland plateau was 2400 metres (7874 feet). They found that the ice sheet in this area, although very impressive, was not the highest in Antarctica, as the German expedition of 1938 had claimed. That distinction – the thickest ice in Antarctica, almost 5 kilometres of it – was eventually located inland from Commonwealth

Bay. Mawson had unwittingly established his hut – his 'home of the blizzard' – at the base of one of the longest, steepest slopes on the Antarctic ice sheet, exposing his expedition to the relentless avalanche of gravity-fed katabatics. The force and constancy of the off-shore winds in the region of Mawson's hut also maintain the Mertz Glacier polynya, a site of rapid sea-ice production that is crucial in the formation of densely saline 'Antarctic bottom water' that drives much of the vertical circulation in the global ocean.[38]

The glaciological work of the Norwegian-British-Swedish expedition led to the insight that world sea level is principally controlled by the state of the Antarctic ice sheet. Since there was a much greater volume of ice down there than people had thought, then its corresponding impact on sea levels must be greater; if the southern ice cap melted, calculated Robin, then oceans could rise by 60 metres (197 feet). Robin, who became director of the Scott Polar Research Institute in Cambridge, continued the charting of glacier and ice cap thickness by airborne radio echo-sounding, which led to the under-ice mapping of almost the entire Antarctic continent by 1983.[39]

The great era of over-snow traverses, criss-crossing Antarctica and sounding its depths, had begun. The Russian station, Vostok, 'the Pole of Cold' established on the high plateau in 1958, became a destination for valiant journeys. Men found themselves spending months in the simplified world of the white desert: 'no hills or valleys, no trees, no boulders ... nothing but that endless, glistening whiteness'. Such a diffusion of light that one can see no shadows, no depressions, no rises, no distance. The expedition tractors had to be close enough to see one another but not so close that they skewed the command tractor's compass readings. When violent snow-laden winds set in, the Russians bunkered down, eating mutton, hamburgers and boiled potatoes, baked beans and thick beef soup while the blizzard lasted. When the wind dropped they grabbed shovels, dug out their submerged transport, and resumed their 'back-jarring odyssey into the white void'. The ice sheet at Vostok was found to be the same in thickness as its height from sea level: 3488 metres (11 444 feet). It is so cold there that a billy of boiling water thrown into the air freezes immediately.[40]

In 1962, an Australian team from Wilkes (near the future site of Casey) made the 3000 kilometre (1864 mile) journey to Vostok and back. Travelling in weasels, caterpillar tractors and sledges, they subjected themselves to months of 'bone-shaking torture'. Robert Thomson recalled that 'it was just like having one's eyes six inches from a large white sheet'; it was like floating in a void. One felt not only blind but disembodied as well. The horizon was high and all-surrounding, and ice crystals in the air produced entrancing refractions. Low water clouds appeared and quickly passed; sometimes they suddenly crystal-lised and dissipated in a shower of 'diamond dust'. What appeared to be four or five suns circled you. Time and space, east and west lost meaning. The plateau party noticed a growth of lethargy, a creeping inertia, a loss of interest in food. Antarctica began to envelop them. Thick paint on the weasel blistered and exploded in the extreme cold. It was −63.4 degrees Celsius (−82.2 degrees Fahrenheit), there were kilometres of ice beneath them, and no one knew what lay before them. Waking and breakfasting each day in the metal caravan, where they slept in temperatures of −40 or −45.6 degrees Celsius (−40 or −50 degrees Fahrenheit), produced a rainstorm of condensation. The little heater that warmed the tractor engines had to be started with a string, like a lawn mower. A nightcap of brandy, left beside the bed overnight, froze. When they arrived at the remotest place on Earth, the Russians were not in residence, but the visitors felt like intruders; they wanted to knock on a door. The Australians stumbled down a passage into a dark, large room. Danny Foster held the lamp higher, and his mates squinted into the gloom. 'A figure stared down at us from the darkness. What was this? Worse than a haunted house, I thought. A closer look at the figure; there was no doubt now – it was a painting of Lenin. This was Vostok alright!'[41]

Once the extent and depth of the ice cap became measured and known, research focused increasingly on understanding its history and future. By measuring ice thickness, ice accumulation rates and ice velocities, scientists assessed the mass balance of the ice sheet: is it growing or diminishing, and what changes are due to human industry? It is a complex calculation. A snowflake falling on the high interior of the Antarctic dome might take more than 500 000 years to travel through the ice to the marine edge of an ice shelf, and so changes in climate since the last ice age are still acting on the ice sheet. Furthermore, there is the question of how an ice sheet that size can maintain itself if it is mostly covered by a permanent anti-cyclone with low precipitation. Russian research in the 1950s revealed that in central, high Antarctica, the ice sheet is relatively flat and moves slowly, but that precipitation is heavier around the periphery where cyclonic disturbances occur, and it is there that ice movement is faster. As Phillip Law put it in 1959, 'The Antarctic ice cap is therefore nourished mainly around the edge, not in the centre.' Global warming has increased precipitation over the plateau which, for the moment, balances the rapid melting of ice shelves in the 'Banana Belt', as the Antarctic Peninsula is called.[42]

When Gordon Robin's Maudheim party discovered that the annual layers of snow could be differentiated, they began the archaeology of the ice. Ice accumulates in seasonal layers that can be counted like the growth rings of trees. From the late 1960s, scientists developed a different way to plumb the plateau: the tortuous extraction of deep ice cores. Continuous cores drilled from the ice sheet revealed historic changes in the atmosphere and long-term climatic oscillations. Australian scientists studied the accessible Law Dome (less than 200 kilometres (124 miles) wide, near Casey station) as a microcosm of the whole East Antarctic ice cap. Over several seasons in the late 1980s and early 1990s, glaciologist Vin Morgan and his team worked with instrument workshop manager, Andrew Fleming, and diesel mechanic, Russel Brand, to develop a modified ice drill which enabled the capture of a deep ice core from Law Dome Summit with a record of 120 000 years of atmospheric history through the last ice age and beyond. At the same time, Russians were steadily drilling beneath Vostok: four hours of drilling a day producing a three-metre (9.8 foot) long cylinder of ice, representing over a hundred years of snowfall. Another day, another century. By 1998, they had extracted more than 420 000 years of Pleistocene history from a hole 3623 metres (11 886 feet) deep. They stopped drilling because they found below the ice, nestled in the bedrock, the largest of over 70 subglacial lakes in Antarctica and one of the largest lakes in the world: 224 kilometres (139 miles) long and 48 kilometres (29.8 miles) wide. As I write, the drill is poised above it. This is an historic pause, in awe of what new life may be found so long isolated from the rest of the world, and of the impossibility of studying it without contaminating it. But the Russians have decided to proceed. On the ice surface, visitors to Vostok station find a sign in English declaring 'Welcome to Eastern Beaches of Lake Vostok'.[43]

By the late twentieth century, Antarctica had emerged as a sensitive barometer of global health. It was like a 'space platform monitoring the planet below'.[44] Recent, observable changes in the ice cap became anxiously assessed. As Stephen Pyne put it, by the end of the twentieth century 'the preservation of Antarctica's ice sheets joined the protection of its penguins and whales as a test of human character'.[45] The confirmation of global warming due to human influence comes not only from the behaviour of the ice sheet, but more clearly and dramatically from the air bubbles trapped in those ice cores. The level of carbon dioxide in the historic air bubbles has leapt since the industrial revolution. Before Antarctica was even seen by humans, it was recording our impact.

## Wednesday, 25 December

**64 degrees south,
110 degrees east**

It is a white Christmas indeed. On the ice floes this morning to greet us were penguins – Emperor and Adélie – and seals too, great slugs lolling about nonchalantly, basking in their swimwear on the shelves. As we move cautiously through a great jigsaw of meringue, nudging aside platforms of ice, we disturb hosts of Arctic terns that then settle on neighbouring floes, and we send crabeater seals slipping into the blue-green depths below their floating lounge-rooms. There is a Kodak frenzy on the bridge.

My *ANARE Field Manual* warns us not to stand too near the edge of sea ice, 'particularly if you are wearing a dinner suit'! Leopard seals patrol the edges looking out for anything that might pass as a penguin.

As I write this, the *Polar Bird* occasionally shudders from stem to stern as it breaches a floe. The day and the sea are calm, the air is sharp. Our red ship is a tiny embellishment on the wedding cake.

We were initiated last night, having well and truly crossed the line of 60 degrees. The ritual used to involve harsh humiliations such as drinking something disgusting, having batter poured in your hair, and perhaps even a dip in the brine (a 'ducking'). In 1985 Stephen Murray-Smith and other initiates were abused by King Neptune and mustered into a net ready to be slung into the sea. Stephen was quite disconcerted! But ours was a tame ritual: it was simply a dab of vegemite on your forehead from a broad paintbrush dipped into a great metal pot, and then we were presented with a certificate from King Neptune declaring each of us to be 'of sound but watery character'. I was officially confirmed as a South Polar Sea Dog. We then had a great barbecue in the between-decks area, a medieval feast, a spit roast of pig with hot gluhwein in mugs which we held gratefully in our hands as the smoke from the BBQ mingled with the condensation from our breaths. Today there was more feasting, and Santa came! He found us out here amidst the ice, working through his branch office at the South Pole.

It has been a time of ceremonies and rituals, of declarations and anointing. 'Crossing the line' (usually the Equator) is an ancient ritual of voyaging, meant to be unsettling: so it was right that Stephen was perturbed. Crossing the line ceremonies were also satires of power. I can't help seeing the vegemite in that tradition, as a play on modern Antarctic politics. The Antarctic Treaty governs the world below the line we crossed, 60 degrees south, and we marked our arrival in this vast, international realm with a risible signature of nationalism.

# PLANTING FLAGS
## Claiming the ice

On 13 January 1930, Sir Douglas Mawson and a party of men landed on a small rocky island off Eastern Antarctica, climbed to its summit, built a cairn, erected a flagpole, raised the Union Jack, read an official proclamation at noon, placed a document in a canister in the cairn, attached a tablet asserting sovereignty, took off their caps, gave three cheers and sang *God Save the King*. This carefully scripted theatre, clumsily performed with rock, pole, plaque, cloth, paper and voice on an island connected only by ice to the mainland, and recorded immediately in maps, diaries and on film, vested all territories between longitudes 73 degrees east and 47 degrees east, and south of latitude 65 degrees in His Majesty King George the Fifth His Heirs and Successors for ever. The black rock, striated with ice, was named Proclamation Island and the inscribed tablet on the flagpole was made to face south towards the other pole. Although everyone cheered, not everyone was impressed. The ship's captain, John King Davis, awaited the return of the shore party with anxiety: 'They have no food or gear but seem to trust to Providence entirely.'[1] The young zoologist, Harold Fletcher, wrote that night: 'We claimed a huge tract of land which we had not landed on and had not even seen all of it – the idea being to get in early. And then the Norwegians are abused for doing less things than that!' Three years later, the *Australian Antarctic Territory Acceptance Act 1933* (Cth) formalised the transfer from Britain to Australia of sovereignty over all this ice. In this interwar period, national territoriality became more assertive in Antarctica. The awesome, undifferentiated ice cap – the 'Big Pav' – was getting sliced and apportioned, and there was a rush to the table.

In 1929, Robert Falcon Scott's old ship, the *Discovery*, had been fitted out again to go south, this time with an expedition led by Sir Douglas Mawson. Davis, the ship's captain, had served with Ernest

Shackleton and Mawson on the *Nimrod* in 1908–09 and Mawson on the *Aurora* in 1911–14, and again with Shackleton on the *Aurora* relief expedition of 1916–17, which included Fred Middleton. Because Antarctica was extreme and remote, its society was incestuous and its material culture often recycled. Mawson, Davis and the *Discovery* were all aging relics of the heroic era, miraculous Antarctic survivors, and their combined *mana* was important to this new expedition. For it was a voyage with an explicitly geopolitical purpose.

As the ship awaited Mawson's arrival in Cape Town in October 1929, local journalists wondered at the sudden congregation of 'the most experienced polar explorers of the day' in southern seas. Norwegian whaling ships were busy in Antarctic waters, the Norwegian captain, Hjalmer Riiser-Larsen, had headed south on the *Norvegia*, the United States (through Richard Byrd) had established a camp in Antarctica called Little America at the beginning of 1929, the Australian Sir Hubert Wilkins was already surveying the continent from the air, and now Mawson's ship was in port and about to weigh anchor. Why the urgency and excitement, and why just now? 'Have the Norwegians a Secret?' asked the *Cape Times*. 'Can they have carried home a report of something seen or something suspected which it was worthwhile to secure? And can some whisper of that report have leaked out to inspire others ...?' Even by the late 1920s, the mysterious icy continent could still nurture feverish dreams. Commander Byrd was reported to believe that Antarctica's unknown interior might harbour deep sheltered valleys in which the summer sun, shining through the whole 24 hours, melted the snow and fostered new forms of life, perhaps even an undiscovered people. Certainly there were other riches to be found: fossils, coal, minerals and whales.[2]

In the early twentieth-century Antarctica had been an additional site of European colonial rivalry, the place for one last burst of continental imperialist exploration, which had been such a trademark of the nineteenth century. The heroic era of Antarctic exploration was 'heroic' because it was anachronistic before it began, its goal was as abstract as a pole, its central figures were romantic, manly and flawed, its drama was moral (for it mattered not only what was done but how it was done), and its ideal was national honour. It was an early testing-ground for the racial virtues of new nations such as Norway and Australia, and it was the site of Europe's last gasp before it tore itself apart in the Great War. The 'long nineteenth century' was particularly long in Antarctica and ended there with Shackleton's death at South Georgia in 1922 while pathetically attempting another glorious adventure.

In the 1920s, a more pragmatic geopolitics quickened in Antarctica, and romantic, masculine heroics morphed into harder-edged territorial theatrics. There was, as the *Adelaide Advertiser* declared in April 1929, 'A Scramble for Antarctica' that might echo the famous 'Scramble for Africa' amongst European powers in the late nineteenth century.[3] Commercial whaling was intensifying along the edges of the ice. Britain had established the Falkland Island Dependencies in 1908 and the Ross Sea Dependency (from New Zealand) in 1923 in order to regulate – and profit from – whaling in those sectors. It was this new regime that Bob Murphy's captain, Benjamin Cleveland, had bridled against so noisily in 1912. In 1924 the French responded by claiming Adélic Land on the basis of Dumont d'Urville's sighting of that part of the continent in 1840. Adélie Land was in the middle of what Australians had come to consider *their* quadrant of Antarctica – the land directly to its south – and so it suddenly became urgent to formalise that understanding. Mawson urged the Australian prime minister, Stanley Melbourne Bruce, to challenge the French claim and act with Britain to secure Australian administration of its Antarctic lands.[4] The voyage of the *Discovery* at the end of the 1920s was a belated response to this geopolitical anxiety, and part of a plan approved at the 1926 Imperial Conference to assert British dominion over all of Antarctica; to paint the whole continent red, as one British official confidentially put it in 1928.[5] The British, Australian and New Zealand Antarctic Research Expedition (or BANZARE, a title and acronym that Mawson considered 'absolutely ridiculous') inverted the priorities of Mawson's first

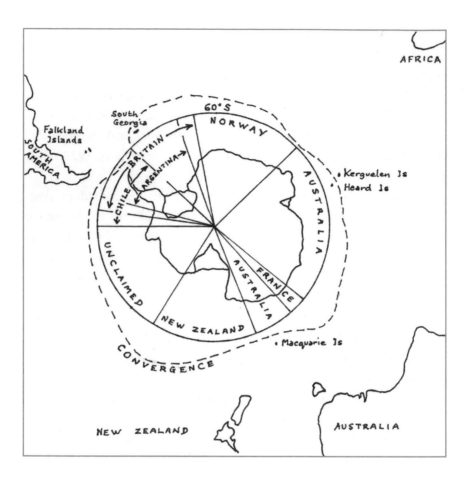

*Antarctic Territorial Claims*

expedition of 1911–14, for it was 'firstly, political; secondly, economic and commercial; thirdly, scientific'.[6] Mawson's secret instructions from the Australian prime minister were to 'plant the British flag wherever you find it practicable to do so'.[7] When Mawson, during one of his many arguments with Captain Davis asserted that, by comparison with the Norwegians theirs was a scientific expedition, Davis replied: 'that was all eye-wash, we were out to grab land'.[8]

In the late nineteenth century, established by African precedents, the then generally accepted way to demonstrate sovereignty was by 'effective occupation'. But how did this doctrine apply to polar regions where long-term occupation was so difficult? On the ice, the formal rituals of taking possession and the bureaucratic and symbolic confirmation of ownership, necessarily orchestrated from a distance, became especially important. Flags, plaques, paper and post offices were crucial. Imperial powers had to demonstrate some degree of control, and desire.[9]

Mawson's discomfort with the primarily political goals of his latest expedition – however much he believed in its aims – was palpable. There are echoes of Scott's unhappy resolution of the goals of science and the pole almost two decades earlier. In 1911, Mawson had settled that tension decisively in favour of science in the planning of his own expedition at Commonwealth Bay. Now in 1929 and again in 1930 he steamed south to claim land, plant the flag and secure economic prospects, and his science was officially the last priority. A deep unease pervaded the BANZARE voyages. Mawson and his old captain, Davis, bickered endlessly and bitterly on the first voyage (KN MacKenzie captained the second). Partly it was the problem of divided command between the leader of an expedition and the captain of a ship, exacerbated on a voyage with no sledging journeys, no hut and few landings. But it was also because they missed the idealism and genuine adventuring of their earlier expeditions, and felt oppressed by their overtly political purposes. Davis grumbled in his personal journal about Mawson's 'lack of plan' and constant changes of mind, and declared that 'If one is to have a scientific expedition one must not hope to do the flag raising business too.'[10] Davis found Mawson difficult to reason with, and Mawson considered asking the doctor to declare Davis insane. The wider politics were corrosive, and ate away at the inner men.

Mawson certainly believed in the strategic and commercial objectives – in fact he had personally campaigned for them – yet he was to jeopardise the expedition's territorial achievements by doing science in defiance of the political timetable. He was under pressure to get south fast and to plant the flag before the Norwegians, but could not stop himself exploring sub-Antarctic islands and dredging for marine specimens along the way. He was, in Davis's words, always 'too busy messing about with the contents of the trawl to attend to his own job'.[11] Mawson's dredged marine organisms are reminiscent of Scott's sledged rocks; both were symbols of scientific commitment on journeys overwhelmed by other, less idealistic purposes. Science could be both a disguise and a solace.

Mawson refused to race for glory or territory. And like Scott, he was haunted by Norwegians. In 1929, Mawson wrote that 'apparently they are out to race and are working in secret, as [Roald] Amundsen did when he beat Scott to the South Pole. However I prefer not to discuss that matter. We are going to carry out a very carefully planned scientific programme.'[12] He insolently suggested that the Norwegians would not. Norwegian activity in the Antarctic waters south of Australia had prompted the BANZARE expeditions. There was a feeling that Australia had neglected its backyard, and the Norwegians had seized the opportunity to extend whaling into these waters. 'It had come about that this very important industry had been allowed to fall into the hands of others', explained Mawson.[13] By the end of the 1920s he was exploring an Antarctic coastline that was thick with the traffic of Norwegian whalers. On a single day the *Discovery* encountered three chasers and two factory ships in the morning, then met Commander Riiser-Larsen's 'flagship of the Sinister Presence' in the middle of the day, and after tea-time was passed by four more chasers going east.[14] It became quite difficult to assert their status as intrepid explorers: 'If that is a factory ship we cannot let her be further south than we are', wrote expeditioner Stuart Campbell in his diary.[15] When the *Discovery* was running low on coal, Commander Riiser-Larsen offered supplies from a nearby whaling ship, but Mawson declined.

Sir Macpherson Robertson, businessman and confectioner (the maker of Cherry Ripes), donated £10 000 to the BANZARE voyages; hence Mawson's naming of Mac. Robertson Land in Eastern Antarctica. Robertson viewed the 1929–31 voyages as a thorough scientific investigation 'with a view to the exploitation of any economic possibilities'. He compared Antarctica with Alaska, a region Britain had declined to purchase from Russia because it was seen to be uninhabitable. Yet the United States had acquired it and 'had extracted many million pounds worth of mineral and animal products from the country'. 'Australia was now facing almost a parallel case with respect to the Antarctic', he argued. 'No effort should be spared in seeing that the colossal mistake was not repeated.'[16] Robertson's portrait stared down at Mawson in his cabin on the *Discovery*.

Mawson explained the aims a little differently. There was a misconception, he said, that they were investigating mineral resources or seeking to establish a settlement. Neither was on their agenda, and 'the idea of a civilization at the South Pole belongs to the realm of dreams'. But there was a resource they would survey with the aim of commercial development, and that was the marine life of Antarctic waters.[17] Australia had the opportunity to participate with profit in the whaling industry down south. But good science would precede any kind of commercial development. Whaling might need to be controlled to ensure its sustainability as an industry.[18] Meteorology as well as oceanography would concern the expedition. A great stretch of coast line had yet to be surveyed, and it was as near to Australia as Hobart was to Fremantle. Any commercial development there was 'Australia's birthright'. Mawson was 'filling in the gaps in the chart' left by his 1911–14 expedition. Antarctica remained mostly a mystery. 'More was known about the surface of the moon', he reflected.

Claiming something as slippery as ice was fraught with frustration and laced with comedy. It was not an easy task even to see the land, let alone get onto it. RG Simmers, the meteorologist on board, presented the following doggerel on New Year's Eve, 1929:

> Oh! This is the song of the B.A.N.Z.
> > On the *Discovery*,
> The Antarctic coastline seems totally fled,
> > From the *Discovery*,
> Bay ice and bergs and penguins galore,
> But no bloody sign of the mythical shore,
> But it's New Year's Day today, so let us all say,
> > Here's to *Discovery*.[19]

Mawson had planned no landing parties. They would go ashore when they could but there would be no land camps or bases. Their business was with 'the great waters'; dredging, sounding, and mapping coastline. But they *did* have to plant the flag, wherever and whenever they could. Mostly, according to Davis, they were 'blindly groping about in a maze of ice'.[20] The Proclamation Island occasion in early 1930 was momentous because it was a rare foothold on rock, albeit an island.

On 25 January 1930, Mawson flew from the ship in a Gipsy Moth biplane with his pilot, Stuart Campbell. When they were over land-ice, Mawson read the usual proclamation out loud, his lips miming the legal formalities, for his words were lost in the wind and engine noise and even Campbell could not hear them. Then Campbell induced the aeroplane to stall so that its nose dived dramatically and, as they began to plummet, he threw a Union Jack on a wooden pole from the cockpit. The Gipsy Moth recovered speed and the two men confirmed that the flag was indeed lying on the ice surface.[21]

On 13 February the following year, 1931, at Murray Monolith, they searched in vain for a safe place to land their small boat in the dangerous swell and were forced to touch an oar against a rock while they read the proclamation and raised the flag in the boat, and then threw the document in a canister amongst boulders above the beach. They also threw a wood plate with a copper inscription and the flag on a pole, but both struck rocks and fell back into the sea.[22]

Five days later, there was a ceremony in a valley at Cape Bruce in Mac. Robertson Land. Simmers recorded that:

> Dux [as they called Mawson] jumped ashore and ran up the valley waving a flag and looking as pleased as punch – and so he was because at last he was in a position to make a legal, complete and entire observance of claiming land, which we forthwith proceeded to do, building a cairn, hoisting the flag, reading the same proclamation as on the 13th, God Save the King, and cheering. In all other attempts (except, of course, at Cape Denison) there has been some little thing wrong – 'Proclamation was only an island; the first proclamation on the 13th floated out to sea and the second on the 13th had no board or proclamation, but this time things have been done properly – mainland, cairn, board, proclamation, several people, lusty singing – lusty because in the confined space of the valley our voices seemed very loud and cheerful compared with previous efforts, which have been on the tops of hills where the voices seem thin and are easily lost in space.[23]

Here, finally, the human presence seemed momentarily resonant. Simmers reported that the ceremony was made 'doubly correct' by the pouring of a little champagne over the cairn, and the rest down their throats. And the empty bottle was also left there, as another kind of proclamation.

Antarctic scholar Christy Collis has superbly analysed the cultural processes through which six million square kilometres of Antarctica were made Australian territory.[24] She focuses on the Proclamation Island moment, the symbolic heart of the BANZARE voyages, a scene that became the centre of Frank Hurley's film of the expedition and which was later reproduced on the cover of an official history of the Australian Antarctic Territory and an Australia Post Office stamp. The British had acquired the Australian continent with the approved rituals under the doctrine of *terra nullius*, because it was regarded as belonging to no-one, and now settler Australians themselves had gone south with the same law and ceremony to acquire a polar empire. And in Antarctica, Australians found a territory that could be claimed without moral complications, unlike their own Aboriginal lands. There was something redemptive about the continent's unalloyed whiteness. Collis reminds us that a shocking massacre of Aboriginal people took place at Coniston in Central Australia in 1928, just as BANZARE was preparing to go south to claim genuinely uninhabited lands. The *Discovery* sailed with a plain white flag.

The physical presence of the explorer – especially one such as Mawson in a ship such as *Discovery* – was essential to the legal and emotional force of the sovereignty rituals.[25] It was Mawson who had to set foot on the island first and Mawson who had to touch the oar to the rock. It was Mawson who read or recited the proclamation (occasionally with prompting). It was Mawson who, through documents he carried, was legally invested by the Crown to claim land. Thus, writes Collis, 'a ponderously worded text transformed Mawson's body into a legal vessel'.

Mawson had not carried such legal power south in 1911, so the BANZARE voyages – as well as 'filling in the gaps' – also revisited the sites of Mawson's earlier expedition. There is a self-consciousness and discomfort in all legal formalities, especially amongst Australians, but the BANZARE voyages had a heightened sense of history and irony. As well as a voyage of legal enactment, it was also one of historical re-enactment. They revisited the earlier expedition's hut on Macquarie Island. Frank Hurley found a reproduction of one of his own World War 1 photos on the wall of a nearby sealer's hut abandoned in 1919. On celebratory occasions they sipped 'Discovery port', laid down in 1902 to commemorate the ship's first voyage under Captain Scott. They anchored in Commonwealth Bay, raised the flag at Cape Denison, and re-entered the old expedition hut there. Mawson was already a tourist to his own history. He was proud that the hut was still standing, even though the ice had penetrated it. Inside it was like a 'fairy cavern', and his fellow expeditioners finally saw the place they had heard about and imagined so vividly. When they made tea outside, they could have lit the fire with wood, for there were plenty of dry planks lying about, but the hut was already historic and so was the timber. In any case, Mawson insisted on firing up the Nansen Cooker because that's what real explorers did and 'nobody knew how to work them but himself'. They took souvenirs from the hut. When expedition members had their group photo taken, Mawson insisted that they 'all dress up for it and make it look cold'. 'He loves a bit of effect like that', observed Campbell.[26] One experienced expeditioner lived with one hand constantly on Mawson's book of the 1911–14 expedition, *The Home of the Blizzard*, 'ready to produce the answer to any question at a moment's notice'. 'These old explorers die hard', reflected Campbell. 'I wonder if we'll all live in the past like this in the years to come. I hope not.' Here they were on Earth's final frontier, already steeped in the past.

Stuart Campbell affectionately recorded in his diary the activities of 'the O.M.' (the Old Man). Approaching the age of 50, Mawson emerges from this private record as likable, even lovable, as boyish in his excitement, immensely knowledgeable, completely lacking a sense of humour, needing his men to be confident and positive, subject to frequent changes of mind and huge mood swings 'like a pendulum', but indubitably the emotional and intellectual focus of the ship: 'He's got thousands of faults and yet he's undoubtedly "The Man" on the expedition.' The Old Man's body, inscribed as it was with the trauma of Antarctic endeavour and plagued by sciatica and lumbago, was their most precious piece of history, and their most effective legal instrument.

Near the old hut at Commonwealth Bay, on 5 January 1931, he 'Proclaimed Land'. Campbell recorded the occasion:

> We formed a mass around the foot of the mast and the OM slightly embarrassed read the Proclamation. Hurley filmed it. There was some difficulty at first because the instructions said to form a hollow Square but owing to the nature of the ground this was impossible. Still one can only hope and pray that our omission of this very important rite does not arouse any international complications. Then after a lot of argument as to

which wire should be pulled the captain hoisted the flag ('Caps on or off Dux?' queries Johnstone). Then sang *God Save the King*. Another photo. We then cheered again for the cinematograph and sang *God Save the King* for the cinematograph again. Then sang *God Save the King* again for the cinematograph. And thus were thousands of square miles of virgin ice clad land claimed for His Majesty King George V by his dearly beloved servant Douglas Mawson (what a bloody farce).

The men felt not just amusement at these proclamations, but contempt. Following the ceremonies, where even Mawson seems 'embarrassed', they retired to their bunks and laughed or exploded into their diaries.

'Rumor Says Mawson's Expedition Was Flag-Planting Excursion' declared a headline in *Smith's Weekly* at the time of the return of the first BANZARE voyage of 1929–30. These expeditions achieved, in Mawson's words, not just 'the landing and hoisting of the Union Jack on new land' and ultimately the establishment of Australian Antarctic Territory, but also the charting of coastline extending through about 40 degrees of longitude, the definition of the Antarctic continental shelf, the discovery of new whaling grounds off the Enderby Land coast, a meeting with the Norwegians that resolved some tensions over territoriality, and a growing conviction amongst Australians that their country should have a presence in Antarctica.

Once Scott and his four exhausted companions had turned their backs on the South Pole in the late summer of 1912, no-one was to stand again at 90 degrees south for 44 years. But in 1929, a North American did plant a flag on it, from a great height.

If Norwegians ruled the Antarctic seas in the interwar period, it was the North Americans who colonised the ice. North Americans were less traumatised by World War 1 than Europeans, and led the return to Antarctica. And it was Richard Byrd who became the dominant Antarctic figure of the late 1920s and '30s. He was a handsome, charismatic Virginian who had established his fame as an adventurous airman. In the mid-1920s, Byrd had found himself – like Scott 15 years earlier – in a race to the pole with Amundsen, but this time it was to the North Pole, by air.

Historians, like explorers, find themselves voyaging to unknown and unexpected destinations. And even more than novelists, they can be surprised by the characters in their own stories. The people you are trying to understand often take you where you don't expect to go. The French writer, Gustave Flaubert, once said that he had to furnish the whole house before he could describe just one room. Writing history is like that. Sometimes you need to look in the attic before attending to the basement. Here, we need to pay some attention to the North Pole before we can return to the South.

Amundsen and the North American Lincoln Ellsworth had flown to within a few degrees of the North Pole in 1925, and Amundsen was preparing another assault, by airship, in May 1926. Suddenly he found he had a competitor. The North American Richard Byrd and his co-pilot Floyd Bennett were also attempting to fly to the North Pole, leaving from the same place. When Byrd and Amundsen first met the year before, their companions had been struck by the contrast between the two men; Amundsen 'shaggy-haired and rugged and seasoned as an oak-mast', and Byrd 'immaculately groomed and slender and every inch the cultured gentleman'. Amundsen was half a head taller than Byrd and, sensing the American's discomfort, invited everyone to be seated.[27] When they met again in Spitzbergen, apparently racing one another to the pole, Amundsen made sure his rival got his flight in first and he also supplied Byrd and Bennett with survival gear in case their plane was forced down.[28] Scott's death still shadowed Amundsen's life and he wanted no further resentment or controversy. On 9 May 1926, Byrd and Bennett returned from a 15 and a half hour flight in their Fokker Trimotor plane with the news that they had indeed become the first to fly over the North Pole. Amundsen graciously welcomed them back and warmly embraced them. The news of Byrd's feat was greeted with some satisfaction in England, for many saw it as Scott's revenge. Three days later, Amundsen himself flew over the pole. The true significance of that event remained, for the moment, hidden.

Byrd's North Pole triumph catapulted him to heroic status. Aviators were idols of the 1920s and a polar aviator was as exalted as a movie star. In America, these were the 'ballyhoo' years when celebrities, big business and mass media began to discover their combined power. Byrd was both a manipulator and beneficiary of this emerging corporate chemistry. When his eyes turned south, they did so with a mixture of motives. He was attracted to the Earth's last vast realm of *terra incognita*. He felt that this great silent continent was still sleeping and that he would help awaken it. He would be its prince. Byrd certainly became its 'mayor', which was – appropriately – the civic term he chose to describe the relationship he would build with Antarctica.

At first glance, Byrd was a fitting heir to the heroic age. He was an admirer of Mawson, a rival of Amundsen's in the north, and he became celebrated as America's 'last explorer'. Of Scott he declared: 'In dying, in failing, he gave the world something much more worth while than success. He gave it an intellectual experience worth the attainment of a dozen South Poles.'[29] In going to Antarctica, Byrd planned laboriously for success, but he too was seeking an intellectual experience, even a spiritual one. In his personal magnetism, romanticism and daring, he shared much with the legendary figures of Antarctica's first phase of continental exploration. But, as a systematic coloniser of the ice, he was also a bridge to the latter half of the twentieth century and the era of permanent occupation of Antarctica. He aimed to transport and implant a society in Antarctica, not just a lone hut in the wilderness, but a 'city', a mini-civilisation. He became an advocate for permanent colonisation as the way to secure national possession.

He would awaken Antarctica with technology, and so his expeditions were novel for their scale and mechanisation. His three chief instruments were the radio, the aeroplane and the aerial mapping camera.[30] He also carried the Stars and Stripes and made unofficial territorial claims on behalf of the United States. His chosen base was deliberately on the edge of areas still unclaimed.

In 1928, he took 82 men south and established what he called a 'colony' of half-buried huts on the Ross Ice Shelf. He named it 'Little America'. It was situated on the Bay of Whales, just a few kilometres from Amundsen's old base, 'Framheim'. On 30 December 1928, while seeking a site for the base, Byrd's advance party pitched camp on 'the Barrier', as the ice shelf was known, and became 'the first members of an American expedition to sleep on the Antarctic Continent'. In his diary that night, Byrd wrote: 'It is as quiet here as in a tomb. Nothing stirs. The silence is so deep one could almost reach out and take hold of it.' That silence would, for the next year, physically embrace them, forever tightening its ice-like grip. Half the party stayed for the winter and literally dug in, embedding themselves in this floating glacier, living in a village of huts connected by tunnels and haunted by the occasional rumble of an ice-quake.[31]

The hum of technology marked the distinctive character of this expedition. When winter blanketed their colony, all that could be seen of Little America from a distance were the radiotowers, the American flag and the smokestack. Mawson's expedition of 1911–14 had pioneered the use of wireless in Antarctica, and had battled to erect and repair the transmission masts against the katabatic winds. But Mawson's publisher had warned him that frequent communication might diminish the romance of exploration. When Byrd's first radio mast was erected on the Ross Ice Shelf, he too reflected that although it ended the isolation of Antarctica, it could also 'destroy all peace of mind, which is half the attraction of the polar regions'.[32] In November 1920, perhaps one US household in 500 possessed radio equipment.[33] By 1929, even at Little America, they had daily press bulletins and every Saturday there was a two-way broadcast with the US, in which the men could talk with their families and friends back home.[34] In the early Antarctic spring of 1929, Wall Street collapsed and the men at Little America heard it happen by radio. Byrd recalled that they 'watched Chrysler go down with the thermometer'.[35]

The *New York Times*, whose circulation had surged through its promotion of Robert Peary's claim to the North Pole, sent a journalist, Russell Owen, south with Byrd.[36] In the early stages of the expedition, Byrd decided to withhold Owen's dispatches from his fellow expeditioners. 'I had a definite reason for this', he wrote. 'Publicity is the worst disease that a weak man can get. It nourishes dissatisfaction, even jealousy. It can rend the most compactly organised group of men. Down here, I had hoped to get away from that evil. If it were possible, I wanted to create a single attitude – a single state of mind – unfettered by the trivial considerations of civilisation.'[37] Byrd enforced a division between what the outside world knew and what his men knew. He was a creature of the new media, a professional manipulator of information.

In a career devoted to discovery by air and an expedition dedicated to showing what modern technology could do for Antarctic exploration, Byrd's flight to the South Pole became the fitting centrepiece of this mission. It was the bookend to his North Pole flight three years before. Aviation in Antarctica in this period was perilous. The Australian Hubert Wilkins had made the first Antarctic flights from Deception Island in November and December 1928, the second an 11-hour flight across the Antarctic Peninsula and covering 2100 kilometres (1305 miles). On a long flight into the unknown, a single mechanical fault, navigational error or weather event might ground you forever in remote, inaccessible terrain. Byrd was certainly conscious of these dangers; some say that he was wracked with fear and inclined to panic. He was not a great pilot himself and not even a particularly skilled navigator. Nor was his sextant always at his side on his momentous flights: one historian alleges that 'Byrd's sole navigational instrument during the [South Pole] flight was a bottle of cognac.'[38] But he chose his co-pilots well.

There was at least one of Byrd's companions on the ice during that long winter of 1929 who knew his dark secret. Just as Scott was eventually debunked, so was Byrd's reputation savaged within a couple of decades of his death. Doubts that had dogged his life hardened into public challenges. Had he reached the North Pole on that 1926 flight? Even at the time, attentive observers had wondered if his plane had been in the air long enough to cover the distance, especially as it returned with a leaky engine. It's possible that Byrd was initially mistaken in believing he had reached the pole. But he had also been under immense public and financial pressure to make that urgent flight successful, and as well, Amundsen had been there awaiting his return, like a giant shadow. Byrd's only sextant fell and was damaged on the home journey, so his surprisingly rapid return flight had not been plotted. Inexplicably, he had failed to drop over the pole any of the 100 small and several large American flags he had carefully packed in the plane, territorial signatures which Amundsen might have sighted and confirmed on his own flight just days afterwards.[39] Byrd's family impounded his papers and expedition records for three decades after his death, so it is only relatively recently that historians have scrutinised his flight record, which some have judged to be doctored. Aging compatriots condemned Byrd as cold, ambitious, deceitful, even cowardly. Stories emerged that he had rolled out of the plane drunk on his return to Little America from the 1929 South Pole flight. Good friends, conscious of their debt to him, sometimes chose a judicious silence. In fact, for a while Byrd was almost forgotten because his defenders preferred to stifle research and writing on him than be party to a complex portrait.[40] Even his contemporaries said of Byrd: 'He is very well known, but nobody knows him very well.'[41] The man was contradictory: he could inspire love and loyalty as well as envy and resentment; he was egalitarian yet egotistical; he made a business of personal fame yet craved solitude. We need to try to understand Byrd's temperament as best we can, because – as we will explore in *Solitude* – these tensions in his character were to place him in a spectacular predicament in the winter of 1934 during his second expedition to Antarctica.

One implication of this revisionist historiography is that it seems possible that the South Pole was attained before the North. It also belatedly awards Amundsen a rare honour. If Byrd did not fly over the North Pole in 1926, then Amundsen reached it first. If – as also now seems likely – neither Frederick Cook nor Robert Peary made it to the North Pole across the ice in 1908 and 1909 respectively (they too may have fabricated their claims), then Amundsen was the first there by any means, and therefore the first to both poles of the Earth. He was never to know this, for in 1928 he died in a plane crash during a rescue mission in the Arctic. It also seems that Amundsen may have been unusual in telling the truth, and in assuming it of others.

The lives of Byrd and Amundsen illustrate how, in the early twentieth century, the poles of the Earth were considered one realm. For a while their shared extremes, and their joint role as a final earthly frontier, obscured their real physical and cultural differences. When flags were planted at each end of the globe and explorers looked for new quests, the unconquered Mount Everest – another domain of ice and snow – became known as 'the third pole'.

The reasonable doubts about Byrd's flight over the North Pole in 1926 did not apply to his South Pole flight of 1929. He and his pilot, Bernt Balchen, photographer Ashley McKinley, and relief pilot and radio-man Harold June, did fly over the featureless wasteland of 90 degrees south, or as near as they could estimate. They had to throw out their two emergency food bags on the journey so that their Ford Trimotor airplane could lift itself over the Transantarctic Mountains and gain the polar plateau. And when they thought they were over the pole (Amundsen's tent of 1911 could offer no verification for it had long been swallowed by the ice), Byrd opened the trapdoor of the plane to drop his carefully prepared claim. It was a small American flag weighted with a stone from the grave of Floyd Bennett, his co-pilot on the North Pole flight who had died of pneumonia in 1927. Floyd had kept the secret of that trip, or so Byrd thought.[42] The news of the South Pole triumph was immediately radioed from the plane (which was called the *Floyd Bennett*) direct to Little America and there relayed to the *New York Times* station and announced by loudspeaker to milling crowds in Times Square. The American flag, recorded Byrd, had been advanced 1500 miles farther south than ever before. 'And that', he concluded, '... is all there is to tell about the South Pole. One gets there, and that is all there is for the telling. It is the effort to get there that counts.'[43] His pilot, Bernt Balchen, wrote that he was glad to turn the plane back towards Little America. 'Somehow our very purpose here seems insignificant, a symbol of man's vanity and intrusion on this eternal white world. The sound of our engines profanes the silence ...'[44]

In early 1939 an expedition from Adolf Hitler's Germany bombed Antarctic ice with hundreds of cast-iron swastikas, each carefully counterbalanced so that it stood upright on the surface. This rain of metal spears was reinforced by the landing of a shore party which planted the flag 'to secure for Germany her share in the approaching division of the Antarctic among world powers'.[45]

In the early 1940s, Britain and Argentina established their own – and removed each others' – symbols of possession on the Antarctic Peninsula and nearby islands. Stamps, post offices, maps and films were weapons of war in a region which, depending on your nationality, was known as Tiera O'Higgins, Tiera San Martin, Palmer Land or Graham Land. In 1939, the British Ministry of Information identified the story of Scott's last expedition as an appropriate subject of propaganda, and the Ealing Studios' film, *Scott of the Antarctic*, was released in 1948, in time to strengthen disputed southern claims with patriotism. In February 1952, Argentine soldiers fired machine-guns over the heads of a British geological party trying to land at Hope Bay on the Antarctic Peninsula. The following summer, in retaliation, British authorities deported two Argentines from the South Shetlands and ordered troops to dismantle Argentine and Chilean buildings on the islands. But British and Argentine crews still had good enough relations at Deception Island to hold soccer matches, although they disagreed as to who was the home team.[46]

Immediately after World War 2, in 1946–47, America's Admiral Richard Byrd led the largest ever expedition south (his third), called 'Operation Highjump'. Just as Mawson had been mobilised for Australia's shows of sovereignty, so was Byrd assigned south when the United States wanted to assert its presence. The 'operation' involved 4000 personnel, a dozen icebreakers and an aircraft carrier. One newspaper declared that it was the beginning of an international race for uranium. 'The Byrd show was mainly a Naval manoeuvre', wrote BANZARE veteran Stuart Campbell, 'to test out flying gear & operation thereof in Polar regions and for obvious reasons they thought the tests could be carried out with less international comment in the South than in the North.'[47] An official Navy directive of 1946 identified the expedition as a means for 'consolidating and extending United States potential sovereignty over the largest practicable area of the Antarctic continent'.[48] Operation Highjump attracted international criticism for its unambiguously military character. In 1950, the USSR announced its renewed interest in Antarctic exploration, occupation and sovereignty. The Cold War and other conflicts had found their way to the coldest part of the planet.

Since before the war, Douglas Mawson had been lobbying the Australian government to consolidate its Antarctic claims and to continue the job that his BANZARE voyages had begun. So, the Australian National Antarctic Research Expedition (ANARE) was established in 1947 and, as Campbell recorded, 'Mawson … wants me to run the show'. The fact that other nations were also strengthening their postwar presence in Antarctica had quickened Australia's commitment.

The Australians had another reason to hasten. They had a secret plan to forestall an American claim to Heard Island, a bleak, ice-covered volcano in the 'furious fifties'. An official Australian landing on Heard Island ('Island X') had been confidentially requested by the British government to prevent the Americans from claiming it on the basis that it was they who had discovered it in 1849. The British and Australian plan to go to Heard 'was all very hush-hush in the early part' recalled Phillip Law, ANARE's newly appointed senior scientific officer. Law was a 35-year-old physics graduate who knew that 'they used science as a pseudo reason for going – as all the nations did – in order to cloak the real reason'.[49] The necessary displays of sovereignty were planned. There was to be a flag-raising ceremony conducted by Campbell at Heard, and the Postmaster General's Department had organised a special postmarking of envelopes to be mailed from the island.[50] This first summer of ANARE's activities also comprised plans to establish a meteorological and scientific station on Macquarie Island, and a reconnaissance of the coast of the Australian Antarctic Territory to determine the site for a future scientific station on the continent itself. In 1946, with the huge American expedition in progress, the Australian government had received legal advice that discovery followed by annexation was insufficient for claiming territory unless followed by some continuity of 'effective occupation'.

Although the wharfies pilfered their grog and emergency chocolate rations before they sailed on 16 November 1947, the expedition to Heard Island encountered its most serious challenge in the wild seas of the roaring forties, where their ship, the *LST 3501*, a relic of World War 2, developed a crack in its steel deck a metre long.[51] It yawned alarmingly on the crest of a wave, but the crew – with exquisite timing between waves – bolted a plate to secure it. The sinuous twisting of the *LST 3501* through a southern swell became legendary: a distant figure walking the long passageway below deck would disappear from view as the ship twisted from port to starboard.[52] When Campbell landed at Heard Island in mid-December 1947, he shooed away the elephant seals at the door of the tiny Admiralty Hut, stepped inside and found emergency rations and a Union Jack left by Mawson's BANZARE party in 1929.[53]

On 26 December 1947, under instruction from the Department of External Affairs, Campbell conducted an official proclamation ceremony on Heard Island. In spite of his tiresome exposure to these rituals on Mawson's BANZARE voyages, he was unsure exactly what he should do and 'looked through books and books and books and a huge volume of External Affairs papers on how to claim land'. They raised the flag, buried a capsule with the proclamation, built a cairn of rocks and signed a declaration claiming Heard Island and the adjacent McDonald Islands for the Commonwealth of Australia.[54] Leaving a party of 14 men for the full year at Heard, the *LST 3501* voyaged again in February 1948 and left another party of 14 men for the winter on Macquarie Island.

Meanwhile, the reconnaissance voyage to Antarctica had met with serious difficulties. Their ship, the *Wyatt Earp*, an old sealer purchased by the Australian government from Lincoln Ellsworth in 1939, proved to be hardly seaworthy and was forced to return twice to port for refitting. It was a small wooden sail-assisted ship with limited ice-breaking abilities and when it did finally reach the Antarctic coastline, the season was well advanced, the ice was heavy and there was little opportunity or capacity to make a landing or even to survey. Because of the lack of a suitable ship, Australian plans for establishing a presence on the continent were stalled for several years, and ANARE concentrated on developing scientific programs at their two island bases.[55]

But in 1953 the Australian government chartered the Danish ship, *Kista Dan*, and ANARE was again voyaging to Antarctica, this time under the leadership of Phillip Law, who had succeeded Campbell as director in 1949. Law suffered from chronic seasickness, yet he was to shape a career that maximised his crossings of the Southern Ocean. As director of ANARE from 1949 until 1966, he made 28 voyages to Antarctic and sub-Antarctic regions and mapped 6000 kilometres (3728 miles) of Antarctic coastline.[56] On this voyage in the summer of 1953–54, he had some flag-planting to do.

The Australian minister for external affairs, RG Casey, compared Phillip Law's 1954 mission to establish a permanent Antarctic base with Mawson's 1911–14 and 1929–31 expeditions. Casey, a strenuous advocate since the 1920s of Australia's presence in Antarctica, had persuaded Cabinet to support this initiative against the opposition of the prime minister, Robert Menzies. 'For strategic reasons', announced the minister, 'it is important that this area, lying as it does so close to Australia's back-door, shall remain under Australia's control.' Meteorology, coal, uranium, aviation routes and food resources were other incentives to Australia's commitment south, as well as the challenge Antarctica offered to the nation's pioneering spirit: 'For us to neglect the Antarctic could be as serious as if our forefathers had confined themselves to a small strip of coastal settlement in Australia and left it to others to develop the resources of the rest of the continent.' Antarctica was an icy extension of Australia's inland frontier.[57]

This expedition was the culmination of Mawson's sledging and flag planting, and the new permanent station was to bear his name. He had been crucial in Australia's return to Antarctica in the post-war period and the establishment of ANARE, and he and Phillip Law had a generally constructive relationship, although Mawson felt that Law lacked sustained field experience. Law was acutely conscious of the Mawson mantle and all that it meant. They were both ardent champions of science. Mawson was privately (and occasionally in public) a critic of Law and the directions in which he was steering ANARE.[58] 'The Old Man' believed that ANARE should do more to pay its own way and he continued to champion the alliance of exploration and exploitation, especially the development of Antarctic fisheries to Australia's economic benefit. He was keen finally to see an official station established on the ice. Leader of the 1954 wintering party was Bob Dovers, son of George, one of Mawson's men from 1911–13.

The *Kista Dan* proceeded to Antarctica via Heard Island, to relieve the Australian base, and then Kerguelen Island, to pick up two men (André Migot and Georges Schwartz) from the French station, who had been invited as 'observers' for the southern journey and its claiming rituals. While at Kerguelen, where the French had also been busy reaffirming sovereignty, the Australians tried out their poor language skills 'with mixed success', but Fred Elliott recorded that, in any case, 'The "Entente Cordiale" was much to the fore and not much work was done.' The Antarctic continent was sighted on 2 February 1954. Elliott later reflected on the emotion of first seeing 'the land of Mawson, Scott and Shackleton' gradually rising above the horizon beneath the palest of blue skies.[59] Their objective was a small outcrop of rock that Phillip Law had picked out from the aerial photographs taken by Byrd's 'Operation Highjump'.

At 5 pm on 13 February 1954, Law gathered his men on the small sloping area of granitic rock on Horseshoe Harbour and officially named the station 'Mawson'. Today it remains as the oldest permanent station on the continent. Another ceremony was about to be inserted into a hectic schedule of clearing cargo heaps. It was a moment of necessary formality when men, fumbling symbols with cold hands in an Antarctic wind, would give enduring expression to the seriousness of their government's political purpose. They 'self-consciously' sang *God Save the Queen* and raised the flag. However, there were some problems unfurling it. The mast was lashed to one of the sledge-caravans and the flag was furled at the top of it, ready to be released by the pulling of a cord. But when Law gave the nod, Dick Thompson pulled the cord and nothing happened. Thompson pulled again but still with no effect. So he pulled harder and the whole mast came down. 'General guffaws from the troops. Phil was not amused', remembered Elliott. Exactly a year later, Phillip Law and Fred Elliott stood at the same spot to commemorate 'the first anniversary of the raising of the flag at Mawson', and one of them was still trying to be grave and the other irreverent. 'I found it hard to stop laughing' wrote Elliott in his diary, 'as he [Law] tried to make it a historical occasion.'[60] Comedy stalks the self-consciously momentous, and so Antarctic history is full of pompous public moments subverted by derisive private ones. Phillip Law's 'troops', just like Mawson's, felt their vulnerability and insignificance most keenly at these moments of imperial display on scraps of rock on the edge of the Great South Land.

## Thursday, 26 December

**66 degrees 17 minutes south,
110 degrees 32 minutes east**

We've arrived. The continent is outside my window. We anchored this morning about 10.30 am in Newcombe Bay. It's stunning to cross such a wilderness of ocean and ice and then to find buildings here – they are both a reassurance and an obscenity – and then to find people emerging from them, and then to find that they have broad Australian accents and that their first words bellowed from the Zodiac as it pulled alongside the ship were: 'What the hell took you so long?'

The mist cleared in time for us to behold Casey in sunlight as we approached, and I was up on the bow looking over the edge into the icy waters. Our gentle final progress was heralded by leaping penguins swimming just ahead of our prow. We seemed to slice the silence of a year. The skittish and playful way those Zodiacs came out to greet us, speeding out like little flies to play around the big metal beast as it prepared to drop anchor, was expressive of excitement and celebration.

As well as excitement on board this morning, there is relief too. 'No more play-acting now', declared Peter ('Bloo') Campbell, the new plant inspector, as we stood together on the C Deck balcony looking at his home for the next 14 months, those primary-coloured buildings on the grey-brown creamy rock and white cornices. Bloo, who comes from the hot hamlet of Humpty Doo in the Northern Territory, recalled that it was exactly a year since he applied for his job (or first enquired about it), and his training began last August. He could tell me about every building on that rock and ice outcrop even though he had never set eyes on them until this moment. From photos and maps he had learned about the power and the piping of every steel structure. The time he had to wait between our anchoring and his getting ashore was perhaps the hardest hour of his long anticipation.

The ship is still for the first time in ten days and everything is sparkling, the blue waters of the bay are shimmering, and the station (appropriately known as Legoland) is a series of red, green, blue, yellow and orange buildings on the hill a kilometre away. I'm told it is 6 degrees Celsius (42.8 degrees Fahrenheit) right now! I've happily had my cabin porthole open. Small distant icebergs catch the sun incandescently like beacons. The ship's artist has set up her easel outside at the top of the ship, on the port flying bridge, and I've just been taking photos of her in her magnificent outdoor studio.

It's been fun to watch the mechanics and barge operators − people I've been having breakfast with every day − swing into action with such gusto and relief. It's like watching a good dancer; the ease with which they move with their machines, the skip in their feet after ten days of waiting. The man on shore who oversees all unloading operations is known as 'the beachmaster', which is also the term for the dominant breeding male seal who maintains his territory through sheer strength and display. The team unloaded the barges into the water using the cranes built on the port side of the *Polar Bird*, and started their engines. A fuel line was connected from ship to shore. When they tested the fuel pipe with air, they found a hole in it four metres from the ship. Once this was patched, the essential 650 000 litres of fuel needed for the next year started flowing to shore. Zodiacs move to and fro along the fuel line, watching for breaks, patrolling passing bergy bits and lifting the line over them when they threaten to drift in and sever the flexible pipe. The rest of us get to be useful too. On two of my six days I am rostered to work in the store on shore, helping with the unpacking; an essential bit of fieldwork for any historian interested in material life.

The ship swings on its anchor in a slow arc, creating a hazard for the fuel line. The Zodiac crews are working hard, cradling the precious pipe – a station's lifeline – over their tiny stern, trying to balance the rogue movement of the ship, striving always to maintain tension and clearance. Antarctica is glistening at the moment, lulling us into trust. I'm told that if you see mist on the moraine line (the rocks along the horizon here), then you have just a few minutes, perhaps 30, to find shelter.

# COLD PEACE
## Reds down under

The launching of the Russian spaceship, *Sputnik*, on 4 October 1957 seemed to many in the West a threatening symbol of escalating Cold War rivalry. But in Antarctic skies it was welcomed as the culmination of a huge, cooperative human endeavour. *Sputnik* was the most visible efflorescence of the International Geophysical Year of 1957–58 (known as IGY), and the appearance of the new satellite (followed by a US one some months later) made it 'the Year of the New Moons'. IGY was a great turning point. It was the single biggest cooperative scientific enterprise ever undertaken on Earth, a hugely successful intervention of science into politics, and it was centred on Antarctica. IGY cut through the increasing cacophony of post-war territorial rivalries down south.

International Polar Years had previously been declared by the scientific community in 1882–83 and 1932–33, but they had focused almost entirely on the Arctic. In the 1950s, the idea emerged that there should be a third polar year, at a time of an expected peak in sunspot activity, and the plan quickly grew to include coordinated scientific observation of the whole globe – an International Geophysical Year. It was agreed that Antarctica would become the priority, as well as those other regions now made newly accessible by technology: outer space and the ocean floor. Nearly 30 000 scientists from 66 nations took part at locations across the globe. In Antarctica, 12 countries were involved: Argentina, Australia, Belgium, Chile, France, Great Britain, Japan, New Zealand, Norway, South Africa, the United States and the USSR. Both Chinas also applied to join IGY, but neither would tolerate the presence of the other. During IGY, the number of stations rose from 20 to 48, and the wintering population increased from 179 to 912. The summer population in Antarctica reached almost 5000.[1] Free exchange of data between nations was part of the agreement, and there was a constantly expressed intention to put science before politics. As Phillip Law put it, for 50 years the main motive for Antarctic work had been territorial conquest and 'scientific work was, in general, of secondary importance. The IGY changed all this.'[2]

The American presence in Antarctica had increased immediately after World War 2 through a series of military-inspired expeditions that hardly even paid lip service to science. Polar military experience became essential to the US because it now faced a superpower rival across the Arctic. But 'big America' had so far resisted bringing 'Little America' under its legal jurisdiction, and it refused to recognise any Antarctic claims by other nations. Doubts had always existed about the legitimacy of any US claims made on the basis of Charles Wilkes' wayward mapping of ice, cloud and land in 1840, but Admiral Richard Byrd's 'colonies' did provide the grounds for a possible claim. But the United States had always declined to claim Antarctic territory, partly out of fear that the Soviet Union might do the same, and partly to reserve access to all sectors of the continent. Douglas Mawson was one of many who, in the mid-1950s, still hoped that America would make a formal territorial claim so that it could more readily sympathise with the geopolitics of claimant nations.

In 1948, the US government proposed a different kind of Antarctic politics. It formally suggested an international trusteeship for Antarctica, proposing that the seven claimant states, Argentina, Australia, Britain, Chile, France, New Zealand and Norway, together with the United States, should form a condominium. It would, however, require the renunciation of claims. The apparent idealism of this *terra communis* was tempered by the proposal's immediate goal of excluding the Soviet Union from the power bloc. The claimant nations rejected the proposal for internationalisation because they were reluctant to cede sovereignty, perhaps Australia most of all. The Chileans, however, replied with a modification which became known as the Escudero Plan. This proposed that sovereignty be suspended rather than renounced. By 1950 Chile and the US had agreed to pursue this revised plan for internationalisation, and their rare unity of purpose came from the common goal of excluding the Soviet Union. But by 1954, the accelerating momentum of IGY was already changing the political scene, for the Soviet Union had emerged as one of the 12 nations committed to the cooperative scientific campaign in Antarctica. As the geopolitical historian Klaus Dodds has argued, 'The emergence of the Soviet Union as a dominant presence in the Antarctic was the decisive element in the IGY period.'[3]

After the death of Stalin in 1953, Soviet scientists began to join more freely in international organisations and the USSR became a member of the Special Committee for Antarctic Research, which led the Antarctic component of the IGY. The USSR had hitherto shown little interest in Antarctica, although the early explorations of Fabian Gottlieb Thaddeus von Bellingshausen, in the sloops *Vostok* and *Mirny* in 1820–21, had given it grounds for territorial claims should it have wished to press them. In 1946, Soviet whaling flotillas entered the Antarctic ocean for the first time. Now, as part of IGY, the USSR proposed an expedition, including the establishment of three stations on the ice, one of which was to be – of course – at the South Pole. But the Americans had also proposed, as part of IGY, the establishment of a base at the South Pole, which is the one place in Antarctica where a single base can occupy every possible territorial sector. To avert superpower jostling at 90 degrees south, the Soviets graciously agreed to occupy two other poles, the South Geomagnetic Pole (or 'auroral pole') and the Pole of Inaccessibility in East Antarctica (that is, the place furthest from any coast, or the 'thermic pole'). The Russians did not regard these poles as second-best. Rather, the challenge of colonising such remote and unknown regions of the ice cap would be an even tougher demonstration of polar expertise.

Australia was the claimant nation most anxious about the Soviet Union's presence in Antarctica. In 1954, the year that the Australian government established Mawson station, diplomatic relations between Australia and the Soviet Union were suspended because of allegations of communist espionage in Australia. On the eve of the 1954 Australian federal election, Prime Minister Robert Menzies announced the defection of a Soviet diplomat, Vladimir Petrov, and his wife Evdokia. Petrov revealed that he had received information from a communist spy ring in Australia, and the government called a royal commission to investigate the allegations. Fear of 'the reds' was rife. Three years earlier Prime Minister Menzies had narrowly lost a referendum to ban the Communist Party. The Petrov affair and subsequent royal commission fuelled hostilities towards the Soviet Union in Australia and ensured that 'the red scare' was the substance of hysterical party politics.[4] By 1955, when suspicion of Soviet activity was at its height in Australia, it became clear that the USSR planned to be a full participating member of IGY and would be establishing a base in the centre of the Australian Antarctic Territory. Australians were very distrustful of Russian intentions and of how they would use the guise of science. A Soviet presence south of Australia was regarded not only as a threat to Australian Antarctic sovereignty but also as a threat to Australia itself. 'We do not want the Russians to mount installations in the Antarctic from which they could drop missiles on Melbourne or Sydney', explained the minister for external affairs, Richard Casey, to a meeting of Commonwealth partners in London in 1957. Sovereignty in Antarctica was a kind of buffer to protect sovereignty at home, just as Australia's inland deserts were a buffer against Asian imperial ambitions ('the yellow peril') in the north. Australians felt that these vast tracts of space on the map – the white desert and the red heart – needed to be occupied to be defended. Law stressed to his political masters the necessity for Australia to confirm her southern claim with an active presence: 'If she did not occupy it some other nation would', he argued, using words that Australians more familiarly applied to their own inland and north. And since Australia could not stop the Soviets from undertaking scientific activity on Australian Antarctic Territory, it was important at least to assert a 'spirit of possession'. Casey spoke to Cabinet about Russian intentions and issued a statement welcoming the Soviets to Australia's sector of ice, for 'As we can't stop them, we'd better take it with good grace.'[5]

The first Soviet Antarctic base, Mirny, named after one of Bell-ingshausen's ships and located in the middle of Australian Antarctic Territory's coast, materialised on the ice with awesome speed in early 1956, under the supervision of architect AM Afanasyev. The unloading of 10 000 tons of buildings, stores and equipment was done under dif-ficult conditions, with hurricane-strength winds breaking the ice sheet between ship and shore so frequently that the road across the ice barrier had to be renewed five times in two weeks. Erected along Lenin Street were 19 bungalows, each of 69 square metres (743 square feet), and five further buildings made of steel shields which housed the electric station, aerological pavilion, bathhouse, laundry and warehouse. A log cabin was built for the magnetic pavilion and a seismic pavilion was cut into solid rock. The Russians also erected a radio station, power station, a mess room to feed 90 people at once, kitchen, medical centre, warm foodstore, telephone exchange and mechanics' shop. Within a few weeks, the snow was criss-crossed with tractor tracks, and aerials and telephone poles had sprouted from the ice. By 1957, 189 people were wintering at Mirny. These people knew about ice and snow and many had been toughened by Arctic experience. Their Antarctic work was done out of a new arm of the well-established Arctic Research Institute.[6]

From Mirny that first winter, the Russians also established the first station in the interior of Antarctica, Pionerskaya, at an elevation of 2700 metres (8858 feet) and 375 kilometres (233 miles) inland towards the South Geomagnetic Pole. It was opened on the day the sun no longer rose above the horizon, 27 May 1956. Oasis station, in a coastal region free of glaciation was opened on 15 October 1956. Komsomolskaya station, 850 kilometres (528 miles) from the coast and at an elevation of 3540 metres (11 614 feet), was opened on 6 November 1957, and Vostok station (named after Bellingshausen's other ship) was established near the South Geomagnetic Pole the following month. Sovetskaya, 3700 metres (12 139 feet) above sea level, 1420 kilometres (882 miles) from Mirny and near the Pole of Inaccessibility, was opened on 16 February 1958. The Soviets had staked out Australian Antarctic Territory.[7]

In spite of official diplomatic tensions, relations between the two countries were friendly and cooperative on the ice. Phillip Law visited the site of the first Soviet station in the summer of its establishment. After initial inhibition on both sides, the men got down to business, and the two leaders were impressed by one another. The leader of the first Soviet Antarctic Expedition of 1955–57 was Mikhail Mikhailovich Somov, a respected oceanographer with Arctic experience. Somov told the USSR Academy of Sciences that he appreciated Law's helpful, practical advice on weather conditions and equipment, and found him 'a very thoughtful, observant, highly erudite and at the same time a very straightforward cordial man. This impression was strengthened at each new encounter with him, three times in Australia, his home country, and twice in the Soviet Union.' Mirny opened on 13 February 1956 with a flag raising ceremony. Law reported to Casey that Mirny was 'larger and more elaborate than any station in Antarctica' and that the expedition had an impressive and comprehensive scientific program and 'appears to be civilian and non-military'. Law found the Russians frank, friendly and welcoming of his visit and interest. The Soviet journalist, Yuhan Smuul, who met Law in Antarctica a couple of years later, commented on Law's 'close affinity with the Soviet scientists due to the commonality of their scientific interests'. Mikhail Somov, looking back with satisfaction in 1966 at the first ten years of Soviet Antarctic activity, celebrated the warmth of international cooperation on the ice. He was referring not so much to the willingness of nations to help one another during emergencies, but to 'the most commonplace mutual aid, which lacks any romanticism or heroics but is full of good, warm humanity'. Somov wrote of the 'intimate and friendly association' among scientists: 'Here, the atmosphere is one of true friendship and selfless mutual help which, unfortunately, has not yet been attained in the other areas of our planet.' He especially recalled the assistance he received in that foundation season from the nearest station, the Australian base of Mawson, 1500 kilometres (932 miles) from the chosen site for Mirny. Within a few days of the Russians contacting Mawson station, explained Somov, the leader of Australian operations, Dr Law, 'sought permission' to pay a 'courtesy call'. Somov readily agreed to the visit for he was keen to learn from a nation that had already experienced two summers and a winter nearby. Within six days of Somov's first message, the *Kista Dan* lay anchored beside the *Ob* and a meeting of the two expeditions took place 'in an atmosphere full of natural, and therefore undisguised, curiosity on both sides and unusual simplicity'. Their exchange of information on a variety of levels was open and supportive. Somov concluded his article by observing that 'Antarctica remains the paradox of our globe; in the most under-developed continent are practised the most advanced ideas in the world regarding friendship between nations. In the coldest continent are to be found the warmest human relationships.'[8]

After Law's visit to Mirny in January 1956, he was informed by the assistant-secretary of the Department of External Affairs, JG Kevin, that 'any display of undue friendliness' on the part of the Soviets was to 'be played down' in any statements Law might make about the visit, and that 'any suggestion that Australian officials were *invited* ashore on Australian Territory by the Russians should [be] avoided'. Furthermore, Law was urged to avoid even mentioning the station's name, Mirny, because it recalled Bellingshausen's ship and thus reinforced the recognition of very early Russian interest in the Antarctic.[9] The Department of External Affairs would have been disturbed by Mikhail Somov's description of international goodwill on the ice and especially by his report that Dr Law 'sought permission' to visit Mirny.

But Law and Mawson continued to reassure the government and the public about the genuine and open motives of their fellow scientists from the Soviet Union, much as Edgeworth David had done over 40 years earlier when the first Japanese expedition to Antarctica had wintered at Parsley Bay, Sydney in 1911. Under suspicion as secret service agents of an imperial Japan, Lieutenant Nobu Shirase and his men on their ship, the *Kainan Maru*, had warmly thanked Professor David for allaying the concerns of Sydneysiders: 'you were good enough to set the seal of your magnificent reputation upon our bona-fides, and to treat us as brothers in the realm of science', wrote Shirase to David as his ship departed. 'Whatever may be the fate of our enterprise, we shall never forget you.' When the *Kainan Maru* weighed anchor in Sydney harbour and sailed south again (where they sledged to 80 degrees south in the same season as Amundsen's triumph), the Japanese recorded that suspicions of the local populace had so subsided that 'We felt that we were leaving home all over again.'[10] Mawson had also visited the Japanese ship in Sydney in 1911, and now it was the Russians he welcomed at Port Adelaide in 1956 on their return from establishing Mirny. 'Some Australians hold that theirs is a pretense [sic] at peace whilst hatching a deep plot', Mawson wrote to Davis. 'My view is not so. I think they genuinely desire peace and are attempting to put their house in order after the shocking rule of Stalin and his clique.'[11] Mawson was politically conservative and a supporter of Menzies but he recognised and admired the Soviets' genuine scientific commitment. And they, in turn, were honoured to have the attentions of a living Antarctic legend.

The common language of science – and the solidarity of pioneering – continued to break down the political and cultural barriers between Australians and Russians on the ice. 'How absolutely spontaneous in the Antarctic is goodwill towards strangers!' declared John Béchervaise when he first made contact with Russians at Mirny in 1959. 'It scarcely mattered into which sphere of observational work I entered at Mirny, I found equipment, functions, and intentions which closely paralleled our own.' The Australians had many happy times with their Soviet neighbours that year. A particularly joyous spring gathering with the Russians was nicknamed 'the October revelation' by the Australians for, after days together, they discovered how much they had in common, socially as well as scientifically.[12] The scale of Russian infrastructure astonished them. John Bunt, an Australian biologist at Mawson, remembers visiting Mirny and the Russian ship, *Ob*, on his way south in late 1956: '*Ob* ... left me dumbfounded. *Ob* was literally a floating, entirely self-contained institute of oceanographic and geochemical research. Nothing was lacking, from the most sophisticated analytical instruments to a range of deck gear that would make it easily possible to retrieve samples from the most abyssal depths of any ocean in the world. By comparison, my own capabilities were extremely modest ... The mere sight of Russians strolling about *Ob* in lounge suits, stewardesses hurrying back and forth in spruce uniforms, even glimpses of crowded couples dancing in night club settings made it difficult to realise that we were in the Antarctic at all.'[13]

The rocket program of IGY – especially as represented by the Soviet launching of *Sputnik* – is what the world most remembers. The other central event of IGY that gained world attention was the dramatic crossing of the continent by Vivian Fuchs and Edmund Hillary, the final realisation of Shackleton's trans-Antarctic dream. This achievement – in true Antarctic tradition – managed to have a 'race to the pole' unexpectedly embedded within it. Fuchs and Hillary were approaching the pole from opposite directions, and Hillary's role was to blaze the trail for the crossing party to follow after they had reached the pole. But Fuchs was delayed and Hillary, a New Zealander still glowing from his successful first ascent of Mount Everest ('the third pole') a few years before, could not resist making a dash for 90 degrees south, thus becoming the first person to reach it by land since Captain Scott; albeit this time by tractor. To the annoyance of Fuchs, it was this drama that captivated the media.[14]

But the real achievements of the IGY were scientific and political. Scientifically, the greatest advances were made in glaciology through increased understanding of the size and stability of the Antarctic ice sheet. Another example of the long-term benefits of the global scientific assault on Antarctica during IGY was the beginning of an investigation of the ozone layer – the thin skin of $O_3$ in the Earth's upper atmosphere – from the British base in Halley Bay on the Weddell Sea. Because of these studies in the 1950s, scientists in the 1980s were able to measure the dangerous thinning of the layer, which normally filters the sun's ultra-violet rays.[15] Politically, IGY was such a resounding success that it cried out to be continued and institutionalised. The Soviet scientists needed IGY, or something like it, to give their science status at home and themselves some freedom abroad. The United States had formally sought the internationalisation of Antarctica since 1948, originally with the aim of excluding the Soviet Union, but now it was clear that any political solution had to include the USSR.

Intensive diplomatic activity following IGY culminated in a draft for an Antarctic Treaty, the main object of which was to promote the peaceful use of Antarctica and particularly to facilitate scientific research in the area; in other words, to perpetuate the cooperation that had been achieved during the IGY. It incorporated the compromise of Chile's Escudero Plan: that territorial claims should be frozen for the period of the treaty but that nothing in the treaty should be interpreted as depriving any party of a claim or, on the other hand, as recognition of a claim. Military activity and testing of any kind of weapons were to be prohibited within Antarctica. The USSR prevailed in its insistence (initially against US wishes) that the treaty should ban nuclear explosions and the disposal of radioactive materials. Information was to be shared and inspections of other nation's bases allowed at any time. Australia argued especially for non-militarisation, freedom of scientific research and the freezing of territorial claims, and Richard Casey helped to persuade the Soviets to accept this last, crucial provision. The Russians were initially opposed to any mention of claims at all, but a meeting between Casey and Nicolai Firubin, the Soviet deputy foreign minister – held near a Queensland beach in March 1959 – brought agreement. A conference was held in Washington in October and November 1959, and the Antarctic Treaty was signed by the 12 nations that had participated in IGY. The International Geophysical Year had indeed achieved the unexpected. Science as an international social system had never before revealed itself to be so powerful.[16]

Politically as well as geophysically, Antarctica seemed 'the last place on Earth, the first place in Heaven'. Perhaps, as Alan Henrikson puts it, it was at the threshold of two ages: one of competitive nationalism and the other of cooperative internationalism. Antarctica could be seen as the site of both the latest phase of imperial partition and the first expression of planetary awareness. Its treaty regime became the model for management of the sea and outer space.[17]

It was fortunate that these crucial meetings and negotiations about the Antarctic Treaty took place in a brief period of reduced East–West tension. A major determinant for success in the treaty negotiations was that neither the US nor the USSR, having established presences on the continent, was prepared to withdraw and leave the ice to the other.[18] The treaty was signed on 1 December 1959 and came into force on 23 June 1961. In the months following signature of the treaty, Soviet and American relations disintegrated dramatically when an American U-2 spy plane was captured over Russia. Soon afterwards, the world held its breath during the Cuban missile crisis. The truce between the US and the USSR was over. The Cold Peace, one might say, was established during a brief interglacial in the Cold War.

At the end of IGY, the North Americans and Russians continued the new era of cooperation by initiating an exchange program for scientists, as did many other nations. In 1959–61, US scientist Gilbert Dewart joined the Fifth Soviet Antarctic Expedition. Dewart was based at the main Soviet station, Mirny, and was also a member of a four-month summertime traverse inland to Vostok. Dewart was studying the geophysics of the solid earth of the region. But even more urgently, he was learning the culture of a nation that was inaccessible to most of his fellow-citizens, and one that was shrouded in secrecy, suspicion and fear. His trip was not only to Antarctica but also to Russia; for he was planted into a 'revealing microcosm of its society in this distant outpost'. The essence of Soviet culture was concentrated down south, reflected

Dewart, 'much the way an ancient ship concentrates for archaeologists the artefacts of a lost culture'. Before embarkation, Dewart narrowly avoided an approach by the CIA, and on arrival at Mirny he was universally expected to be a spy. The only question amongst the Russians was who he was spying for: the CIA or the KGB?[19]

On board the ship voyaging south in late 1959, Dewart had in his possession a draft copy of the Antarctic Treaty, which had then just been signed. The US Department of State had sent Dewart a copy translated into Russian as a gift for his hosts, who had not yet seen a copy, and so, when an opportunity arose, he casually plucked the document from his jacket pocket and presented it to his new friends. They were immediately wary. 'What are you giving us this document for? ... Are you afraid we might violate the provisions of the treaty?' Dewart insisted that it was just for their information; it was a courtesy. 'They wanted to be sure that what I was doing was not an official act but a personal gesture, and I had to reiterate that nothing I was doing there was an official act.' Dewart reflected that 'The American bearing gifts was genuinely suspicious to the Russians.' What obligations did such unsolicited gestures entail? Also, Russians were highly sensitised to the official flow of information and the protocols it demanded. Printed papers and informality were mutually exclusive.

But Dewart also discovered that Russians enjoyed making fun of their superpower rival. The leader's quarters at Mirny were known as 'The Pentagon'. And when one Russian protected himself from sunburn by fashioning a pointed hood out of a white sheet, with holes for his eyes and mouth, his countrymen instantly dubbed him 'the Ku-Kluxer'. But some jokes did not easily translate. At Wilkes station in Antarctica, where Dewart had been based during the IGY, expeditioners had a name for the pale circles around the eyes caused by wearing dark glasses in the icy glare: it was known as the 'panda effect'. Dewart introduced this term into the lexicon of Mirny, but when he first tried to describe the 'strange pudgy creature that lurks in the mountains of China', he thought his hosts suspected him of 'making some kind of crack about Mao Tse-tung, who was still a hero in the Soviet bloc'.[20]

Gilbert Dewart was impressed by the Russians' living arrangements. Their quarters had 'a much more individualistic and homelike quality than the sterile dormitories of the American stations', he wrote. 'The décor was Victorian with a Slavic accent, the furnishings those of a comfortable though antiquated hotel, with traditional Russian touches like oriental carpets on the walls. It was, indeed, more like a settlement or a village than an institutional station.' However, in the dining and conference hall or *kayut kompaniya*, there was little decoration apart from the large portraits of Nikita Khrushchev and Vladimir Lenin hanging in conspicuous positions. Seating for meals was strictly hierarchical. Leaders sat at a separate table and enjoyed the refinements of a tablecloth and Russian tea glasses in metal holders. There was another cluster of tables for people of intermediate status and another area for the humblest (where Dewart wisely chose to sit). 'Individualistic' and 'hierarchical' were not adjectives an American expected to use of Soviet life. Meals were chiefly of meat, potatoes and cabbage, cooked with garlic, pickles, peppers and chopped eggplant. Green vegetables were rare. But 'the glory of the meals was the soup', *borshch*, of course, but also bean, fish, rice, potato and black mushroom, all with freshly baked bread. On Sundays there was a leisurely brunch of *blini* (small pancakes) with sour cream and huge quantities of caviar.[21]

Days began with the Soviet national anthem played over the base's public address system, a station weather report and a tape of the latest news from Radio Moscow. Dewart enjoyed the regular movie nights, for the Russians had a much better and more international range of films than the Americans did at Wilkes station. The philosophical bent of the Russians made them 'great, one might even say obsessive, conversationalists'. The Americans and the Russians discovered many more similarities between their two countries than differences. But history hung heavily over the Russians. Dewart learned of the shocking shadows cast across his companions' lives by Adolf Hitler and Josef Stalin: 'their ghosts hovered over the land and the survivors like poisonous clouds'.[22] During Dewart's time with the Soviets, a tragedy erupted in their midst. Fire – always a hazard in Antarctic stations – engulfed a building at Mirny in August 1960, and the inferno, fed by a gale with wind velocities of up to 200 kilometres (125 miles) per hour, trapped and killed eight meteorologists.

There were new rituals for the American to learn. Drinking, especially vodka, was a serious custom. The question of drinking with the Russians, recorded Dewart, 'was not a trivial matter' and sipping or nursing a drink was 'on the same level as cheating at chess'. The daily ration of vodka was 100 millilitres (which had to be taken standing up and in one gulp) and you had a strong obligation to stay the distance with your comrades. Baths were a major social event, and they took place every second Sunday. The *Russkaya banya* involved vigorous scrubbing in hot water with coarse brushes, followed by time in the steam room, including perhaps some self-flagellation with birch branches, then a run outside and a naked dive into a snow bank. Afterwards, you lounged around the bath house drinking a cold fermented fruit drink or wine.[23] And the biggest celebration of the year was May Day, when the hall was hung with bright red political banners. The men toasted the success of the forthcoming Paris summit conference of the Big Four: Nikita Khrushchev, Dwight Eisenhower, Harold Macmillan and Charles de Gaulle. Détente was in the air. Khrushchev, upon assuming power in 1958, had repudiated aspects of Stalin's regime and had been willing to travel to the US. There were now plans for Eisenhower to visit Moscow. Dewart had seen more of Khrushchev (during his visit to the US) than the Russians ever had, and so the American entertained them with accounts of their own leader: his penchant for jokes, his success with the American media. On May Day 1960, they raised their glasses to peace and friendship.

But on that same day in the northern hemisphere, something happened that was to unravel these hopes. An American spy plane (called a U-2) was shot down in Soviet air space, and the pilot, Francis Gary Powers, parachuted safely and was captured. Although the United

States government initially denied knowledge of the flight, the U-2 was indeed spying for the American Central Intelligence Agency. The Russians had already suspected they were spying (such flights were supposed to have been suspended), and of course they had spy flights of their own. But fate had delivered into Khrushchev's hands an international media moment he was not going to pass up. There were public exhibits of the plane wreckage. Eisenhower refused to apologise for the spying and Khrushchev stormed out of the Paris summit.

At Mirny, the news of the spy plane broke a few days after May Day via a bulletin on Radio Moscow. The PA system announced that a very important open party meeting would be held in the *kayut kompaniya* that night. To the surprise of many, Dewart decided to attend. 'It was definitely not a pleasant evening', he recorded. Taped speeches of Soviet leaders denouncing the US government were played, and the expedition leaders spoke in the same terms. A resolution 'expressing solidarity with the Fatherland' was passed with acclamation. At the end of the evening, though, Boris Kamenetski turned towards Dewart and urged: 'But let none of this affect our relationship with our good comrade Gil. Whatever problems may exist between our governments, let our personal friendship live on.' Gil reflected that 'Now I had a new sense of responsibility. Antarctica seemed to be the only place in the world where Russians and Americans were still on speaking terms.'[24]

While on the over-snow traverse from Mirny to Vostok in the summer of 1960–61, Dewart and his companions continued to conduct what he called 'the Great Trans-Antarctic Russian-American Seminar'. Hunched together in their tractors on the great white plateau, they pondered their place in the world. They heard of the ratification of the Antarctic Treaty when they arrived at Komsomolskaya, and they talked about its implications for a long time. Could Antarctica avoid the destructive national rivalries that were tearing the rest of the world apart? One of the Russians voiced his doubts: 'The Antarctic Treaty should work very well, at least until something of real commercial value is found down here to set us at loggerheads again.' Outside, a crystal desert surrounded them, as dead in high summer as it would have been in deepest winter. What had brought them here? This 'vast, cold, white country' had somehow seduced them to pay this scientific pilgrimage. 'But was it the wasteland that we really sought', they wondered, 'or was it the integrating effect that this menacing environment had upon the social consciousness of our little community?' Was it, paradoxically, the peculiar society they cherished even more than the remarkable environment? In coming to Antarctica, were they really seeking 'the heightened sense of humanity that grows from a communal struggle against hardship and peril?'[25]

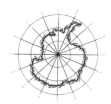

In 1963, Charles Swithinbank became the first Briton to serve with a Soviet Antarctic Expedition. He did so for two years, at the height of the Cold War, soon after the Cuban missile crisis of October 1962 had gripped the world. Getting on a Russian expedition took a lot of organising, and his old expedition mate, Gordon Robin – by then director of the Scott Polar Research Institute in Cambridge – came in handy. Swithinbank was another 'exchange scientist', and his opposite number was Dr Garik Grikurov who travelled to the British base at Stonington Island. When Swithinbank clambered aboard the Russian ice-breaker, *Ob*, there were 225 people on the ship but only one, he reflected, had a home west of the Iron Curtain. And 30 of them were women (who stayed with the ship). When the *Ob* came into port at Le Havre and Fremantle, on its way south, the passengers were allowed ashore provided they walked in groups of three and one of them was a full member of the Communist Party. Swithinbank was a glaciologist studying the rate of deformation of the ice sheet, and Mirny was built exactly at the boundary between a floating ice shelf and the grounded ice sheet, a critical transition begging to be studied. As with Gil Dewart, all the Russians assumed Swithinbank was a spy.[26]

And also like Dewart, Swithinbank was impressed by the quality of life at a Russian base. He soon discovered that he was paid much less than his fellow scientists. 'On some British expeditions', he recalled, 'men lived in conditions resembling those of stabled horses – though a horse is given a place of its own, while the men were stacked in bunks.' By contrast, the Russians at Mirny had writing desks, cupboards and hot-water radiators, and even their rooms were wall-papered and hung with landscape paintings, and their floors carpeted. The scientific instruments were also excellent. 'In short, it was the best-equipped Antarctic research station that I had seen.' But that portrait of Nikita Khrushchev hanging in the hall during Dewart's time had to come down on Swithinbank's watch. The news came through in October 1964 that Khrushchev had been deposed and so his portrait quietly but swiftly disappeared from the dining room, leaving just an unfaded patch on the wall. Swithinbank later happened upon the portrait in the clothing store, 'preserved, I guessed, in case he should ever be rehabilitated'.

In his account of his experience with the Russians, called *Vodka on Ice*, Swithinbank recorded a slightly more austere diet than Dewart had experienced. Vodka, of course, was still essential, although in its absence, Eau-de-Cologne was an esteemed substitute. Occasionally dinner consisted only of boiled potatoes. When Swithinbank arrived home after 18 months of this high carbohydrate diet, he found that high protein foods made him ill. The doctor diagnosed his ailment as 'prisoner-of-war syndrome'. At Novolazarevskaya station, they had been using paper napkins at meals, but when they ran out during the winter, they were issued with pages torn from the works of Lenin. Old copies of *Pravda* eventually replaced the toilet paper.

In the station library at Novolazarevskaya, there were 15 books by Lenin and a number about him, and books by Friedrich Engels, Karl Marx, Josef Stalin, Nikita Khrushchev, Cho En Lai and Fidel

Castro. The political literature alone spanned more than two metres of shelfspace. 'Nothing', Swithinbank wrote in his diary, 'is left to the imagination.' But he also found, translated into Russian, the collected works of William Shakespeare in eight volumes, as well as works by Charles Dickens, Mark Twain, John Galsworthy (15 volumes), Jack London, Jules Verne, Anatole France, Nikolay Gogol, Stendhal (15 volumes), Thomas Mann (10 volumes) and Somerset Maugham. The chosen foreign authors, observed Swithinbank, 'wrote of the class distinctions that were held to be inherent in capitalist society'. On the voyage home, when the ship stopped at Abidjan, one of the Russians bought three books banned in his country. They were Ian Fleming's *From Russia With Love*, Aldous Huxley's *Brave New World* and George Orwell's *Nineteen Eighty-Four*.

Swithinbank, an experienced polar campaigner, concluded that the Russians were 'the most congenial community I had ever lived with in the Antarctic'. He made treasured friends such as Nikolay Yeremin, the leader at Novolazarevskaya. Yet he also found it a very stressful year because of the 'mental isolation and immersion in an alien faith'. 'I began to understand how total immersion in any religious tradition – and I see communism as one – can overwhelm the mind.' At Molodezhnaya station, Swithinbank attended Communist Party meetings and was struck by how much they had in common with Church services back in England, and he found them just as boring.

On the voyage home, Swithinbank recorded that 'an event occurred that might, at a stroke, have destroyed my faith in the integrity of the Soviet Antarctic Expedition, or at least some members of it'. One of the ship stewardesses was told to clean out the laboratory where Swithinbank had left his diary and field notes in a drawer. She emptied all the drawers into a waste bin and was about to heave the contents over the side when a Russian scientist stopped her. 'What upset me immeasurably', reflected Swithinbank, 'was the realisation that, if she had disposed of everything that I had recorded in the last 15 months, I would never have believed that it was due to a misunderstanding.

Instead, I would have concluded that it was an official act to deprive me of my records, while explaining this away as an accident. Such was the mind-set of the Cold War in 1965.'

On returning to England, Swithinbank could not write freely to his new friends and their own messages became stilted and formal; he received mass produced postcards with just their signatures. It was another 30 years before he could meet any of his Russian comrades again. They held a reunion in St Petersburg in 1995. When Swithinbank finally renewed his conversation with his long-lost companions, he discovered that the Cold War had once reunited him with one of them in a rather different way. While Swithinbank was serving on a British nuclear submarine under the ice of the Arctic Ocean, his old friend Nikolay Yeremin had unwittingly been tracking him.

In 1991, as the *Professor Molchanov*, the first Soviet tourist ship, approached America's Palmer station near the Antarctic Peninsula, a voice came over the radio with the demand: 'Will the white ship approaching Palmer station please identify itself. We don't recognise your flag.' The Soviet Union had just fallen, and two days earlier the crew of the *Professor Molchanov* had anchored in a bay, pulled out the vodka, lowered the Soviet flag, chiselled the hammer and sickle off the stern, and raised a hand-sewn Russian flag.[27] The collapse of the Soviet Union in 1991 led to dramatic cuts in Antarctic and Arctic programs. By that time, the USSR had succeeded in building a ring of bases totally encircling the continent, representing the largest national operation in Antarctica. Molodezhnaya had become the largest base, with a winter population of well over 100.[28] But from the early 1990s, Russian life in Antarctica became precarious. Stories of winters past in Vostok can be frightening.

At times it was a community desperately close to the edge of survival, just a few fuel drums above freezing. There was much anxious arithmetic: how many fuel drums should be left for heating food and melting water, how many could be used for drilling? It was the old tension between science and survival, the same dilemma faced by Scott, Mawson and Byrd. But science *was* survival. The drill was the primary justification of Vostok; the ice core illuminated the whole history of humanity and the drill edged closer to the vast, hidden lake kilometres below them.

In 2000, Novolazarevskaya had the air of a struggling mining town, reported a grateful Australian visitor, Alexa Thomson. The communication equipment was antiquated, the buildings shabby and run-down, and the walkways between buildings were splitting and rotting. But the sauna was still in good working order. Thomson referred to the station's 'jumble of tatty buildings and auto-wrecker ambience'. Spartan ingenuity and frontier improvisation reigned in a community of people uncertain just when they might be able to return to temperate lands.[29]

In 2006, Russians celebrated 50 years of continuous Antarctic presence and research, and did so by renewing their commitment down under. Decree Number 713-r of 2 June 2005 declared the aim of 'maintaining and reinforcing the Russian Federation's position in Antarctica', and increased the funding for 2006 by 47 per cent. There are five permanent stations and two summer field stations, a new expeditionary vessel is under design, and plans have been announced to reactivate the Leningradskaya and Russkaya stations as well as to substantially upgrade Progress station (near Australia's Davis station), which will become Russia's main base.[30] At Vostok, the drill is turning.

## Friday, 27 December

**Casey station,
Antarctica**

I do feel both sensitised and empowered by being here
and experiencing Antarctica. The past is in the present
all around me. People think historians look mostly for the
changes over time, which are of course important, but I
am perhaps even more interested in the continuities. For
example, I am myself experiencing both the boons and
constraints of benevolent paternalism on a large scale.
Understanding that predicament, that culture, is a key to
writing about Antarctic exploration and science, especially
as it has become more institutional.

This is on my mind today because I've just visited
a huge bit of Australian infrastructure on the ice, and
because I (and other expeditioners) are constantly treated
with a touching (if occasionally constraining) care by the
Antarctic Division. I can see why people can get addicted
to it: three meals a day cooked by someone else, the world's
greatest wilderness outside the window, the best gear and
finest machines at your disposal, and challenging work that
feels connected to some kind of national mission. Some
must feel lost when they leave it.

In Australian Antarctica, scientists are like the
penguins; they come back with the sun. Not one of the
17 men who will winter down here this year is officially
classed as a scientist. This is a legacy of the massive
building investment of the 1980s that created a secure,
comfortable base (some say too comfortable) but also
generated constant maintenance demands. Compare this

with the 'light footprint' of New Zealand, which has made a minimal investment in buildings and maximum in science and mobility. In the 1980s, when building workers massively outnumbered scientists on the Australian ships and bases, two of their common sayings were: 'We're not here for a long time, we're here for a good time', and 'We're not here to make friends, we're here to make money'. Stephen Murray-Smith landed here at Casey in the midst of a minor coup when building workers isolated and humiliated the station leader. Some of those tensions now seem to have subsided (although they remain as undercurrents) and everyone I've met is here to make friends, and a difference. And the great summer influx of science – like an exuberant blossoming of life – is made possible by the quiet work of the winter contingent.

But you must imagine a fine, still morning, with the palest, sweetest blue low on the horizon just above the line of snow-covered hills, and a barge with 30 people on board dressed in yellow and orange gear and life-jackets heading away from the big red ship and into shore. I gather up my black backpack and step over the edge of the barge and on to the great white continent itself. My foot is greeted by grey industrial rubble from the making of the road that leads from the station to the wharf. But a few more steps, and I am on the snow, deep and crusty, just like in the Australian Alps. The drifts of snow, though, are deeper. And you lift your eyes and the water of the bay has a steely polar glare to it, and in and beyond it are icebergs, outrageous presences.

As we walk towards the station, about 500 metres up the hill on a flagged walking path, we pass five Adélie penguins playing. We have been warned not to approach them too closely because they may become stressed, and 'scientific research' (with implanted electrodes) shows that their heart rates go up when approached by humans, particularly if you are tall and they are nesting. 'Simple reflection', however, reveals that their heart rates go up even more if they are captured and have electrodes implanted. But these five naughty, cute, funny, wonderful little penguins haven't read this particular research paper (it probably wasn't in their preferred international refereed journal) and they are coming over to look at ME! My heart rate goes up.

The centre of the station is the Red Shed. It is also known as the Casey Hilton. I will tell you about it, although I fear it will make my 'exploration' seem less heroic. It's like a hotel. From the outside it is a very big red shed, and the walls are of two metal plates with six inches of foam between them and the windows are triple-glazed. It looks industrial, modular, prefabricated. Inside it has comfort and, if not character, certainly dignity. There is a mess where you can feed yourself silly at almost any time of the day or night. And there is a comfortable lounge, with a sunken sitting area looking out across the bay through huge windows; this is known as the Wallow, which is what sea elephants lounge around in. There is gentle canned music and it is warm and picturesque and there are lots of pictures on the walls and a good library upstairs, and someone opposite me is reading a faxed copy of yesterday's Hobart *Mercury*. Fancy faxing the *Mercury*, of all papers! Tasmania's slice of imperialism.

Opposite the Red Shed is a tall pole sprouting signposts indicating the direction of cities and Antarctic stations, and their distances from Casey: London 15 966, Bucharest

14 211, Sydney 4471, Adelaide 3934, Hobart 3424, Macquarie Island 2870, South Pole 2637, Mawson 2029, Davis 1396, Vostok 1383, Dumont d'Urville 1298, Mirny 778. I challenged Bloo Campbell, the plant inspector staying for the winter, to see if he could give his tiny home town, Humpty Doo, the dignity of its own sign on that celebrated pole.

The station leader, John Rich, gave me a tour of his village and so I learnt a lot today about the workings of a station, and the life of people in it, and I spent some time in the library reading old station logbooks and looking up every now and then and gazing out the window at the ice, taking the historian's delight in reading archives *in situ*. The scientists underestimate the gold to be found in these documents. An hour or two of fossicking amongst the logbooks enabled me to tell the waste management engineers and biologists over lunch quite a lot about how and when their tip sites evolved. They were incredulous, as if I'd just performed a piece of suspect magic.

My first official job in Antarctica was to collect a huge tub of vanilla ice cream and carry it down 'Main Street' to the station kitchen. The full symbolism and irony of this action have yet to be teased out, but I sense it is full of promise.

But the best fun of the day was in the store, the Green Shed, a huge warehouse with walk-in freezers and ranks of 20 metre (66 foot) high shelving. There is also a Warm Room in the store where some food is kept at refrigerator temperatures. We had to wait around a while for the food container (or one of them) to be unloaded from the ship and brought up the hill and lifted by crane to our door, and the new chef (Trent) was getting impatient, but it finally arrived and we threw the door open and started organising and unpacking. We unpacked cardboard boxes of every food item you can imagine fitting in a freezer. I can guess the menus of expeditioners for the next year. There were straight-cut chips, shoestring chips and seasoned wedges, carrot rings and baby carrots, brussels sprouts (small number), broad beans (even smaller number), broccoli, mixed vegetables, mango cheeks, raspberries, blackberries and boysenberries, avocado mush and banana cream, huge amounts of poultry, porterhouse steaks and eye fillet steaks and T-bone steaks and lamb fillets and veal fillets and salmon and bream and beef ravioli and chow mein and potato gems and lots and lots and lots of ice cream.

# WINTERING
*Surviving the polar night*

The deep Antarctic winter and the long polar night can play with your sanity. In the final years of the nineteenth century, humans first discovered the perils of wintering in Antarctica. Two expeditions surrendered to the ice and darkness, and returned with painful testimony.

The first was the Belgian Antarctic Expedition of 1897–98, captained by Adrien de Gerlache. Its members included a young Norwegian named Roald Amundsen who offered to sail unpaid for the experience (as second mate), and an American doctor, Frederick Cook, who had accompanied Robert Peary on his first North Greenland expedition. On the voyage south the expedition was almost shipwrecked, the life of a young sailor was lost, and they slipped weeks behind schedule. The precious summer light was already dwindling by the time they crossed the Antarctic Circle. The lateness of their voyage forced de Gerlache to abandon his plan to land a wintering party, and as the *Belgica* cruised along the coast of the Antarctic Peninsula, the captain discussed with his crew the possibility of wintering in the ship. His proposal was rejected. Yet, on the last day of February, he ordered the ship south again towards open water. Within days, the *Belgica* was trapped within the pack ice and many on board suspected that de Gerlache had meant to be caught.

These 18 men, hijacked by their captain's gamble with the ice, became the first humans to experience an Antarctic winter. They were hardly handpicked for the ordeal. There had been serious disagreements amongst the crew well before they reached southern latitudes. Even language divided this smallest and most isolated community on Earth. Some on board the *Belgica* spoke Norwegian, and the rest spoke French except for Cook who spoke English and a little German. During the long winter, one of the Norwegians came to believe that the French word for 'something' actually meant 'kill', an unfortunate misapprehension that made everyday conversation dangerous.[1]

Cook, the doctor, wrote a grisly account of the voyage in his book entitled *Through the First Antarctic Night*, a study of what he called a 'new human experience in a new, inhuman world of ice'. He aimed to record those things 'usually suppressed in the narratives' of polar explorers. In particular he documented the winter degradation of body and spirit. He tells how they farewelled the sun and bunkered down for the coming of the night 'with its unknowable cold'. Cook wrote that, as the sun wandered northwards, so did their hearts and spirits. 'The curtain of blackness which has fallen over the outer world of icy desolation has also descended upon the inner world of our souls.' Their energy waned with the light. 'I can think of nothing more disheartening, more destructive to human energy, than this dense, unbroken blackness of the long polar night'. Cook, who had experience of exploration in Greenland, was appalled by the very different pole in the south. He missed the company and instruction of the Eskimos and, stranded as he was in a sea of ice, he also missed real, solid land, 'not the mere mockery of it, like the shifting pack that is all about us here'. Could there be a more melancholy, a more maddening, or a more hopeless region than this, he wondered?[2]

'We are under the spell of the black Antarctic night ... We have aged ten years in thirty days.' They grew physically weaker and their hair turned grey. They became puffy around their eyes and ankles and their skin grew oily and took on a pale and greenish hue. 'We were all reasonably good-looking when we embarked', wrote Cook, 'but we were otherwise when we returned. The long night offered a radical transformation in our physiognomies.' He generalised these symptoms as 'polar anaemia' and administered 'the baking treatment' to the worst victims. This involved long periods exposed in front of the fire.[3] In Cook's account, the crew of the *Belgica* slipped into a netherworld that was pale, attenuated and agonising. Two sailors went mad. One tried to walk back to Belgium. The popular Lieutenant Danco died of a 'weak heart' (more probably scurvy) on 5 June 1898 and the crew dropped him through a hole they cut in the ice. Cook was haunted by the image of Danco floating about under the ship in a standing position.

Scurvy was an even greater threat to the expedition than anaemia and depression. The classic symptoms of swollen gums and stiff joints became apparent and Cook belatedly insisted on the consumption of fresh penguin meat. He became fascinated by the effect of the winter on the hearts of the crew. Their hearts expressed their vulnerability and edginess; they beat more weakly, they became irregular, they surged at the least provocation. Mitral murmurs became audible, resounding loudly in the silence of the polar night. The sun, decided the doctor, 'seems to supply an indescribable something which controls and steadies the heart. In its absence it goes like an engine without a governor.'[4]

By June their nights had dipped to a temperature of −30 degrees Celsius (−22 degrees Fahrenheit), but the deepest cold was to stalk them in early September when the thermometer registered −43 degrees Celsius (−45.4 degrees Fahrenheit). But it was light they missed more than heat. 'Oh, for that heavenly ball of Fire!' exclaimed Cook. 'Not for the heat – the human economy can regulate that – but for the light – the hope of life.' On 22 July, a month after midwinter's day, some of the crew climbed to the crow's nest to catch the first glimpse of the returning sun. For some minutes they could not speak. They felt life stir inside themselves again. Soon they were lying on the ice in the few hours of light, sunning themselves 'like snakes in spring', feeling the blood quicken in their arteries. By late November, they were 'feasting their souls', as Cook put it, on direct sunshine.[5]

Getting through, and recording, the first Antarctic night might have been Frederick Cook's finest moment. He was praised by the crew for his personal strength and medical care.[6] He forced upon them a life-saving regime of fresh seal and penguin meat. His energy and skill even helped them force their eventual escape from the ice. But a decade later, he had become a controversial global figure because of his claim to have reached the North Pole ahead of his mentor and rival, Robert Peary. Cook, using Peary's methods, stole a march on him and forestalled him. Peary unleashed a virulent campaign against Cook that escalated into a war between the newspapers that had supported each explorer, the *New York Times* (Peary) and the *New York Herald* (Cook). The *New York Times* won and its circulation soared, the *New York Herald* went into decline, Peary became revered, and Cook was disgraced and discredited (although he never retreated from his claim). The Eskimo men (and one African man) who accompanied the white explorers to their 'poles' were not allowed to share in the honour of the achievement, nor were they deemed reliable witnesses. It also emerged that Cook's claim to have climbed North America's highest mountain, Mount McKinley, was false. The irony was that Peary's own attainment of the North Pole now seems likely to have been faked.[7]

Roald Amundsen, the demonstrable conqueror of both poles, one on skis and the other by airship (as mentioned in *Planting Flags*), was a fellow-expeditioner of Cook's on the *Belgica*. He greatly appreciated Cook's Greenland experience and took every opportunity to learn from him, and through him from Peary. Years later, when Cook's claim to have reached the North Pole was attacked, Amundsen defended Cook. He was speaking up for an old shipmate, but also out of admiration for Cook, and indebtedness to him. But, more vitally, Amundsen was speaking out against the idea that anyone – neither a Peary nor a Scott for that matter – could own a pre-emptive right to a pole. In standing beside Cook in the controversy, Amundsen was defending a stealer of the pole, a man who (whatever one says about the truth of his claim) had the temerity to march unheralded into a territory that Peary had publicly owned for years. In the very days that Cook's controversial northern adventure was publicised to the world and Peary's campaign against him was unleashed, Amundsen was himself taking the first steps to go secretly south and launch his own race for the pole from a base in 'British' territory, the Ross Ice Shelf. In the drama of Amundsen's belated decision to turn south in 1911 and to winter under Robert Falcon Scott's nose, it is often forgotten that the Norwegian had first wintered down there 14 years earlier when Scott had scarcely even imagined Antarctica.

While the men on the *Belgica* were still wondering if their ship would escape a second winter trapped in the ice, another ship was nosing towards the Antarctic coastline with men who hoped to winter on the land. This ship, the *Southern Cross*, carried a British expedition led by Carsten Borchgrevink, and its members believed the *Belgica* to be lost. One of them, a young Tasmanian named Louis Bernacchi, had expected to join the *Belgica* during its summer resupply. But the ship had not returned, and so he had opportunistically signed onto the *Southern Cross*, which was sailing south in 1898.

Bernacchi was haunted by the ice.[8] He had grown up in Tasmania, for his parents had arrived in the colony when he was aged seven, and his father had leased Maria Island, a rugged island off the east coast. This short, stocky, red-haired child had the run of the island in his formative years, an island that was exposed, as he put it, to 'the great seas of the "Roaring Forties"'. And nearby, where he went to school in later years, was Hobart, the haunt of whaling ships and men, a rich source of gossip and folklore about the far south, even those mysterious realms of ice. No wonder that Bernacchi, after some years working and training in meteorology and magnetism at the Melbourne Observatory, travelled to London to join an Antarctic expedition, led by Carsten Borchgrevink. Borchgrevink was Norwegian but, like Bernacchi, had spent years in the Australian bush, and the expedition he led was British. They were both sniffing the air for the familiar whiff of eucalyptus as their ship rounded the south-west coast of Tasmania, and Bernacchi looked lovingly on the familiar shores of 'the prettiest and most genial isle in all the world!'[9]

Hobart has a great tradition of welcoming and celebrating polar voyagers, perhaps with one famous exception. In March 1912, Roald

Amundsen anchored off Hobart on his return from Antarctica, having just efficiently bagged the South Pole. He came ashore alone and without fanfare, anxious to telegraph the world with news of his stunning achievement. Dressed in his peaked cap and blue sweater, he booked in at Hadley's Orient Hotel. On the eve of his greatest fame, Amundsen noted in his diary that he was 'Treated as a tramp ... given a miserable little room.'[10] Today you can book into the Amundsen Suite at Hadley's and they proudly boast that it is the very same room he stayed in that historic night. (You must pray that it is not.) But Amundsen did quickly receive generous recognition from Tasmanians who also took pride that his momentous news had been released in their city.[11]

While Bernacchi's *Southern Cross* was in port in December 1898, Borchgrevink took the opportunity to pay tribute to Hobart's polar heritage by laying a wreath at the foot of the statue of Sir John Franklin. Amongst the distinguished guests at this 'graceful and spontaneous' occasion was Sir James Agnew, the former Tasmanian premier and senior vice-president of the Royal Society of Tasmania, who remembered Franklin welcoming the great Antarctic ships of almost 60 years earlier: the French ships of Dumont d'Urville, the *Astrolabe* and the *Zélée*, in 1839 and '40, and the British ships of James Clark Ross, the *Erebus* and the *Terror*, in 1840 and '41. The balls and festivities that greeted these ships became legendary. Amidst such adventurous visitors, Franklin could transcend for a moment the petty politics of the convict colony and again play the famous explorer. The arrival of the British polar ships, wrote the governor's wife, Lady Jane Franklin, 'added much to Sir John's happiness, they all feel towards one another as friends and brothers, & it is the remark of people here that Sir John appears to them in a new light, so bustling & frisky & merry with his new companions'.[12] Within seven years, Sir John had taken the *Erebus* and the *Terror* to the Arctic and was dead.

Similar festivities welcomed the *Southern Cross* in 1898. Bernacchi wrote that 'The whole town appeared to us to have but one occupation

– entertaining the members of the Expedition.'[13] Perhaps it was because, as Bishop Montgomery declared at their farewell service, 'Take a good look at these men for they will never be seen again.'[14] Hobart residents would have been conscious of the fact that little official exploration had been attempted of the Antarctic coastline since that remarkable voyage of James Clark Ross almost 60 years before. Also the latest ship to go south, the *Belgica*, had not been heard of since it sailed towards the ice a year before. There had been a few landfalls on the Antarctic Peninsula earlier in the nineteenth century, but the first recorded landing in East Antarctica had taken place just three years before the *Southern Cross* sailed, and it had involved Borchgrevink.[15] He was one of many in the party who later claimed to have been the first to stumble ashore at Cape Adare. Now Borchgrevink was taking the *Southern Cross* back to that same bit of the continent, this time to build a hut and stay the winter. The state of knowledge about Antarctica was poor. Was it a continent or just an archipelago of islands connected by ice sheets? Could humans survive a winter down there? What kinds of animals might they find? The Hobart *Mercury*, in explaining to its readers the need for Borchgrevink's expedition, speculated that 'there is reason to believe that the two Poles of the earth do not correspond'.[16] The small *Southern Cross*, just 45 metres in length, was equipped with shotguns and ammunition to deal with polar bears or other threatening land animals.

They arrived in the middle of February 1899 and ten men and 75 dogs settled into a prefabricated hut on stony ground. They played a lot of cards and chess and were battered by 'the din of the roaring elements'. The silence also assaulted them: 'The silence roared in our ears', wrote Borchgrevink, 'it was centuries of heaped-up solitude'. When they had the chance to walk outside, they always walked towards the north, 'towards brightness and sunshine and life and home. Never towards the silent and cold south.'[17] They farewelled the sun in May. 'An absolute sterility prevailed', wrote Bernacchi.[18] He generally rose early each day to read the meteorological instruments, but most of the others slept in to 11 am, which was breakfast time in winter. 'Suppose they will sleep most of the winter', wrote Bernacchi in his diary. 'It would be rather amazing if one fine day we slept on for a century and then awoke and returned to civilization', he mused. 'We would most probably be entirely forgotten. If we searched among musty volumes in some old library we might perhaps find the bare mention of the fact that an expedition went south in 1898, a hundred years ago. And if we made our identity known who would believe us?'[19] They felt so remote, beyond the known world, and even outside time.

But they were unbearably close to one another. Inside the hut, Bernacchi reported that by the beginning of June, 'A strange spirit of irritation prevails among members. Scarcely bear [the] sight of one another. Some crusty and others morose, [making the] most unpleasant remarks they can possibly think of to one another.' 'What children of the light we are', reflected Louis. Even the dogs seemed affected by the darkness, snarling, fighting, killing and devouring one another 'for no apparent reason'.[20] The commander, Borchgrevink, had by this time isolated himself from most of the men, partitioning off a room and taking his meals only with the two Finnish men of the party, who offered him an uncritical audience. Borchgrevink was drinking heavily. The zoologist, Nicolai Hanson, wrote that 'Such oppressive feelings is reigning within our four walls, that everyone looks as if he is half dead.'[21] Bernacchi had lost all respect for his suspicious, idle and unstable leader and mocked him, sometimes to his face. As they approached mid-winter, Bernacchi confided to his diary:

> Wish the God I had never joined such a numskull and his expedition. Getting to positively loathe the sight of all. For a handful of men to live together day after day, see nothing and hear nothing but one another, is agony … What beasts we all are and how astounding our conceit. One would think that here at the extremity of the globe one would at least be spared these contemptible human passions. But it is quite the contrary for I have never before seen them in such nakedness.[22]

A few days later he recorded his certainty that his leader was now insane. Bernacchi could not even stand the bread his commander occasionally baked. Borchgrevink called it an 'Australian damper', no doubt with some nostalgia for his days surveying in outback Australia, but Bernacchi and others regarded it as inedible. It became 'emergency food' and no emergency was ever desperate enough to warrant recourse to it. Bernacchi wrote later of the commander's damper that, should you need it, 'It is still in the Antarctic regions on the left hand top corner of the shelf inside the hut.'[23] And the hut is indeed still there.

In early October, just as the penguins were about to return for nesting, the zoologist Nicolai Hanson died after a long, agonising illness. Before his death, he entrusted many of his precious scientific notes to his leader, but Borchgrevink either lost or destroyed many of them, possibly out of suspicion for what they might reveal of his own behaviour during the winter. On his birthday in December, Borchgrevink hoisted the British flag in his own honour, but his companions declined to propose his health. At the end of January, after watching anxiously for the return of the ship around the clock, the *Southern Cross* arrived while they were all asleep. The first the wintering party knew of its blessed return was when the captain, with the mail-bag on his back, approached the disturbingly silent hut, opened the door, heard them all snoring, put the bag on the table and called out: 'Post!'[24]

The expedition did not add much to geographical knowledge and, because of the death during the winter of the zoologist, the natural history collections and observations were incomplete and flawed. But the expedition did prove that humans could survive – just – a winter on the mainland, and it did collect meteorological and magnetic observations across an entire year. This had been Bernacchi's chief responsibility and he contributed crucial insights into Antarctic winds and weather.

Bernacchi is such an appealing witness on Antarctica because, as his grand-daughter Janet Crawford has so ably chronicled, he had a love–hate relationship with the continent, one that I think can be seen as typical of the emotional magnetism of the place.

Bernacchi gave voice to both wonder and horror. He was powerfully drawn to the place, and having got there, longed to escape it. Having escaped it, he could hardly wait to return to it.[25] You could love Antarctica, but you couldn't live with her for long or achieve a fulfilling intimacy, so perhaps it was destined to be a relationship of ardent declarations, desperate break-ups and distant longings. Bernacchi signed up as physicist on Scott's first expedition south in 1901, only a year after getting home.

But when he first saw the coast of Antarctica in mid-February 1899, these were his words:

> As we drew closer, the coast assumed a most formidable aspect. The most striking features were the stillness and deadness and impassability of the new world. Nothing around but rock and ice and water. No token of vitality anywhere; nothing to be seen on the steep sides of the excoriated hills. Igneous rocks and eternal ice constituted the landscape … Approaching this sinister coast for the first time, on such a boisterous, cold and gloomy day, our decks covered with drift snow and frozen sea water, the rigging encased in ice, the heavens as black as death, was like approaching some unknown land of punishment, and struck into our hearts a feeling preciously akin to fear when calling to mind that there, on that terrible shore, we were to live isolated from all the world for many long months to come. It was a scene, terrible in its austerity, that can only be witnessed at that extremity of the globe; truly, a land of unsurpassed desolation.[26]

He and Frederick Cook were amongst the earliest observers to find the words to describe the confronting difference of the Arctic and the Antarctic. One was full of life, the other was implacable and silent:

> Enveloped in an atmosphere of universal death, wrapped in its closely-clinging cerements of ice and snow, the one expression of the Antarctica of to-day is that of lifeless silence. If, in the domain of Nature, such another region is to be found, it can only be in the heart of those awful solitudes which science has unveiled to us amid the untrodden fastnesses of the lunar mountains.[27]

This became another theme of Bernacchi's: that Antarctica's proper parallels must be drawn from outer space, from other planets and moons. Antarctica was literally otherworldly. When, in the long winter, he contemplated the moon from Antarctica, he reflected on their alikeness: 'a dead silent world above you, and a dead silent world at your feet'.[28] From the top of Cape Adare, he looked inland across the continent and this is what he saw:

> The silence and immobility of the scene was impressive; not the slightest animation or vitality anywhere. It was like a mental image of our globe in its primitive state – a spectacle of Chaos.

> Around us ice and snow and the remnants of internal fires; above, a sinister sky; below, the sombre sea; and over all, the silence of the sepulchre![29]

Bernacchi felt that he was looking not just at ice-age Earth, at his world in the deep past of the Pleistocene, but also at the bleak future of his planet, spinning cold in space.

'Polar madness has always interested me', wrote Raymond Priestley, who himself survived one of the most extraordinary experiences of winter isolation. 'If it does not include depression to the point of suicide it appears to be curable.' Priestley was one of the six men of Scott's northern party who spent a winter at Cape Adare and then were stranded for another winter in an ice cave at Terra Nova Bay. That experience of maintaining sanity for more than six months in a cave three metres (nine feet) by 3.5 metres (12 feet) made Priestley an authority on polar madness. He knew men who had gone over the edge. Bertram Armytage committed suicide in the Melbourne Club after the first Shackleton expedition. Edward Nelson shot himself after his return from the Scott expedition in 1913. George Abbott, who shared the cave with Priestley, went raving mad on the voyage home and recovered months later. Dennis Lillie, a biologist with Scott, did not recover his sanity. Hjalmar Johansen, who had quarrelled with Amundsen, shot himself in the streets of Christiania on his return from the *Fram* expedition. Priestley heard that at least three of the crew of Shackleton's *Endurance* were 'mentally deranged' before Elephant Island was reached.[30] Did Antarctica make these men 'go mad' or did it attract people with a certain extremism in their personalities, not just looking for the edge but already near it? Whatever your state of mind, Antarctica can be destabilising, it can be life-changing. 'Its primordial simplicity, vast bulk and preternatural silence throw nearly everyone slightly off balance', explains Barry Lopez.[31]

Douglas Mawson was marooned with a madman. When Mawson staggered back to his Commonwealth Bay hut in early 1913, he faced an unexpected extra winter. In retrospect, he came to believe that it may have saved his life. Weakened by exhaustion, malnutrition and vitamin A poisoning, he may not have survived a rough voyage on the Southern Ocean and a swift return to the hectic schedule of civilisation. But a long, dark second winter in the hut held its own threats. Sidney Jeffryes had arrived with the *Aurora* in January and was one of the six men selected to wait in hope for Mawson's return. He was the wireless operator and undertook what Mawson called the 'tedious and nerve-wracking' work of establishing and maintaining two-way contact with the men at Macquarie Island, and hence with Australia. But soon after mid-winter, 'though curiously logical at times', Jeffryes began suspecting that his companions were plotting to murder him, and he believed that Mawson had cast 'a magnetic spell' on him. He threatened to expose these machinations on his return to Australia. On 27 July, Jeffryes tendered his 'resignation' from the expedition. Determined to isolate the madness in case it might catch, Mawson assembled his men around the table and publicly isolated and ridiculed Jeffryes with a carefully written speech, pointing out that 'the accommodation houses are few and far between in the Antarctic'. Jeffryes' madness waxed and waned. They needed him, for he was the only one who knew how to operate the wireless, although even this task was taken from him when one day in September he was caught transmitting a false message in Mawson's name. Jeffryes was admitted to an asylum on his return to Australia in 1914.[32] Looking back on 1913, Mawson confessed to his fiancée, Paquita, that 'Most of my time during this winter was occupied in keeping myself and others sane.'[33]

What weight does one give to the evidence of a 'madman'? One could argue that Jeffryes' madness made him a privileged witness to the hauntings of a winter hut. In the year of his return from Antarctica, Jeffryes wrote a lucid letter to a friend, Miss Eckford, from the Ararat asylum. 'When I tell you that my reason, memory, and all my faculties of mind, are as good as ever they were; and yet I am insane, it will seem paradoxical, but it is really so.' Jeffryes remained, as Mawson had put it, 'curiously logical at times'. The trouble started, according to Jeffryes, when he defended Mawson from Cecil Madigan who 'made a scurrilous insinuation against Mawson, with regard to Mertz'. Whatever we think of this testimony, it does identify a likely taboo, a subject of disquiet in the polar night in a hut where Mertz's bunk lay empty and Mawson still lived.[34] Once suspicions, however groundless, had been voiced, there was no escaping them. They infiltrated the very fabric of the hut and, like ice in the crevices, set hard.

In 1967 at Mawson station, the doctor disintegrated. In March he announced that he would not do any work other than attending to patients. By May he was in a 'wild state' and was officially put on sick leave. One night, he picked quarrels with everyone, constantly complained of being assaulted and then finally had his fears confirmed when he was thrown out into the snow. The officer-in-charge (known as the *oic*) reported to the Antarctic Division that the doctor was becoming more withdrawn and had 'occasional violent outbursts'; he was threatening to cut people's heads off or to break their necks, which probably discouraged them from seeking his medical assistance. The oic sincerely hoped that 'he gets relief by saying and not doing'. By September, the doctor was sending 'lunatic incomprehensible' cables home.[35] His companions were concerned that he was secretly drinking some of the methylated spirits kept in the paint store. With the coming of summer, his behaviour became unremarkable again.

There are rhythms to the year, and to the Antarctic experience. Seeing the ice for the first time is a great adrenalin burst; the sea becomes calmer and one's soul exalts. The ship leaves in late summer and the real adventure begins. Mid-winter comes more quickly than you think because you forget that you have not yet reached the thermic depths of the year. There is the excitement and release of the mid-winter dinner and pantomime, the mirror moment of the year when you play at opposites, cross-dressing and subverting authority for a day. 'Cinderella' is the traditional pantomime. Whether the station leader is cast as Cinderella or an ugly sister may be significant. In Bloo Campbell's year, a memorable event was 'the rat night in the workshop' when a red-bearded Osama Bin Blooin turned up to sabotage the evening. Bloo marked his mid-winter with a special celebration. He sent me a photo of himself in the half-light, up a pole with spanner in hand, proudly displaying the new sign that pointed across the world from outside Casey station: *Humpty Doo 6141.*

Mid-winter is a moment of celebration because the sun begins its long journey back to you. But the worst is yet to come. The polar winter deepens, the hours seem to move more slowly, night is indistinguishable from day, the blizzards enclose you, the cold enters your bones, there is the weird scintillation of the aurora, and you hear the sighing and sobbing of the ice pack and the gurglings, blowings and snortings of seals; or was it of something else?[36] Nature, if one could call it that, can be heard but not seen. The inertness of Antarctica begins to claim you. August to October are the danger months; the post-midwinter blues can set in. In late spring there is the excitement of returning penguins and of quickening summer activities and plans, but also the gnawing anxiety about release: Will the ship come? When will it come? The ship brings the flu, and the invasion.

There is a famous story of the director of the Australian Antarctic Division, Phillip Law, arriving at Macquarie Island in 1950 to relieve a wintering party and finding everyone speaking to one another with theatrical nineteenth-century gentility. The men had survived the winter by repeatedly working through their small film collection, and the group's favourite was *Pride and Prejudice*. Once they tired of watching it, they turned down the volume and acted out the voices themselves. This ventriloquism easily tipped over into daily relations, and soon men were bowing and holding doors open for one another, and addressing their colleagues with sweet and elaborate civility. 'Such affability, such graciousness – you overwhelm me' they could be heard saying to one another.

At Maudheim in 1950–52, the men of the Norwegian-British-Swedish Expedition decided that an international group had fewer frictions than a national one. Charles Swithinbank remembered that 'At Maudheim, each one of us, to the best of our ability, leaned over backwards to suppress our national prejudices and preconceptions. That surely was the key to our success. We made allowance for the differences in cultural background; whereas on a national expedition it is all too easy to assume that men brought up in the same culture should think alike. Finding that this is not so can lead to conflict.'[37] Members of the *Belgica* expedition, with its language problems, may not have agreed.

Frenchman, André Migot, was a member of a wintering party on Kerguelen Island in 1953. In his book, *The Lonely South*, he provided a perceptive analysis of life in a small, isolated community. He reflected on the strangeness 'of the team we had come to relieve'. Clearly a year of isolation where 50 men were 'shut up in a sort of multiple solitary confinement' had a strong effect. He commented on 'the concentration camp atmosphere of the base' and 'the suffocating promiscuity of camp life', by which he meant the tendency to dwell on negative gossip. They found that some people would communicate with one another only by registered letter. The newcomers felt that they were observing a flawed social group and that they would be different. Unhappily, Migot discovered that they were not.[38]

By late winter, Migot wrote: 'Our daily work is tedious, and the diet monotonous; the time-table is inflexible, the camp ugly and the landscape unvarying.' Cliques had formed. There were the administrators, the radio men and meteorologists, the soldiers and workmen (who formed the largest group), and the scientists (the smallest group). The scientists were not only fewest in number, but the products of their work were invisible, and so they were persecuted.[39] Winter tensions between scientists and tradesmen became a feature of Australian Antarctica in the 1980s, when Australia launched a massive building program on the ice.

When Stephen Murray-Smith arrived at Casey station in the summer of 1985–86, he found a dysfunctional community. The previous officer-in-charge had been isolated by his fellow-expeditioners and there had been an 'effective takeover' of the base during winter. Murray-Smith reflected:

> It can happen so easily, it seems. A wrong word here, a suspicion of too much clubbiness among a few there, a decision to go out bivouacking in the old Antarctic way, then when the weather blows up a bit a call over the radio to be collected in the search-and-rescue Hagglund. Suddenly authority has crumbled – the oic's authority can only be based on his personality, anyway – and the tough guys are getting up in the mess at dinner time announcing that there will be a public holiday tomorrow, while the oic eats by himself, shunned even by his friends, in a corner.[40]

Portraits of the Queen and Lord Casey hanging in the station were damaged and taken down. The annual report of the isolated oic did not mention any of these incidents or even acknowledge the crisis, but commented that 'Casey is traditionally considered to have lower morale than any other station', because of the tunnel design of the old station (now superseded and removed) and because it is the least scenic of the Australian stations. He added: 'It is nice to stay in bed when there is a blizzard but this privilege is confined to DHC personnel [building workers] and others have to fly balloons, keep skeds [schedules] or whatever.'[41]

I was able to follow some of the drama in the station logbooks in the library at Casey station. Logbooks are kept by the oic and have become less frank and more bureaucratic over the years as confidentiality has proven a chimera. The 1969 Casey logbook, available on the open shelves, includes a note at the start (pasted in later) which states: 'Please be aware that this log may contain sensitive or contentious issues. At the time it was written, logs were considered to be staff-in-confidence documents.' The incoming oic at Casey in November 1985, Barry Martin, inherited a social disaster. Within a few weeks he had written in the logbook in capital letters: 'D.H.C. '85 WINTERERS A PROBLEM!' The Department of Housing and Construction (DHC) provided the personnel for the massive building program in Australian Antarctica in the 1980s. The undercurrent of tension between building workers and scientists caused much of the unhappiness at Casey. Boffins and builders were not only out of sympathy with one another; they worked for different institutions and did not share a mission. 'Saturday nights in The Club are a real problem', noted Martin, 'with 85 winterers playing long and hard – 86 crew avoid it in the main because of the unpleasant atmosphere.' On Christmas Eve, Martin recorded that '86 folk are by nature more conservative and milder than 85'ers and are being intimidated by the hard-drinking, foul-mouthed crew who gather in the mess every evening. The main offenders are DHC'. These men, he later concluded, 'seemed malicious in intent'.[42] The decision was made to send home all those building workers not attached to the new wintering party, and to despatch them on the next available ship; which was Murray-Smith's *Icebird* (the *Polar Bird* in its earlier guise). The ship's volley-ball court, where the beer is stored, would be locked. There was discussion of permanently closing the bar. The troublemakers were split up among cabins. Back at Casey, liberated from these winterers, the oic recorded that 'it was as if the sun had come out'. But on the voyage home, Murray-Smith, always politically sympathetic to the worker, grew to respect these 'hard cases' or 'animals', as they had been called. 'I was in the army as a private soldier', he wrote, 'and I know how they feel.'[43]

The oic who arrived to relieve the Casey crisis, Barry Martin, reflected on it in his annual report a year later. 'This was an interesting year', he wrote, 'with both extremes being reached, very high and very low. The collective station morale is a very fragile thing, and often varies from almost day to day with no discernible stimulus.'[44] But in this case there had been a serious, sustained breakdown both in leadership and 'the normal controls of self-respect and peer pressure'. The men Murray-Smith enjoyed talking to on the ship home were not the same men who had rampaged at Casey. Something strange had happened to them down there. It was, Barry Martin concluded, a direct result of people staying too long and known trouble-makers being allowed to make more than one trip to Antarctica. Throughout the sustained building program, the Antarctic Division did not have the final say on the selection of DHC personnel. The oic at Davis station in 1986, Robert Easther, made similar criticisms of the selection policy, although morale during his year had been mostly good. Two unsatisfactory expeditioners in his wintering team had both previously had 'very negative OIC reports' and 'should never have been re-selected', he reported. One even boasted openly about the ineffectiveness of the OIC report in preventing him from returning. In another case, the Antarctic Division Personnel Section had told Easther that, although an individual was known to be unsatisfactory, he 'was going to have to "wear" him'.[45]

Ten years later, in 1996, Easther was summoned south as an emergency mediator in another Casey coup. At the end of the long winter – well after Mid-Winter Day but before summer life had quickened, traditionally the lowest moment of winterers' morale – a letter from Casey station curled off the fax machine at the Australian Antarctic Division in Kingston, Tasmania. It was signed by all the expeditioners except the leader and declared that they would no longer heed his authority. As the chosen mediator, Easther was given three weeks to pack his books on conflict resolution and join the first ship south for the summer. The media loved it: 'Hot words in frozen south', 'Antarctic staff freeze out leader of mission', 'Antarctic staff blow their cool', 'Mayday! Ice-bound insanity', 'Stir-crazy in Antarctica', 'Mediator sent to resolve cold snap', 'Mediator cools hot heads in Antarctica', 'Heated row at ice base', 'Hotheads crack as winter drags on', 'Hell in a very cold place', 'Winter of discontent'. The hype was heightened by the news that FBI agents had just been sent to the US base at McMurdo Sound to investigate another winter eruption where one cook had attacked two others with the claw end of a hammer.[46]

'Polar madness' can erupt without warning. *You can get it when someone sits in your chair. You can get it while they are combing their hair. It can come at any time ... dishing up chow or telling them how! As a matter of fact, some have it right now.* It might be as trivial as the way you dress. It might be as provocative as being served a glass of chilled urine at dinner. It might just be the sound of your voice. All these vexations have caused conflict in Antarctic communities, especially the usurpation of favourite chairs and the slurping of soup. In the 1950s an Australian had to be kept in a storage room for much of the winter after he had threatened people with a knife. In the 1960s a Soviet scientist killed a colleague with an axe because he was cheating at chess. In 1983 the doctor at Argentina's Almirante Brown station burnt it down to force an evacuation home. But most of the harassment and hurt we don't even know about because if it can be suppressed, it will be. Open conflict is too damaging. When sledging, Raymond Priestley knew 'to confine your remarks to the afternoon' and 'avoid controversial subjects like you would the devil'.[47] On the ice, minor disagreements can easily snowball. The never-ending polar night, the claustrophobia of a small community, the boredom of isolation: such an environment of forced intimacy can make the personal habits of your companions irritating and unbearable. The *Belgica* and *Southern Cross* expeditions had to deal with the physical as much as psychological traumas of wintering, but 'Red Shed Fever' has its own pathology. Tensions build until the atmosphere is as explosive as a tinder-dry Antarctic power plant. The merest spark can ignite it. Coming barefoot to dinner was a major gripe against the isolated leader at Casey in 1996.

Easther, a former station leader trained in psychology, was astonished by what he found at Casey in the spring of 1996. When the leader sat at the dining table, everyone else would leave. When he needed field equipment, it suddenly wasn't available. They even stopped cooperating with him on that most serious of Antarctic rituals, the fire drill. It was Borchgrevink all over again. If Bernacchi could have done so, he would have faxed Kingston. Casey's besieged leader had come to Antarctica in search of space and peace, and to escape grief, for in the previous five years he had lost his wife to a car accident and his sister to suicide. He believed in rules, pinned notices to boards and liked working alone. Tensions on the station had all come to a head when a traverse team was caught in a blizzard in August. A woman on the traverse was in radio contact with her husband back at base, but there was still anxiety because the team was a week overdue. At last the lights from the traverse vehicles could be seen. In the excitement of the reunion, the husband disobeyed the leader's orders (and modern Antarctic regulations) and drove out alone to greet them. The following day the station leader had a humiliating and abusive shouting match with the husband in the dining room. The fragile pretence of civility was shattered. When Easther arrived two months later, he aimed to restore some basic politeness just by getting people to say 'good morning' to the leader. He failed. But he did succeed in getting the leader to wear socks to dinner.[48]

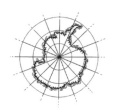

Since 1993 Australian Antarctic men and women have participated in research for the US National Aeronautics and Space Administration (NASA) because the Australian stations, not yet accessed by regular aircraft, experience some of the longest and most isolated winters. Antarctic wintering is regarded as a revealing laboratory for students of medicine and psychology and an ideal analogue for long-term space exploration.[49] Louis Bernacchi would have recognised the relevance of his lunar mountains and his winter hut to this other voyaging. George Palmai was a doctor at an Australian station in 1960 and investigated the psychological adjustment of an isolated group of 14 men.[50] He kept a psychological log and noticed general lethargy, a rise in blood pressure and the lowest ebb of morale from July to September. Depressive symptoms such as poor concentration and increasing irritability, decline in personal hygiene, increased anxiety about home, greater demands on the doctor all assumed 'almost epidemic proportions during the winter months'. In the spring, when the men began anticipating their own return to a sexually active society, their favourite pastime was watching the breeding behaviour of the elephant seal. Later studies showed that testosterone levels in men declined throughout a year spent in isolation from women but remained stable in the one man who had a female partner in Antarctica. Testosterone regularly attained its highest levels in the station leader, which is consistent with the biology and behaviour of other primates.[51] Casual observation suggests those who have wintered in high southern latitudes subsequently have more daughters than sons.[52]

Reading the medical and psychological studies of life in isolated communities, one can't help sensing the limits of faceless, nameless, clinical accounts of deeply personal and cultural matters. In the name of objectivity and rationality, real people are gutted and meaning ebbs away. Such studies, fully referenced, can deliver mundane insights of meaningless generality, informing us, for example, that 'there is experimental evidence that work is an important human need'. History, by contrast, spills over with illuminating, verifiable examples that you can argue with. This person did that here, then, because. History's commitment to context, contingency and particularity has often been seen to weaken its usefulness. But to understand the rigours of wintering, we need stories of winters past, we need some actual experiences to think with: Borchgrevink mistaking power for authority, Mawson humiliating Jeffryes in order to contain the infectious disease of madness, boffins and builders forming cliques in an era of divided command, a station leader himself shattering the pretence of civility. John Rich's crew at Casey celebrated the 2002 mid-winter with a reading of Bernacchi – 'What children of the light we are' – which must have been both patriotic and salutary. Scott's northern party survived six months in a cave by *imagining* space and privacy. They drew a line in the ice and declared that what was said on the other side could not be heard. The reports of Antarctic station leaders over the years are a kind of meditation on these matters. Has anyone ever studied them as a genre, as a compendium of practical advice, or are they trapped within the year of their accounting, each as discrete as an air bubble in its annual layer of snow? Are we condemned to an endless roundtrip? I am arguing that history is a survival manual. Tim Bowden's official history of ANARE, *The Silence Calling*, delivers on this promise of history by recording the rich voice of experience.

The isolated station leader at Casey in 1996 argued in his annual report that 'Stations don't have difficult years – they have difficult periods.' He cited other years and stations with problems, and referred to the violence at McMurdo in the same year by listing as a 1996 achievement: 'Casey wintering residents celebrate 1 year at Casey (without taking to claw hammers).' He also criticised the Antarctic Division for dealing directly with the rebels and sending a mediator because it undermined his authority. He felt that a station leader was an easy 'scapegoat' for a breakdown in group dynamics and concluded that 'I don't think that any of the wintering expeditioners will truly understand what motivated our actions, nor grasp the sources of emotion we displayed during the conflict'. He is telling us it was unfathomable. Or at least that it was not the leader's fault. For himself, the year was 'one of tremendous personal and professional growth'.

One of the reasons Stephen Murray-Smith was in Antarctica was his interest in the social life of isolated communities. He was, after all, the Emperor of one. Every summer since 1966 (except for this icy one) he would lead a small group of family and friends to his camp on Erith Island, a tiny, remote outcrop of land in stormy Bass Strait that could only be reached by fishing boat. He knew about 'incestuous island patter'. 'Although the community was egalitarian, Stephen was its patriarch, magisterial even in nakedness', writes John McLaren.[53] The Tasmanian government had appointed Murray-Smith warden of the islands. The Emperor (as they sometimes called him) might have had no clothes but he didn't need to have it pointed out. Murray-Smith (1922–88) was a distinguished historian, literary editor and practical intellectual. He had studied the history of islands, of lighthouses and their keepers, and now he turned his questioning gaze on Antarctica. He therefore viewed the 1985 Casey coup with something of a professional eye.

The year he went to Antarctica, Murray-Smith had completed a long essay on 'Three islands: A case study in survival'. The three remote communities that he put under the microscope were Tristan da Cunha, Pitcairn Island and Cape Barren Island. Each of them, he noted, was democratic in temper but dependent on individuals with 'authority of character and vision'. Was Murray-Smith thinking of himself and his own island realm? He certainly appealed to the experience of 'Anyone who has taken part in a commune, been on an Antarctic base during the winter, or even attempted to spend some weeks in utter isolation with a selected group of friends'. His academic and practical interest (with Murray-Smith they were always fused) was in the emergence of a moral order in such communities. He told their stories as parables for living in any small, isolated society, and he saw their unexpected successes as 'a precious imaginative resource to others, and a most significant historical lesson to all human beings striving to come to terms with the broken societies produced by the industrial era'. He wanted to know: What is it that caused some communities to self-destruct?

'Everyone has been familiar with the general situation since the mutiny on the *Bounty*', commented Murray-Smith as he sailed into the Casey coup of '85.[54]

The seed of the famous mutiny on the *Bounty* in 1789, argues historian Greg Dening, was Mr Bligh's bad language.[55] It was not that Bligh was physically violent, the flogger of legend. Rather, it was that he constantly violated the boundary between public and private on a small ship; he was ambiguous in his command. Dening reflects on the difficulty for historians of recapturing language in its full expression; for it is not just a matter of words, but also timing, an inflection, a look in the eye, a turn of the lip. It is not just how it was said but how it was heard. Dening's analysis of the mutiny is a sparkling evocation of implicit codes and expectations in a total institution. His book, *Mr Bligh's Bad Language*, should be recommended reading for all leaders at Antarctic stations. A station in winter is rather like a ship voyaging at sea. As Richard Byrd said, 'everything that one does or says, or even thinks, is of importance to one's fellows. They are measuring you constantly, some openly, others secretly – there is so little else to do!'[56] The station, like the ship, is isolated, inescapable and all-encompassing. In the absence of walls, privacy is achieved and honoured through behaviour. Under Finn Ronne's tyranny at Stonington Island, one rebel started calling his bunk space 'Pitcairn Island', the desperate refuge of Fletcher Christian.[57] On the vast ocean, stormy or frozen, some things cannot be said because there is not the space for them. Ceremony and custom can be your saviours. Some people find they have power without authority and others have authority without power, and the successful leaders have both. 'Hegemony', as Dening says, 'is made of trivia'. Bligh's mutiny may have begun with a tease. The little things do matter. It really could be dangerous to dine without socks.

## Saturday, 28 December

**Windmill Islands,
Antarctica**

I never expected this trip to give me experiences such as
today. I have been out on a Zodiac miles away from Casey,
exploring the inlets and islands and ice cliffs of the coast. I
have heard – for the first time – the silence of Antarctica.
Everywhere else I have been within the throb and thrum of
engines – even the ship at rest, at anchorage, is constantly
drumming, cranking and sighing in order to sustain the
community on board. Last night, at 11 pm, when the air
was still, I went on deck and leaned over the rail for a while,
gazing across a peaceful, dark sea towards a sky of gentle
mauves and pinks and light blues. The lone windmill on the
hill at Casey was not moving at all; it was a steel statue. I
felt the silence but could not hear it. Suddenly I craved it, I
wanted to escape the ship and the station for a little while
and hear the ice creak. And today I did! But even better than
I could have hoped. It was a day full of otherworldly beauty,
I gasped constantly. It was such a contrast to yesterday,
when I negotiated and added to the human infrastructure
of Antarctica. Today we were fragile creatures way, way out
of our biome, space travellers temporarily visiting another
planet. We had to be constantly aware of the danger of this
environment, the margin for error is small. What is the
wind doing, where are those clouds coming from, how firm
is this sea ice, this overarching cliff, how cold am I getting?
There was a time when we were preparing for our trip when
I thought 'it does not matter about the camera or keeping my
bag dry, what matters is that I stay warm and come home'.

We skidded over the polar waters, between floating

bits of ice, and under great rearing cliffs of ice four or five storeys high. We stopped beside floes populated by penguins and some of them swam beside us, leaping gracefully like dolphins in arches from the water. We followed a channel between two islands and nosed our boat into an ice crack, and I had to jump out and tether it to this great plain of thin sea ice that probably won't even be there next week! Walking around this area you had to step over the tide cracks which were like small crevasses, and once the ice suddenly gave way up to my knee and I had to scrabble for a handhold. We had lunch here in the company of three Weddell seals sprawled on the snow, hardly batting an eyelid at our presence. And the little Adélie penguins came perkily up to us to investigate and one or two walked virtually between our legs. One waddled up to me in a friendly fashion and regarded me from only a metre away with respectful interest. It was the most beautiful setting, remote and potentially dangerous. Yet we were having a picnic! I really did hear the ice creak.

Antarctica is like a kingsize playground for intellectual physicality. The marriage of ideas and bodies, of the most abstract aspirations and the most basic biological needs is close. Everyone you talk to here – whether they are a diesel mechanic or a physicist – is driven by curiosity and passion of some kind. This is a community of wonder. Although I am a contemptible roundtripper, I have not been patronised or disdained in any way. On the contrary, oldtimers seem to share the pleasure of your innocence, and thereby happily relive their own.

# SOLITUDE
## *An experiment in loneliness*

The silence is palpable in Antarctica. It is a presence, not merely an absence. As Richard Byrd wrote after his first night on the continent, one could 'almost take hold of' the silence. On the windiest continent, it is the silence that people remember. Douglas Mawson lived for two years in 'The Home of the Blizzard' yet he felt 'The Silence Calling'. Byrd's geologist, Laurence McKinley Gould, wrote in praise of the Antarctic silence:

> When there is no wind this is a land of unparalleled quiet. But it is a different quiet than one feels back home. I have stood in the woods at home when the world seemed dead. There was no kind of sound. But in that world where a variety of sound was the rule rather than the exception such a silence was oppressive if not ominous. Not so here – this is a land of silence. One stands in the midst of it without any feeling of oppressiveness. It is an expanding sort of silence. It is inviting. It is the natural state here and I like it – I have come to feel at home in the midst of it.[1]

Morton Moyes, an Australian meteorologist on Mawson's 1911–14 expedition, spent ten weeks alone in Queen Mary Land in 1912–13, awaiting the return of sledging parties. He documented the silence in his diary. November 7: 'All alone here now and the silence is immense.' December 20: 'The Silence is so painful now that I have a continual singing in my left ear, much like a Barrel Organ, only it's the same tune all the time.'[2] When his companions returned he stood on his head for joy.

In Antarctica, it is dangerous to be alone. Antarctica often draws people who seek solitude, only to condemn them to an intense human intimacy. Gould identified this as the chief paradox of Antarctic life. He distinguished between 'group isolation' and 'individual isolation'.

Describing Byrd's first expedition in the winter of 1929, Gould wrote: 'More complete geographical isolation than ours could scarcely be imagined; but it was a group isolation. Individual isolation was about the most unattainable thing in Little America. There were 42 of us living in very compact quarters – much more intimate than any of us had ever lived before for any such period of time.' As a more recent expeditioner put it, 'It is the greatest irony: claustrophobia in Antarctica.'[3]

André Migot, the French doctor at Kerguelen in 1953 and Mawson in 1954, was committed to solitude. He had sought it mountain climbing in the European Alps and he had sought it in the Cistercian Order. He had made journeys to China and Tibet, and found peace in a Buddhist retreat at Shangu Gompa. He longed for 'cosmic unity'. So he went to Antarctica. After a year at a French sub-Antarctic base full of unhappiness and tension, he joined the foundation party of Australians at Mawson station in 1954. He relished the pioneering spirit of a group that was establishing a base on the ice, but he also welcomed what he described as the 'British' or 'Anglo-Saxon' reserve of his new companions. Their social style and expectations – and no doubt his status as an outsider who spoke a different language – gave him the right mixture of space and society. 'I must admit', he wrote, 'that for once a communal life was almost better than a solitary one.'[4]

It has been said that those who survive a polar winter best – or a concentration camp – tend to be the shy and withdrawn people, those not utterly dependent on the social group for their identity and energy.

What happens, then, if you are a creature of the limelight and yet also a romantic seeking that cosmic unity? What happens if you are a leader who decides to be alone? What happens if you make solitude a stunt?

In 1934, a famous North American conducted an awesome experiment in personality on the ice. As an explorer of the last frontier, Admiral Richard Byrd combined a modernist determination to use new technology with utopian and primitivist purposes. He wanted not just to transplant civilisation to the ice, but to create a new society, one purified by its isolation and simplicity. He explained that 'If it were possible, I wanted to create a single attitude – a single state of mind – unfettered by the trivial considerations of civilisation ... In a word, we are trying to get away from the false standards by which men live under more civilised conditions.'[5] And Byrd's most gruelling social experiment was an anti-social one: he decided to spend a polar winter on the ice alone.[6]

The Bolling Advance Weather Base was established on the vast, flat ice of the Ross Barrier at 80 degrees south in April 1934 and Byrd became, for four dark months, its sole occupant. It was the first inland station ever occupied in Antarctica and its purpose was to record continental rather than coastal weather. But, in spite of this scientific justification, the truth was that Byrd 'wanted to go for the experience's sake'. America of the 1920s had rediscovered Henry Thoreau and his celebration of life at Walden Pond. But the spareness of Antarctica made Byrd's nature retreat seem radical and modernist. Byrd considered himself a reasonably happy man – as happy as a famous explorer could be – but he had been so busy exploring, managing, lecturing and raising money for expeditions, that 'a crowding confusion had pushed in' to his life. Byrd was proud that he had flown over both poles, but these were such abstract and ultimately disappointing goals. In the wake of such achievements, he was 'conscious of a certain aimlessness'. There were books he wanted to read, music to listen to, new ideas to learn about. He had to drop out of the world to catch up with it. But why bury himself at the pole, people asked? Because a man in the limelight – someone 'who must go to the public for support and render a perpetual account of his stewardship' – must even be alone publicly. Anyway, he 'wanted something more than just privacy in the geographical sense'. He 'wanted to sink roots into some replenishing philosophy':

> Out there on the South Polar barrier, in cold and darkness as complete as that of the Pleistocene, I should have time to catch up, to study and think and listen to the phonograph; and, maybe for seven months, remote from all but the simplest distractions, I should be able to live exactly as I chose, obedient to no necessities but those imposed by wind and night and cold, and to no man's laws but my own.[7]

Perhaps, he admitted, he was also in search of a more rigorous existence than he had known: 'much of my adult life had been spent in aviation', he remarked. 'The man who flies achieves his destiny sitting down.'

He took this trouble to explain his reasons for the solo experiment, because at the time it was rumoured that he had been exiled by his own men, or that he had gone out there to do some quiet but serious drinking. However, it's true that this, his second expedition, also needed a heroic centrepiece, one that might match the South Pole flight of his first expedition in 1929. Byrd was going out on the ice alone, but he made sure that the eyes of the world were upon him.

Byrd was his own guinea pig in a laboratory of solitude. He was well prepared, but on his first night he lay in bed realising he had forgotten an alarm clock and a cookbook. Although a promoter of technology, he did not much understand it. He had a gasoline-driven generator that provided power and radio contact, and when the engine became erratic due to poor maintenance, Byrd had to communicate to the outside world by cranking an emergency transmitter by hand. This was his only link with base camp at Little America, 160 kilometres (99 miles) away. He inhabited a world underneath the ice that he could span in four strides going one way and three strides going the other, and it was heated by an oil-burning stove. It took him a long time to realise that the queer sickish oil smell it emitted might kill him.

In the autumn, Byrd went for regular walks from his buried hut, taking with him an armful of split bamboo sticks to mark his return path. Once, in a reverie, he walked too far from his last stick. He was alone and now he was lost. With a rising terror, he built a mound of snow and explored from it in ever-widening circles until, at last, he found his farthest stick.

Byrd committed himself to routine and ritual. The silence of the Barrier was enthralling.

> I have never known such utter quiet. Sometimes it lulled and hypnotised, like a waterfall or any other steady, familiar sound. At other times it struck into the consciousness as peremptorily as a sudden noise. It made me think of the fatal emptiness that comes when an airplane engine cuts out abruptly in flight. Up on the Barrier it was taut and immense; and, in spite of myself, I would be straining to listen – for nothing, really, nothing but the sheer excitement of silence.[8]

There were times of mystical revelation:

> I was learning what the philosophers have long been harping on – that a man can live profoundly without masses of things. For all my realism and skepticism there came over me, too powerfully to be denied, that exalted sense of identification – of oneness – with the outer world which is partly mystical but also certainty. I came to understand what Thoreau meant when he said, 'My body is all sentient.' There were moments when I felt more *alive* than at any other time in my life.[9]

But after six weeks, it was death that stalked him. Byrd felt a creeping unease and a growing depression, and he felt his mind slipping in frightening ways. On the last day of May he collapsed under the combined influence of chronic carbon monoxide poisoning from the heating stove and the gasoline-driven radio generator. He was very sick and alone in the polar night, wracked by pain and 'literally shaken by the thumping of my heart'. He lay in his bunk mumbling like a monk, and when his voice stopped 'the silence crowded in'. He had expected that the real threat to his survival would come from outside, but it came from within. He wrote final letters to his wife and children. And as he vomited and stumbled and crawled around his cavern he was ashamed:

> [A]gainst cold the explorer has simple but ample defences. Against the accidents which are the most serious risks of isolation he has inbred resourcefulness and ingenuity. But against darkness, nothing much but his own dignity.[10]

Well, he had now lost his dignity. But not his pride. For although the radio theoretically gave him the means of rescue, he would not use it in this way, not just for pride's sake but for the sake of his men at base camp, for any winter journey was perilous. His geologist, Laurence McKinley Gould, believed that 'Commander Byrd's most outstanding characteristic is his concern for the safety of his men.'[11] Byrd had left firm instructions not to be rescued, even if radio contact was lost. And so there began months of cryptic radio conversation with his men in Little America, during which he fought to hide his sickness and they became increasingly puzzled by his halting oddities and longer silences. In his cavern, as the winter cold deepened, he became fatally bound to his stove. 'This fire was my enemy', he wrote, 'but I could not live without it.' But in choosing between freezing and poisoning, he chose the enemy he could fight. 'Cold I could feel', he wrote, 'but carbon monoxide was invisible and tasteless. So I chose the cold', and he turned off the stove fire for two or three hours every afternoon and shivered in his sleeping bag.

But his men at base camp perhaps knew him better than he did himself, and their radio banter strategically hid their growing anxiety that something was up. They contrived a late winter journey in his direction; not a rescue, mind you, but a scientific expedition. Minutes of staff meetings at Little America in early July 1934 reveal highly fraught discussions about the personal, political and scientific justifications of this proposed trip. Was it for Byrd or for science, was it for rescue or research? Or was it, perhaps, for the media? Byrd's alliance with the media had enabled him to fund, and make money from, large-scale polar expeditions, but it also placed him under constant pressure to be newsworthy. Byrd had signed a contract with CBS radio for live broadcasts from Antarctica. He had also negotiated film rights for the expedition with Paramount Pictures, who demanded a 'novelty' comparable to the South Pole flight of 1929, something 'which had tremendous publicity value'.[12] Before leaving for Antarctica he had hinted to his media partners of a strictly confidential plan, one that would 'have more drama than anything I have ever done before'. Byrd's lonely vigil was to be that event. He risked making himself a public sacrifice.

Byrd's experiment in loneliness had dangerously exposed the contradictions in his personality. He craved publicity, yet he was secretive. He was romantic yet manipulative. He was 'passionately interested in isolation for its own sake' and genuinely yearned for an experience of 'tranquillity', yet by testing himself alone on the ice in the middle of an expedition he had invited the world inside his head. The public nature of his solitude endangered him. And as he crawled around his cavern in mid-winter, dizzy and sick, his fate was being shaped by the needs of the media. By June 1934, Byrd's forthcoming rescue from Advance Base was being touted as 'a news event of the first importance' and had become a bargaining chip in negotiations over renewal of the CBS radio contract. Media representatives in America seem to have known of Byrd's rescue before he did. The media was influencing field operations in Antarctica, shaping the events it reported even at the farthest reaches of the globe. The dramatic story and its release to the world had to be planned. But what of the man's survival? Historian Robert Matuozzi believes that the expedition's dependence on commercial broadcasting arrangements indirectly imperilled Byrd's safety and that of the men who rescued him.[13]

On 11 August 1934, tractor headlights seared the darkness at Bolling Advance Base. Byrd was still alive, but very sick. He did not talk to his men about his ordeal and they never asked him. A curious silence settled inside the hut, a rather different one to that which reigned outside. '[S]omething deep inside me', Byrd recalled, 'demanded that I close my mind to the notion that I had been rescued.' Four years passed before Byrd could bear to write about his experience. In some ways, he never fully recovered.

## Sunday, 29 December

**Beyond the moraine line,
Casey station**

Today we got right up onto the glacier – the plateau
– behind the station. We travelled a little way into the
icy interior in a Hagglund (Swedish-built articulated
over-snow vehicle) and got out in what was hopefully
a crevasse-free zone and tried to absorb some of the
vastness of the ice that climbs inland from Casey station,
the Law Dome. I looked closely at the ice and imagined
dragging sleds across it, imagined pitching the tent on
it, imagined seeing nothing but it for months. I was even
sitting on ice reading *Sitting on Penguins*. We could still
see the coast and had a wonderful view of the inlets,
islands and bays we explored yesterday. And beyond
them we could see the Vanderford Glacier calving into
the sea. Now I see where those icebergs come from and it
*is* scary.

In the Wallow there is a chart just like ours on the ship
that shows the progress of the *Polar Bird* across the southern
seas. Our voyage was carefully mapped as we headed for
Casey, no doubt with both excitement and apprehension
as winterers anticipated the end of their year. I notice that,
according to the chart at Casey, the *Polar Bird's* excellent
progress came to a sudden end on 23 December when it was
swallowed by a sea monster. The great creature is drawn in
some detail on the chart.

I've been working in the archives of the station library
again. This morning I found Stephen Murray-Smith
there, his visit as a ministerial observer duly recorded in

the 1985–86 logbook. The saga of the unhappy station community of that year also unfolds in those pages.

This year by contrast, has been a happy one. The community consisted of 16 men and one woman. John Rich believes that even having just one woman on station substantially changed the tone of the community for the better. Humour and irony have also been saving graces. When the *Kapitan Klebnikov* anchored in Newcombe Bay last October, it was greeted by a huge sign on the front of the fuel farm declaring that, after their long voyage, they had finally arrived at 'DAVIS'.

'Spider' had skilfully covered the 'CASEY' sign with sheets of plywood. The Casey 2002 Winterers have prepared, as is the tradition, a Yearbook. It includes a glossary. A 'True Story', it explains, is 'A story unlikely to be true'. The bar in the Casey Meteorological Office is called 'The Ice-O-Bar' and the bar in the workshop is called 'The Refinery'. 'Back in '32' is the phrase you use to introduce 'true stories' to show previous years did it tougher.

# HONEYMOON ON ICE
## *Love in a cold climate*

Douglas Mawson and Paquita Delprat became engaged before Mawson's Australasian Antarctic Expedition of 1911–14 and then unexpectedly faced a 27-month separation before their marriage. Mawson's sledging tragedy and the failure of wireless communication in the first year lengthened and deepened the silence. Letters – and very few of them – had to sustain their love. There was little hope of conversation, even written, because any letter had to be delivered by the *Aurora*, and even then might not get off the ship, as we shall see. 'It seems like writing to a wall', Paquita declared to Mawson in September 1913, 22 months into their ordeal and without a letter from him for a year and a half. 'There is no reason why you shouldn't like me as much as before. But this everlasting silence is almost unbearable.'[1]

They had met in Broken Hill soon after Mawson returned from the Shackleton expedition of 1908–09, when Douglas was 27 and Paquita 18. Francisca ('Paquita') Adriana Delprat was the youngest daughter of Broken Hill Proprietary Ltd's general manager. They did not meet again until Mawson had returned from a half year in Europe, but in the winter of 1910 he courted her and, in early December 1910, he proposed. A year later, he was to leave for Antarctica again, anticipating a 15-month absence. His future father-in-law approved of the match but encouraged Mawson to abandon his plan for another Antarctic expedition. 'Do you think it is a fair thing to make a woman go through?'

'I don't want to doubt you dear', Paquita continued in her letter of September 1913, 'but I'm afraid of the fascination of the South. All the members [of the expedition who returned in 1913] say they would go again and here is Shackleton off again … I long for our meeting but in a faint way dread it. Do you feel hurt at that sentence, or do you understand it? … I wonder why I feel so sad tonight. Am I afraid there is not room in your heart for the expedition and I? … Oh Douglas don't *don't* let Antarctica freeze you. If I only had some words to go on with.'

Mawson was reluctant to use the precious radio communication for personal messages, especially as leader. And a letter he wrote to Paquita just before his tragic sledge journey was forgotten by his captain, John King Davis, and not delivered when the *Aurora* returned to Australia in early 1913. If the captain had remembered to give it to Paquita, the letter would have told her this:

> we shall get away [on the sledge journey] in an hours time. I have two good companions Dr Mertz and Lieut Ninnis. It is unlikely that any harm will happen to us but should I not return to you in Australia please know that I truly loved you from an admiration of *your spirit*. And should we meet afterwards under other circumstances please know and love me as a brother.[2]

Paquita was heartbroken that there was no letter from her fiancé when the *Aurora* returned without him. But what would she have made of these sentiments had they been delivered? She feared the iciness of Douglas: 'Are you frozen? In heart I mean', she asked in August 1913, in frustration that he would not send her radio messages. 'Am I pouring out a little of what is in my heart to an iceberg? … Can a person remain in such cold and lonely regions however beautiful & still love warmly? You were not in love when with Shackleton.'[3] Her love, he reassured her, melted the ice and the 'frozen, austere solitude'. She warmed her 'proxy iceberg', as he called himself, and he felt the cold less this time.[4] But Douglas, in turn, urged Paquita to be more affectionate: 'don't come to me as ivory – as you sometimes do – I want you warm and – well *you*'.[5] Her reserve during their courtship had been due to sexual innocence and fear. Before their departure she had been constrained by her belief that kissing could lead to pregnancy. Now she tried to tell him that she was transformed by new knowledge and understanding, just as he was by Antarctica: 'I own now I was rather cold before you left through ignorance of everything.'[6] She also wrote, 'when I look back at

myself [at] the time you left it seems a different person – so young &
silly. How could you love me? I must have hurt you, my love, often but
believe me & I know you do it was only in ignorance. I loved you then
as a girl who knows *nothing at all* of life & now – as a woman.'[7]

But her man, too, had changed. That past summer had ravaged
Mawson's body. His hair had fallen out, the soles of his feet had peeled
off, and he felt despoiled and worn. He sent a wireless message to
Paquita: 'DEEPLY REGRET DELAY ONLY JUST MANAGED
REACH HUT EFFECTS NOW GONE BUT LOST MY HAIR YOU
ARE FREE TO CONSIDER YOUR CONTRACT BUT TRUST YOU
WILL NOT ABANDON YOUR SECOND HAND DOUGLAS.'

He worried: '*Can you be happy with me.* I have aged in appearance with
this strain and may not appeal to you now. My body tissues have been
strained and cannot be so good.' He urged her to size him up critically
on his return, and to be honest about any reserve she felt, for 'the merest
indications of splits are apt to widen to become fissures and crevasses'.[8]

The worthiness of his love was directly related to the character of
his achievements on the ice. Paquita's love had helped him get home on
that dreadful trek. But, early in the expedition, when his hopes of even
establishing a land base in Antarctica had seemed in jeopardy, he wrote
to her: 'This morning I wished that I had never spoken to you of my
love – that was because a large proportion of failure appeared to stare
us in the face ... things looked so bad last night that I could do nothing
but just roll over and over on the settee on which I have been sleeping
and wish that I could fall into oblivion without affecting you, darling.'[9]
Paquita was a responsibility, a fetter on his freedom, as well as an inspiration. In bad times, her love could be as much a burden as a comfort.

Paquita and Douglas were separated not just by continents, but by
age and experience, too. Their uncertainty, about the other's love as
well as their own, increased with the separation and silence. Paquita
wrote: 'I have lost confidence, not in you, but in the future.'[10] Mawson,
fighting for his and others' sanity in the hut, meditated on love and evolution. He subjected his feelings to scientific scrutiny. 'Love is unknown

in the lower animals, hardly perceivable even in mammals except in human beings', he wrote to Paquita. And even amongst humans, he felt that 'true love in the ethical sense' was a quality of the higher races, whereas the 'savage races' felt a love that was 'mere passion and instinct'. So there in Antarctica, stranded from his fiancée for another year, it seemed to Mawson that physical separation, an icy setting and a manly society might cultivate a pure form of love. After all, the frontier could be morally improving, and it had been an evolutionary force in human history. Mawson reflected in his diary that 'It seems to me that man and his brain have evolved as air breathing animals and chiefly in temperate and cold climates, because of the vicissitudes there.'[11] In early twentieth-century racial theories, the enlivening influence of cold climates was often compared to the deadening and degenerative effect of the tropics. Mawson's blue eyes and 'Nordic' appearance, together with his achievements, qualified him for the accolade of 'an Australian Nansen'. So, here in the coldest of climates, there might be evolutionary opportunities: 'in this stern country of biting facts ones love gets frozen in deeper'. 'No', he replied to Paquita's concerns: 'I am not frozen in heart you may be sure, this is where the warm hearts are bred.'[12] Antarctica, as the last bastion of frontier manhood, perhaps represented an earlier, essential state of society and evolution, and all were now under threat.

In the late nineteenth and early twentieth centuries, polar exploration narratives were crucial to the social construction of white masculinity. Men battled not only the elements but also their weaker selves to establish their manliness.[13] The polar explorer emerged as an international icon of manliness at a time when the politics of 'the new woman' was gathering momentum. Mawson shared a general concern in Australian society in the early twentieth century that the rising power of women would suppress the birthrate. A decline in the white colonial birthrate had been observed in the 1890s and had registered in the 1901 Commonwealth of Australia census, and was sufficiently disturbing to prompt a Royal Commission in New South Wales.[14] What would 'the demographic transition' – that change from a high birthrate and high deathrate to a low birthrate and low deathrate – mean for modernising, western societies? This was a national, even racial issue, and also an utterly intimate one about the relations between a man and a woman. 'What will be the outcome of it all?' wrote Mawson. 'Will the altered relations of men and women affect love between the sexes?' Douglas explained to Paquita that 'the female represents the passive vegetative state – the male is the active animal state', and this biological classification would remain true 'no matter how much the *new woman* may think to the contrary'. 'The new woman' was a phrase in currency in the early twentieth century as campaigns for women's suffrage and other social and political rights gained momentum. From the sanctuary of Antarctica Mawson declared: 'Woman has her sphere – man has his – between them is the whole future of the race.' His geological sense of time reminded him of the inevitability of extinctions and of the passing on of the dominance of the Earth, from species to species and perhaps from race to race. 'What is it that will overthrow the dominion of the genus homo? Is it woman with whom in such a degree lies the fate of the man of tomorrow?' Mawson feared that the political emergence of 'the new woman' would weaken western society, leading first to a matriarchy and then to the triumph over Europe of the hordes of Asia. It was, he decided, the 'insidious germ of social change' that ended empires.[15]

Antarctica was the twentieth-century's prime site for boys' own adventures, and they were very much their own. The ice was a masculine place, to be defended, where women might be imagined and missed but never seen or held. Brigid Hains, in her entrancing study of Douglas Mawson and John Flynn as Australian frontier heroes, has written of 'the boyish camaraderie of Antarctic exploration' expressed in a love of pranks, nicknames, 'gadgeting' and amateur theatricals. At a time when JM Barrie's *Peter Pan* was popular on the London stage, these lost boys down south were living in their own 'Never-Never Land'.[16] 'We felt like boys again', wrote Robert Falcon Scott's photographer Herbert Ponting, 'and acted, too, like boys'.[17] Belgrave Ninnis rejoiced in his 'second childhood'. The interior worlds of those Edwardian winter huts of Scott, Shackleton and Mawson seemed to capture the innocence of the era, and also a self-conscious spirit of boyish bravado and irresponsibility. In Australia, argues the historian Marilyn Lake, the beginning of the twentieth century saw a contest between men and women for control of the national culture, a conflict which often entered the home.[18] The dominant images of national character in Mawson's country at this time were masculine ones, and they were of nomadic, unattached, rural lives. The wandering bush worker and later, the swagman, epitomised the independence and uncomplicated camaraderie which city bohemians tried to foster in their all-male clubs and cafés. For some bohemians, the family life to which they returned at the end of the night was a source of embarrassment or mockery. Women, though, could be reverenced – and confined – as 'mothers', or as distant, beloved, child-like creatures. Either way, they represented responsibility. Mawson went on his expeditions 'to let off superfluous steam' and he imagined Paquita as his 'engineer, and a *good* engineer can always regulate steam pressure'.[19]

In her study of Mawson's 1911–14 expedition, Hains observes that 'This desire to escape convention on nature's frontiers could be deeply misogynist.'[20] She notes that Mawson's select personal Antarctic library contained Rudyard Kipling's verse, the works of Robert Service (a Darwinian celebrator of the frontier) and essays by Robert Louis Stevenson, one of which laments the dampening effect of marriage on heroic and generous men. In marriage a man 'undergoes a fatty degeneration of his moral being', wrote Stevenson. Mawson's letters to Paquita frankly shared his feelings about wilderness and domesticity (sometimes, she must have thought, too frankly). 'I sometimes think that I am much better out upon a lonely trail for nature and I get on very well together – I feel with nature and revel in the wilds. Here within a gunshot is the greatest glacier tongue yet known in the world – no human eyes have scanned it before ours. What an exultation is ours – the feeling is magical – young men whom you would scarce expect would be affected stand half clad without feeling the cold of the keen blizzard wind and literally dance from sheer exultation – can you not feel it too as I write – the quickening of the pulse, the awakening of the mind, the tension of every fibre – and this is joy.'[21] Young half-clad men revelling on the ice, their skin tingling with the vigour of the cold, an imaginary gunshot ringing in their ears ... The sensuality of Mawson's prose about man and nature could not help but make his protestations of love for Paquita pale by comparison. Of the prospect of marriage, he wrote to her: 'How curious it all will be – to think that I have to bury some 30 years of my life and start on a new line – I wonder how I shall take to it?'[22]

The sheer physicality of Antarctica – its raw elemental embrace – resonated with Mawson's yearnings, for love and death. Douglas wrote to Paquita in mid-winter 1913 about the feeling of literally bursting with love: 'Do you ever feel full like that – have to do something to let it off – go on the ice so to speak – feel as if you would like to stiffen up and die if it would be just useful to somebody. Well men *can* feel that way and I pity those that don't.'[23] He wanted to *do something for posterity*. The sexuality of sacrifice was on his mind. In some ways he envied Scott his famous, eloquent tomb. Mawson wrote from his hut that same year to Kathleen Scott, then the world's most celebrated widow. He fondly remembered their meetings in London before the expedition, and she had recently sent him a message on his return from his trek: 'Love and sympathy come back safe'. Now he told Kathleen (whom Paquita was never to like) that her message 'touches my heart and presses the key of the most tender feelings imaginable ... I cannot refrain from addressing myself to you, now, nearest the scene which has riveted your mind so long ... My own case has enlightened me in the secret feelings of the soul, the willingness to pass into oblivion ... Please accept my love.' Love, sacrifice and redemption might all be found at the edge of empire. Peter Pan had thought that death would 'be an awfully big adventure'.[24]

She was 22 years old, she was a bride, she had never been further south than Florida, she was already missing the hat shops of New York, and she was gazing, appalled, upon the vast whiteness of Antarctica. She felt like an intruder upon a man's world. It was 1947 and Jennie Darlington was about to become one of the first two women to winter on the continent of ice, half a century after the first men.[25]

What was she doing there, at the bottom of the Earth? She had inadvertently married a continent.

The year before, soon after the end of World War 2, she had met over cocktails an American naval officer who had been south with Admiral Richard Byrd's third expedition in 1939–41. In their first conversation, Harry Darlington taught Jennie to pronounce 'Antarctica' with the middle 'c'. And she had made the mistake of asking him 'What was it like up there?'

But she felt his eyes upon her. She was soon to go to Europe and so met Harry again in Washington where she had to pick up her passport. He stole it and would return it only if she agreed to be his wife. She did agree. The only problem was that he was about to go to Antarctica for two years. Their wedding was small and secret. But their honeymoon was to become spectacularly public.

One night, Jennie and Harry dined with the commander of the proposed new expedition, Finn Ronne, and his wife Edith (Jackie). That evening, as Harry pored excitedly over maps, Jennie knew she had lost him. She watched him become alive, assured, purposeful. He was offered the position of third-in-command. 'Already the Antarctic was coming between us', she recalled. 'I was jealous, not of a person, but of a place.'

He was a moody man, her Harry. And a man's man. Antarctica was his separate realm, its difference defined as much by the absence of women as by the presence of penguins. In the aftermath of war, Jennie found that 'men came back from having faced the enemy and could not face themselves'. Many turned south. Harry's inner demons were exorcised on the ice. Like Shackleton before him, Darlington compared the white warfare of the Antarctic to the red warfare of Europe. 'The challenge without the killing', he mused.

Preparation for the Ronne Antarctic Research Expedition continued with excitement and uncertainty. Harry told Jennie that her program was 'to stay out of the way'. She sat on an oil drum knitting socks and ski caps. She felt like an extra rib.

The ship set sail at the end of January 1947, and Jennie Darlington and Jackie Ronne accompanied their husbands, first as far as Panama and then ('a second reprieve') to Valparaiso, Chile. There the expedition was engulfed by crises. The sledging dogs had sickened and more than half had died, prankish looting of the expedition's stores had broken out, and the men were disconsolate. But the biggest crisis was the decision by Finn and Jackie Ronne to continue together to Antarctica. Finn had to persuade Jackie, but she finally gave in. 'I was in love with him', Jackie recalled. 'I would have gone to the moon. It was the moon.' Finn invited Harry to take Jennie as well. But Harry said no – 'of course'.

Jennie knew to stay silent, but inside her something hardened. Then the disconsolate men intervened in the most surprising way. Seven of them (just under half of the expedition) wrote their commander the following letter:

> We, the undersigned, feel that it would jeopardize our physical condition and mental balance if the Ronne expedition consisting of twenty men were to be accompanied by one or more females for that period of time spent in the Antarctic.
>
> Therefore we agree to form a united front to block that possibility. We are all prepared to leave the expedition in Valpo as a group if one or more women accompany it.

But the dissidents did not have a majority, and so they offered a surprising compromise. 'Two women would be better than one. If Jennie goes, we'll tear up the paper', they declared.

Jennie then watched her new husband recoil in horror and masculine angst at the thought that she might accompany him. 'He sat hunched over on the lower bunk, his head buried in his hands. He was trapped, cornered.' If he agreed to take his bride south, the expedition would continue. If he did not, Antarctica would remain as he felt it should be, an exclusively male domain. 'Jen', he said desperately, 'please understand. It hasn't anything to do with you. It's just that there are some things women don't do. They don't become Pope or President or go down to the Antarctic!' This was a man who would find the gender equality amongst penguins shocking.

In their cabin that night, Jennie recalled that 'the atmosphere was as taut as the lines lashing the gear on deck'. Harry refused to meet her eyes, and she said nothing. The next morning, after he had absentmindedly put two socks on one foot, he told her she was going to Antarctica.

The men of the expedition agreed that the main problem with this outcome was that 'It will make it harder to forget women'. The Chileans reminded the Americans that other women had set foot on Antarctica even if they had not wintered before. Jennie's mother cabled with concern to the ship at Punta Arenas in Tierra del Fuego: 'Please be careful about catching cold!'

The Antarctic winter night is more than 2000 hours long and Jennie described it as 'the hardest personality test on earth'. She and Harry had a room at the end of the men's open bunkhouse, a frail and doubtful fragment of civilisation nestled into the ice on the edge of Marguerite Bay on the Antarctic Peninsula. Harry put a door on their partition, 'more for slamming purposes than privacy' he joked, for every word and cough and snore could be heard. What was their first big argument about, down there at the bottom of the world where human existence was constantly imperilled? About Harry refusing to turn off his reading light when Jennie wanted to sleep.

The expedition planned to explore and map what Commander Ronne called 'the world's last unknown coastline', and Harry was head of the aviation program. But relations between the two leaders quickly disintegrated. They were constantly at war with one another, and their wives were caught up in it. The only two women on the continent were unable to communicate during the long Antarctic winter because it may have seemed disloyal to their feuding husbands. In her memoir of this time, Edith Ronne made no mention of the fact that there was another woman with her.[26]

Finn Ronne finally relieved Harry Darlington of all aviation activities and removed him as third-in-command. Jennie therefore found herself managing a man marooned from his calling, who, for the next five months, lived through 'the darkest, most destructive period of his life'.

But there was more yet. 'Fate had another blow to impart', she recalled. 'I was going to have a baby.'

In her voluminous polar gear, she could hide it for a while. But she was anxious about telling Harry: 'Psychologically unprepared for marriage, violently averse to a woman's presence in polar regions, caught in a basic personality conflict to which there seemed no end or answer, my husband could hardly be expected to consider this the most auspicious time to start a family.' Especially as the 1947–48 summer turned out to be one of the rare years when the pack ice did not melt in Marguerite Bay, the ship remained frozen in, and they faced the prospect of another winter – and the first birth – in Antarctica.

'I'm having a baby. I've upset the programme. I'm sorry', she told Harry. But he was jubilant about Jennie's pregnancy although also anxious for her health, especially when, on a sledging trip, she fell through the ice into the freezing waters of the Bay. The baby remained a secret they kept with the doctor. In late February, two icebreaker ships of the US Navy were summoned to penetrate the Bay, and the expedition was able to depart for home. Cynthia, the first child conceived in Antarctica, was born in the northern hemisphere.

Jennie wrote about her year at the bottom of the world in a book called *My Antarctic Honeymoon*. She reflected that she and Harry had grown closer; she felt that she understood him better; she considered that Antarctica (even with its middle c) had equalised their relationship. However she concluded that 'taking everything into consideration, I do not think women belong in the Antarctic'. But if she had the chance she had no doubt she would do it all over again.

The French explorer-historian, Paul-Emile Victor, in his book *Man and the Conquest of the Poles*, summarised the Ronne expedition in one line: 'The expedition ran into all the difficulties ordinarily caused by the presence of women in such circumstances.'[27]

Jennie Darlington's book had an interesting life, too. Harry was just one of a long line of men who found they were unable to work with Finn Ronne, who was a famously poor manager of people. A young physician on the 1947–48 expedition, Don McLean, later confessed to thinking of pushing Ronne off a cliff during a hike from the station.[28] Ronne was a short, smart, vigorous man who could be charming in manner, but who was also insecure, intensely competitive, and authoritarian. He had no ability to negotiate or compromise. He loved setting up games and contests that he could win, by cheating if necessary. In 1957–58, he led a US expedition to Ellsworth station in Antarctica for the International Geophysical Year (IGY). Although IGY made science the primary purpose, Ronne went out of his way to frustrate his scientists, with the result that the entire scientific staff refused to comply with his directives. The men, who felt tyrannised by their leader's defensive ego, found solace in reading and passing round

Jennie Darlington's book, *My Antarctic Honeymoon*. In those pages they recognised their man as he was a decade earlier, and they had fun imagining Ronne (who still had a pinup girl on his wall in 1957) dealing with a young, pleasant and assertive woman.

One day over breakfast, when passing the book between readers (they had it beneath their shirts), Ronne saw them and recognised it. He went 'really mad' and had a shouting match with one of the scientists. He then had the book removed from the station library, and also ruled that none of the enlisted (Navy) men were to be allowed to read or use any of the books in the library. Ronne was unaware that at least two further copies of Jennie Darlington's banned book were circulating in the camp. Later in the year when one of the men prepared a radio message to his wife at home asking her to buy *My Antarctic Honeymoon*, Ronne (who had to approve all outgoing messages) refused to send it. Soon the men knew that any mention of Jennie's name would send the leader wild. One of the scientists, John Behrendt, recorded the many tantrums in his diary:

> Ronne then got off on the subject of Mrs Darlington. He asked why of all people, he [Don] had to pick the one person he was at odds with. Actually Don had done it just as a joke and to needle Ronne a bit, but he told him he thought her book had done a lot for morale at this base. Ronne called her 'that slut'. He said that she had been 'picked up in a bar'.[29]

Ronne met his own wife on a blind date. When they married he promised never to go to Antarctica again. 'Fortunately', recalled Jackie, 'I didn't believe him.'[30]

To a biologist's eyes gazing into the rich life of the Southern Ocean, Antarctica's short summers are erotic. They are an exuberant, opportunistic procreative frenzy under the sunshine of a single long day from October to March. On the edge of the ice, life has evolved to snatch the brief summer moment to reproduce itself. Surrounded by this efficient sexual ecstasy, the human polar communities are condemned, mostly, to dreaming and yearning. The British who shared Stonington Island with the Americans in 1947 (and were forbidden by Ronne from fraternising with them), had heard a story that there were two women at the neighbouring base but wondered for a while if it was just a myth. When Jennie Darlington unexpectedly encountered a British man unloading a sledge from a recent fieldtrip, she described what happened:

> At the sound of my voice the man started. He slowly straightened up from the sledge, turned, stared at me, his eyes going from the ends of my long green and white knitted stocking cap, from under which the pigtails protruded, to the tips of my regulation boots. Then his eyes widened. His mouth dropped open. Astonishment, embarrassment, and a certain confused fear flashed across his weather-beaten face. Involuntarily he reached out, grabbed the shovel, looked at it an instant, and then glanced back at me. Without another word he turned and fled to the British bunkhouse.[31]

Dr Richard Butson, the disconcerted gentleman, later apologised to Mrs Darlington: 'After mucking about on a glacier for several months I mistook you for a mirage.'[32]

John Béchervaise, leader at Mawson station in 1959 and a poet and philosopher, described the internal drifts of snow that insinuated themselves inside the men's huts, sometimes taking form slowly overnight through a tiny aperture. 'I sometimes find myself thinking of inside drifts as female. Their curves are voluptuous, like those of breasts and thighs … I have seen a strange white genie materialise through a nail-hole and, finally, lie recumbent large as life.'[33] Men could wake to find a ghostly female form lying in the corner of their bunkroom, like a dream. The outside drifts, which inhabited the true realm of men, were different; they were 'taut, of chiselled stone rather than of moulded flesh'.

The extremes of intimacy and distance experienced in Antarctica – intimacy with one's fellow expeditioners and distance from those at home – often made people reflect on love. We have seen how Bob Murphy and Grace Barstow built their marriage between these poles of longing and fulfilment. Béchervaise, experiencing a tumultuous year of blizzards and fire at Mawson, 'wrote love-letters from the world's end'. It was a place from which he felt he could write even to his dead mother, for they were temporarily reunited in their habitation of other worlds. He wondered about what human communication and contact meant, was it ever complete? 'Sensitivity, the ability to receive, is as essential as the desire to give. The sensitive may touch through the hem of a garment; the insensitive may be distant in a kiss.'[34]

While Richard Byrd was alone on the ice during his winter vigil of 1934, he reflected on the nature of a continent without women:

> Well, this is the one continent where no woman has ever set foot; I can't say that it is any better on that account. In fact, the stampede to the altar that took place after the return of my previous expedition would seem to offer strong corroboration of that. Of the forty-one men with me at Little America, thirty were bachelors. Several married the first girls they met in New Zealand; most of the rest got married immediately upon their return to the United States. Two of the bachelors were around fifty years old, and both were married shortly after reaching home.[35]

The British Antarctic explorer, Sir Vivian Fuchs, argued the benefits of sexual abstinence in his polar bases, as if he were describing a religious community: 'It is often suggested that the absence of women, and indeed of sex, must be a very real deprivation, but to most of the men it has little importance. They have to put it behind them and are wholly occupied with the life they are leading. It is unusual for the topic to arise, even in conversation.'[36] Fuchs shows a leader's ability not to see what he doesn't want to know. In Sir Vivian's icy outposts, only the British upper lip was stiff.

Like Douglas Mawson and Harry Darlington, many Antarctic men who were in love were attracted to, and vociferously defended, a continent without women. There was something spiritual about male comradeship, something pure about distant yearning and asexual love, and something incontrovertibly masculine about frontiering. The ice was their own inviolable space. In Antarctica, the presence of women could diminish a man. In their absence, one might prove oneself worthy of them. Then there was the fact that women, whatever their virtues, complicated things. Richard Byrd liked the fact that Antarctica was a place 'where men will not strut because there are no women about'.[37] Fuchs argued strongly against mixed communities of men and women in Antarctica. In his book, *Of Ice and Men*, he wrote:

> Should it happen one day that women are included as part of the base complement, problems will certainly arise. To add such stresses, with their consequent jealousies and disputes, to the difficulties of adjustment would surely lead to the breakdown of that sense of unity which is so important to the group.

> This does not mean that women could not compete with the environment – they certainly could – but it might be wise for them too to form single-sex communities, where they could form the same bonds as their male counterparts. It is no part of scientific exploration to provide for family life in isolated groups, where modern facilities could only be built-in at the expense of the work for which they exist.[38]

When interviewed in 1980, he was asked: 'And how do you feel about the fact that women are playing an increasing role down there?' He replied: 'They're not really, their part is *very* small.'[39]

There was resistance to women visiting Antarctica, and then there was resistance to women wintering, and then there was resistance to women getting beyond the coastline and working in the Antarctic interior.[40] The last male bastion was continually rebuilt and defended.

The media reported the arrival of women on the ice as 'invasions' and 'incursions'. A list of firsts gradually accumulated: first woman on the continent (Caroline Mikkelsen, wife of a Norwegian whaling captain, 1935); first women to winter (Jackie Ronne and Jennie Darlington, 1946–47); first women at the South Pole (six Americans jumped hand in hand from the ramp of a transport plane, 1969). But the real challenge was to have women accepted as normal members of expedition parties. The senior male resistance was tenacious and ingenious. Rear Admiral George Dufek, commander of US Antarctic operations, was quoted in 1957 as saying: 'Women will not be allowed in the Antarctic until we can provide one woman for every man.' Two years later, he described the women who want to go south as 'the do-gooders and newspaper girls'. He continued: 'I felt the men themselves didn't want women there. It was a pioneering job. I think the presence of women would wreck the illusion of the frontiersman – the illusion of being a hero.'[41] Throughout the 1950s and '60s, women of many nations unsuccessfully applied to their scientific and Antarctic programs for a trip south. Sometimes their applications were honestly rejected; sometimes the paperwork was lost.[42] To avoid discussion of the emotional and psychological dimensions of having women in Antarctica, male managers offered a simple logistical excuse: 'There are no facilities for women provided in Antarctica.' It all came down to bathrooms. It was as if women were a different species and 'all-male facilities' were a natural, inviolable feature of the ice. Here is how a senior North American naval officer, Rear Admiral James Reedy – a man regarded as sympathetic to the cause of Antarctic women and whose own wife wished to go south – justified the absence of women during the IGY:

> We recognise the fact that one day we're going to have to let women go down there. We'd like to have them, too, for that matter. It would make the place a lot less dreary.
>
> But in the first place there are no facilities for them down there. Every man who is down there is working to capacity on what is essential work. To ask them to turn around and build suitable

accommodation for women when it is not strictly necessary would be asking too much of them. And remember, we can't spare men to do anything other than what is vital to our programme.

Again, space on the aircraft is another thing. Every passenger means less cargo. An average woman weighs around 125 lbs, but by the time we carry survival gear for her, too, it goes up to around 300 lbs.[43]

Admiral Reedy, in spite of his stated sympathy, was doing his best to ensure that Antarctica remained what he called 'the womanless white continent of peace'.[44]

Bars would prove more intractable to reform than bathrooms. It was the social and working spaces that were most stubbornly gendered, especially where military culture prevailed. In the US Navy, women were often labelled a 'bitch or a slut or a dyke', names which focused on women's sexual availability to men.[45] Into the 1980s, the British Antarctic Survey (BAS) continued to prohibit any woman from flying in a BAS aircraft alone with one man (the pilot) on the grounds that, in an emergency, the pair would have to bivouac without a chaperone.[46] Men felt they had to moderate their language in the presence of women and saw this as a curtailment of freedom. Women were offended by the distribution or display of pornographic pictures in public areas. When the British journalist Sara Wheeler toured Antarctica in the mid-1990s, she found that the bar at the British base, Rothera, was still 'occupied territory' and was 'festooned with postcards of women's bottoms'. 'It was', she reflected, 'as if I had entered a time capsule and been hurtled back to an age in which Neanderthal man was prowling around on the look-out for mammoths.'[47] At Rothera, the support staff signed on for two-and-a-half years in Antarctica and felt strongly that women would be destabilising. 'They don't want the complication of a female in such a pristine place', explained another visitor to Wheeler. 'It's visceral.'

In 2000, the first woman wintered over at the Indian base of Maitri. She confessed to some women visitors that although she had good support from the leader, 'a lot of the other men resent my presence down here. They think I'm breaking into their exclusive club. It's worse when they want to watch their porn films.' On such occasions, she had to retire to her room and paint landscapes.[48]

But when the first women went ashore at Casey station in the summer of 1975–76, the men awaiting them at the base sent a welcoming message by teleprinter: 'Men of Casey 1975 delighted to hear you visiting us next changeover stop few only mumbles about invasion of man's last domain but grins give them away and know deep in their stoney hearts they happy to embrace you all on arrival stop'.[49] By the 1990s, husband and wife teams were wintering at Australian stations and double beds were available for people who unexpectedly fell in love on the ice. If your base consisted of just a cluster of tents pitched on blue ice, it was better to make love only when the katabatics were blowing. In his book, *Terra Antarctica*, Bill Fox tells of a guy who fell in love with a woman who worked in a McMurdo lab, and he brought her tokens of affection from his fieldwork: ventifacts from the dry valleys and specimens of sedimentary layers from the Transantarctics. Another woman at the station, experienced in Antarctic ways, observed: 'Just like any good penguin, he brought her stones every day to prove his devotion.'[50] Antarctica began to be reshaped by women's dreams. When Sara Wheeler looked across an awesome field of deep crevasses in 1995, she found herself thinking: 'So often it is the landscapes most inimical to life that are the most seductive. In this respect they are like boyfriends. It doesn't seem fair.'[51]

Expressions of nationalism on the ice soon enlisted women. In 1978, Argentina flew a woman in the late stages of pregnancy to the continent so that an Argentine baby could be indigenous. Emilio Marcos Palma was born at Argentina's Esperanza station. By 1984, the Chileans had matched the accomplishment and had even established a school.[52] Babies joined flags, postage stamps, cabinet meetings, lusty singing, proclamations and science as assertions of territorial ambition.

In his annual report on Casey in 2002, John Rich thanked the only woman who shared their winter. 'The station would have been a much lesser place without her.' The men appreciated 'her maturity, tolerance and good grace'. John continued: 'It is with some sadness that we observe that many wintering parties are now all male ... Unfortunately women are still very poorly represented in the trades.' The Casey 2003 winterers we took down on our ship were all males. Bloo told me later that 'there was not a cross word all winter', and that they moved all the tables together in the mess and ate together regularly. The 2003 station leader, Ivor Harris, reported near the end of the year that the absence of women 'has worked out very well for us. Some of our sources of humour and amusement have been things that have probably worked best in an all male group. I think it just reflects the freedoms we have enjoyed in a "big boys club" atmosphere where we have been able to be just that: big boys!' The camaraderie of the heroic-era huts echoes in those words.

The history of gender on the ice offers a stumbling political narrative of increasingly futile male resistance and a growing acceptance of women as expeditioners, doctors and station leaders, although it's true that wintering parties remain overwhelmingly male because of the dominance of maintenance staff. The history of love in Antarctica has no such progressive trajectory, and it never depended on the presence of women on the continent. Men have fallen in love with one another in Antarctica, and women too, but women and men have more frequently poured their passion into words for distant loves. Just as often, though, people have escaped relationships by going to Antarctica. Then the remote, icy, reflective continent seems uniquely in league with their desires.

## Monday, 30 December

**Shirley Island,
Antarctica**

Not far from Casey station is Shirley Island which you can
walk to over the sea ice for half of the year but which in
summer months you reach by boat. This is where Stephen
Murray-Smith really *did* sit on penguins. We nosed the
Zodiacs into the rocks at the base of the island, tethered
them, and then walked up into penguin city. We positioned
ourselves on rocks about 20 metres away from one of the
colonies and just watched activity in it for about an hour.
We had with us Dr Eric Woehler, who has studied these
and other colonies over about 15 summers, and he quietly
told us about the life of the Adélie.

These are the penguins that run up to you with their
flippers out behind them as if they are about to embrace
you in welcome. These are the ones that line up on the
edge of the sea ice, nudging each other towards the water,
because the first penguin in will test if there is a leopard
seal skulking under the rim waiting in hunger, and once
one has dived and survived they all cascade in afterwards
in beautiful formation. These are the penguins that delight
the crowds by flying perpendicularly out of the water and
landing on their feet on the ice.

We watched them as parents, with one or two chicks
under their feet. Here, the babies are always born on 15
December, give or take a day or so. They have to get their
timing right because there is such a short window for
breeding. The parents protect their precious light brown
fluffy bundles and find some safety in the crowd. Why?

Because also on this island are the skuas – big brown birds with sharp beaks – that nest nearby in cruel symbiotic relationship with the colonies. The tension in the air is palpable. The skuas constantly fly low over the colonies, terrorising them, and the penguins watch them in unison, arching their necks up at the threatening sky and following the cruising bird with the tilt of their bodies. It is like a Mexican Wave. If the skua gets lower and closer the colony becomes shrill in alarm. Sometimes the bird lands right on the edge of the colony and watches and waits with menacing patience. Sometimes it feints an attack on the heart of the colony. And just when you think it might all be theatre, a skua cruises in with noticeable deliberation, pauses in mid-air, drops its neck, falls into the midst of the penguins and swiftly lifts out its prize. We were lucky or unlucky to observe this happen once. The skua took the little limp ball of fluff to a nearby rock and, in sight of the parents, enjoyed a meal. We gazed into the colony trying to observe the effect of this attack on the group or the parents. One of the parents settled on its stomach. Life went on remorselessly around it. If that was their only chick, then those penguins soon return to sea.

Later today we took the Zodiacs over to a very beautiful nearby peninsula to the remains of Wilkes station. This was a US base built for the International Geophysical Year in 1957 and handed over to Australia in 1959. It was abandoned in 1969 because it was constantly drifted over with snow. Wilkes is closer to the rubbish end of the heritage-rubbish spectrum, but there are some aspects of the site that could merit preservation. One of my fellow-roundtrippers, Mark Forecast (an aptly-named meteorologist), wintered here in 1965. Where we saw just a snowdrift overhang on the western tip of Clark Peninsula, he saw the spot where 'Wally used to catch the bar chompers', the tiny, appetising fish that went so sweetly with home brew on a winter's evening. One of Mark's claims to fame is to have driven a motorcycle out on the sea ice and sunk it. It must be true; I found it recorded in the station logbook. It was eerie and also wonderful to spend an hour or two wandering, sometimes alone, around this stunning location amongst abandoned buildings filled with ice. You walk across their rooves. I'm told that when the melt is on you can look down into an ice-cube of a room and see a piano!

# Of HUDDLES
# and PEBBLES
*Life among the penguins*

The most famous scientific story from Antarctica concerns penguins. It manages to combine heroism, idealism and farce. In the first winter of Robert Falcon Scott's last expedition, three men set off on what one of them called 'The Worst Journey in the World'. Dr Edward Wilson, 'Birdie' Bowers and Apsley Cherry-Garrard sledged away from the hut at Cape Evans five days after the mid-winter celebration of 1911 on a 108 kilometre (67 mile) trip to Cape Crozier. It was 'the weirdest bird-nesting expedition that has been or ever will be'.[1] Their purpose was to collect the eggs of Emperor penguins (*Aptenodytes forsteri*). Their embryological quest was driven by a belief that an egg reveals the historical evolution of a creature; it is a developmental museum of former lives. And since the flightless Emperor was regarded as the most primitive bird in existence, its embryology would be a window into a deep biological past. Far south and flightless, and dumbly exposed to winter blizzards, the Emperor seemed the epitome of primitivity. The primitive lurked in the wastes of Antarctica for Edwardian explorers. The wilderness of the ice cap was invigorating, but it also enforced a brutal test of racial vigour that one could fail, as other creatures had failed.[2] The Emperor was thought to be condemned to a cul-de-sac of evolution and was thus expected to be in decline because of its own biological limitations. And so the Emperor's eggs might reveal, as other birds' eggs could not, the origin of feathers and 'the missing link' between birds and reptiles. This was the scientific belief that drove these men out into the polar night. 'We travelled for Science', wrote Cherry-Garrard.[3]

The Emperor penguin breeds on the ice during winter, incubating its egg on its feet. It is the only penguin to remain near the land during the Antarctic night. During Scott's first expedition, Edward Wilson had found a colony of them at Cape Crozier in the spring and had been surprised by the size of the chicks. They must have been born in the winter! He was determined to return to see them, and to secure an egg for 'Science'.

By the time the Emperors saw the three men in that winter of 1911, approaching them surprisingly from the land, the humans looked bent. They did not walk with a steady rhythm and they were burdened. The Emperors knew they were after eggs. Any lone figure on the ice at this time of year needed an egg, either their own or someone else's. An egg on your toes gave you a purpose and a future. Your neighbours helped you with their bodies to survive the winter wind, but where eggs were concerned they could not be trusted. You would huddle together and circulate so that no-one was exposed on the edge for too long – you were one giant organism designed to resist the cold – but someone without an egg was either your returning mate or a thief. The females were not expected back from the sea yet – so who were these?

And this had been a difficult season. The colony had divided because of the uncertainty of the sea ice, so there was anxiety in being so few. There were fewer bodies to shelter amongst, and eggs were at a premium. Some adults nursed hard, cold, dirty balls of ice on their feet. There was some comfort in that, but the moment a real, warm egg was available, it was snatched in exchange. The wind, conspiring with the ice, was the major predator. It was the only thing coming from the land that could threaten you.

The men heard the Emperors before they saw them. Their cries resounded even above the booming of the ice. The men had hauled their sledges for 19 days and Cherry-Garrard could not find the words to describe the horror of that night-time epic: 'They talk of the heroism of the dying – they little know – it would be so easy to die, a dose of morphia, a friendly crevasse, and blissful sleep. The trouble is to go on'.[4] Each 'day' it took them at least four hours to get out of their sleeping bags, fed, packed up and into harness. Their clothes were frozen as stiff as boards and they had to work together to bend them into shape around one another, like armour. At the end of their daily march their balaclavas were so completely iced to their heads that they could not remove them until the primus stove was burning. Great swaying curtains of the aurora – green, orange and lemon – beckoned them on. When they were a few miles from the penguin colony, they built a small stone hut on the lowest slopes of Mount Terror. Giant cliffs and pressure ridges of ice still separated them from the birds and they heard them calling a full day before they could find a way through crevasses to reach them.

There were only a hundred or so birds, instead of the thousands they had observed nine years earlier, and the penguins clustered and trumpeted on the thin sea ice. There were many more birds than eggs. The men were euphoric; they had not quite believed they would survive to see the birds, to complete their mission, and they knew they were the first humans to behold the marvel of such a winter colony. One of the men, being short-sighted, waited above on the cliff while the other two bullied eggs from Emperors, clasping them like gold nuggets in their fur mitts. Then they murdered and skinned three birds because they needed oil for their blubber stove back at the stone and ice igloo they had built. Two of the five precious stolen eggs burst in Cherry-Garrard's mitts as he clambered away from the colony, stumbling in the darkness without his spectacles. He could not even save the yolk for dinner.

That evening the men fired up the blubber stove, which spouted a glob of boiling oil into Wilson's eye. He was in agony the whole night. The next day they packed ice into the cracks of their igloo and pitched the tent beside it. In the middle of the next night the precious canvas was sucked away by a furious blizzard. And the following day, Sunday 23 July, the canvas roof of their igloo was stripped to shreds. 'Birdie', whose beak-like nose earned him this nickname, was their dutiful and conservative meteorological recorder and he logged the storm at Force 11. They lay exposed in the hurricane awaiting death as the snow drifted over them and they slept fitfully.

When, sometime near the middle of Monday, there was a lull in

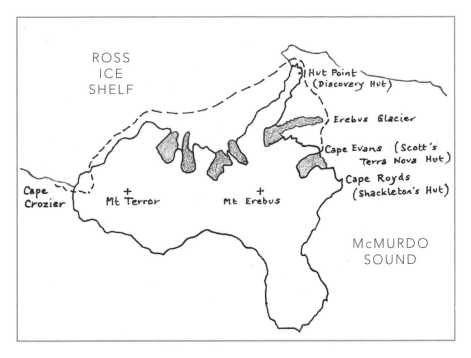

*Winter Journey to the Penguins, 1911*

the blizzard and a little light, they searched without hope for the tent, their lifeline. Sucked up by the storm, it had shut like an umbrella, its bamboos entangled with the outer cover, and so its 45 kilograms (100 pounds) of encrusted ice brought it back to earth half a mile away, and they found it there! It was a miracle that strengthened them: 'Our lives had been taken away and given back to us.'[5] They said nothing. Words were not necessary but their actions were eloquent: they carried that tent with solemnity and reverence and they 'dug it in as tent was never dug in before'. They considered returning to the penguins, but with three eggs already pickling in alcohol on the sledges, they decided it was time to return to the main hut, if they could.

They did make it back to Cape Evans, where they were greeted 'as beings who have come from another world'.[6] But only one of them returned to England, for Wilson and Birdie died with Scott on the polar journey in the summer. 'Science is a big thing if you can travel a Winter Journey in her cause and not regret it', wrote Cherry-Garrard.[7] This lone crusader for 'Science' carried the Emperor eggs home to Britain as reverently as he and his two dear companions had carried that tent. When the expedition had returned to London, Cherry-Garrard wrote to the Natural History Museum in South Kensington and made an appointment to present the sacred eggs.

It is 1913 and he cradles the eggs as he enters the centre of imperial science. He is greeted by a custodian with the words: 'Who are you? What do you want? This ain't an egg shop. What call have you to come meddling with our eggs? … I don't know nothing about no eggs.' When finally he is ushered into the presence of the chief custodian, 'a man of scientific aspect', Cherry-Garrard finds himself rudely dismissed without a word of thanks. 'You needn't wait', instructs the man. But he would like a receipt for the eggs please. 'It is not necessary: it is all right. You needn't wait.' But he will wait, because he wants a receipt, and because he is here not for himself but for his two dead friends, and for the penguins who made the sacrifice, too. So he sits for a long time in a gloomy passage outside, dreaming of murder. People pass to and fro and some ask his business and he always replies: 'I am waiting for a receipt for some penguins' eggs.' The receipt did finally come. But it did not seem to mean very much when Cherry-Garrard returned some time later with Captain Scott's sister, who wished to see the eggs. No such eggs existed, they were told. Nor were they in the museum's possession. But Miss Scott insisted, forcefully, and the eggs were eventually located, and they were even eventually studied. They were found to be no different to an ordinary bird's egg.

It was not until 1952 that Emperor penguins were studied by humans for a whole season.[8] A French expedition wintered at Pointe Géologie in Adélie Land, not far (in Antarctic terms) from Mawson's Hut at Commonwealth Bay. The comings and goings of the Emperors shaped the year for the men. This penguin colony had been found by the French in November 1950, and it was then only the fifth to be discovered in Antarctica.

France was building on the presence it established in Antarctica through the voyage of Dumont d'Urville in the *Astrolabe* and *Zélée* in 1840 and expeditions led by Jean Charcot early in the twentieth century. This new series of explorations, launched in 1948 by Paul-Emile Victor, confirmed the French claim to Adélie Land by landing wintering parties in 1950 and 1951 at Port Martin, 80 kilometres (50 miles) east of Pointe Géologie where d'Urville and his men had hungrily gathered rocks the previous century. This year, a small party was to be landed at the pointe to study the Emperors, and the young science graduate who would lead this quest was Jean Prévost. The French had tried in vain to find 'an ornithologist who was prepared literally to share the life of the birds he was studying'. Prévost had certainly studied some zoology but his training had focused more on agriculture. One day, when preparing for their voyage, Prévost turned suddenly to the expedition doctor, Jean Rivolier, and asked: 'Do you happen to know what these Emperors are? I hear they're some kind of bird.'

The French claim is in the middle of the vast Australian sector, and an official Australian observer was included in the Third French Antarctic Expedition of 1952. Robert Dovers, an Australian surveyor and cartographer who had been a member of the foundation party at Heard Island in 1948 and had also spent six months on Macquarie Island, 'wasn't French' remarked Rivolier. But his companions admired his superhuman energy and decided that he would wear out three pairs of shoes in the time it took them to wear out one. There was a post-war explosion of energy in this man's intensity and vigour: after serving in North Africa and the Pacific, and having fought 'the black evil of

Nazism', Dovers relished 'a clean fight against worthy opponents, the elements of wind, cold and ice'.[9] Bob Dovers had Antarctica in his genes because his father, George, had been a member of Mawson's Australasian Antarctic Expedition in 1911–13. Rivolier was predisposed to like Dovers because of his own brief encounter with Australians on the way south. Rivolier had flown to Melbourne to join the expedition ship, the *Tottan*, and he recalled: 'I have never taken to anyone as I took to our Australian hosts'. Older men would come up to him in the street and begin a stumbling conversation consisting of legendary World War 1 names: the Somme, Verdun, the Bois Bellot. This inherited bond provided a strange common language, the language of his father's generation, but it helped Rivolier to put up with the 'shocking' French accents and 'Mademoiselle from Armentiéres' grammar of these friendly Australians. However, he could not tolerate their endless invitations to see the duck-billed platypus.

When a fire destroyed the French base at Port Martin on 23 January 1952 – which it did in less than half an hour – Dovers was able to remain in Antarctica by being reassigned to Pointe Géologie, where seven men eventually wintered with the penguins. By Sunday 27 January they had constructed the main hut on the Ile des Pétrels near Pointe Géologie and made it habitable. They declared a holiday and had a special dinner 'to commemorate the founding of the republic of Terra Nova' (the tongue of the Terra Nova glacier glistened green and blue nearby).[10] The Proclamation of the Day decreed that everything be held in common and that war be declared on skuas. Dovers fell into the habit of calling Rivolier (the doctor) 'my dear Wilson', and Rivolier responded by calling Dovers 'Scott'.

Bob Dovers wrote that they 'seemed to have found a little Antarctic paradise', a relatively sheltered corner of a famously windy part of the continent. Below them was a stretch of unbroken bay ice, and beyond that was an open bay studded with little islands and fantastic icebergs. The tongue of the glacier loomed on the horizon. Because of the area's more moderate climate and the amount of exposed rock on the island

and capes, it was a stunning site for natural history; there were Adélie penguins, seals, skuas, giant petrels, Wilson storm petrels, snow petrels, and now there were dogs too. As well as his mapping work, Dovers looked after the huskies that had been transferred from the burnt-out station to this paradise for wildlife. This put him in tension with the biologists' purpose even though he shared their quest. For Dovers, the huskies were just part of the undivided nature he delighted in. But he reflected that the biologist, Jean Prévost, 'was more or less patron of the defenceless wild life and I was the captain of the enemy'. Bob's minimalist, laconic journal of 1952 is all about these warring parties, the dogs and the penguins: 'Monday 28 January: Difficulty with dogs. Live seal near dogs. Dogs & Penguins. Blood and gore. Misadventures with dogs.' Every so often one of Dovers' charges got away and wrought havoc. '4 February: Three dogs got loose today. Little giant petrels saved at last moment. Skua breasts for lunch … 6 February: Maru loose killed 200 penguins. Seal brains and liver for tea.' His diary entries offer a spare account of nature red in tooth and claw: the men look after the dogs, the dogs eat the birds, the men study the birds, the men eat the birds, the dogs work for the men, the seals eat the birds, the men eat the seals, everyone hates the skuas. As far as the humans were concerned, Dovers was never troubled by being the lone Australian among the French: 'there was no question of nationality, I remained always one of them'.[11] But he preferred to keep the people out of his story.

As winter approached, all this wildlife abandoned them. Only the Emperors would stay during the darkness, but first they had to arrive. 'For the first time', wrote Dovers, 'men were to be able to watch a complete year of the life of these birds.' That's why the men had come here, why they had decided to stay even though their support station had burned down. They completed their hut, assembled their 'meagre resources' and then awaited the arrival of the Emperors.[12]

With very few exceptions, Emperor penguins do not set foot on land and only feel winter sea ice under their feet. Penguins are not primitive animals but are recent descendants of flighted birds, adapting as Antarctica separated and iced up. They literally fly under water. They can dive to a depth of 300 metres (about 1000 feet). Their life cycle is bound to the ice cycle. What could be the cause of the Emperor's 'aberrant' reproductive cycle, wintering on the ice when all other life (except for a few humans) summered?

On 10 March three Emperor penguins rocketed out of the water onto the bay ice at Pointe Géologie just as the final Adélies were preparing to leave for the outer ice floes. They were early and at first moved about the bay looking for solid, stable sea ice. The men responded by immediately going into a huddle. They were delighted to see the birds, not only because their scientific work was duly launched but because they were also relying on the Emperors for food during winter. They watched them, and the penguins watched them back. For a while the Emperors spent most of their time standing stock-still 'like mystics' for long spells, and the men had no idea what the birds were going to do next, if anything. This could be a very boring winter. By 20 March, there were over 100 Emperors. The men tried to work out which were males and which females. In the next week another 2000 arrived and the noise of the colony could be heard from the men's hut for the first time. The site the Emperors chose was their traditional one; a stretch of ice between the land and the nearest island, although at the end of March a breakaway group moved from the original site to what they thought was firmer bay ice nearby. Some individuals waddled around with large identification numbers in red painted by the scientists on their backs. Courting couples were everywhere, bowing and singing to one another and stretching their golden throats skywards, and the first began copulating on 11 April. The last Adélies left. On cold days in mid-April the Emperors began to cluster in circular groups and on warmer days they would spread out, like a giant creature breathing. The huddles were extraordinary: each bird faced inwards, its beak between the shoulders of its neighbours, its back to the wind. Half a mile from the men's hut there was now a colony of over 6000 breeding pairs. Roast Emperor had become a favoured dish in the French Republic of Terra Nova.

The men tried to capture the life of the birds on film and tape, stumbling around the ice with camera, batteries, amplifiers and wiring. 'We would then go up to one of the Emperors and hold the microphone in front of him in what seemed to us a clear invitation to co-operate', recorded Jean Rivolier who shared the ornithology with Jean Prévost. 'Generally the bird who started to sing was about a hundred yards away.'[13] If you pointed a film camera at a bird it seemed always to stand completely still. They would follow one bird for days, in shifts, and the moods of the human hut would fall or soar with the passion of their chosen penguin. The men identified with the male in his quest for 'a lady' and shared his anguish when he was caught in a quarrelling love triangle. 'He's made it!' ejaculated one of the Frenchmen, bursting into the hut with excitement and pride on the day their man finally made his conquest.

On 5 May the first eggs were laid and immediately passed to the male. On 9 May a few females, looking thin, were observed leaving the nesting colony and marching towards the sea. Groups of 30 left daily, beginning their long walk to food, a journey of over a 160 kilometres (100 miles) to open water across the expanding winter barrier of ice. They were away for over two months, leaving the males to their hungry vigil. Several eggs were found abandoned and frozen on the ice each day, and occasionally a male could be seen beginning a stumbling trek towards the sea, his egg still on his toes, his hunger driving him. The scientists experimented and found the Emperors did not recognise their own egg amongst others. Almost as soon as the chicks hatched and were briefly sustained with a last, hoarded feed from the males, the females returned from the sea, sang until they found their mates, and regurgitated fish into the mouths of their babies. They then tucked the youngsters into their own brood pouches and the starving males were, at last, free to begin their own march to the distant sea for food. This was early July (two months ahead of the cycle at Cape Crozier for this was a more northerly colony), and when the males eventually returned, they too fed the chicks and the relay continued. By 17 August, Emperor chicks were leaving their parents and going for short walks on their own. By 8 September, they were observed forming their own huddles, or 'creches'. The first infants began to depart the colony, instinctively heading towards the sea, from 11 December. On Boxing Day, catastrophe struck. As the sea ice was breaking up, an enormous fragment of an iceberg fell on the breakaway colony of birds that had, back in March, moved closer to the glacier. Five hundred chicks – many of them not yet equipped with their swimming uniforms – were crushed or drowned. In this season of the longest days, the Emperors continued to leave for the sea, which was now all around them. The fragmenting ice briefly carried the memory of their colony; it glowed with a greenish hue and was hollowed where the huddles had stood.

The French were intrigued by the 'communism' of Emperor society and referred to the males without eggs as 'the unemployed'. This subgroup was restless and wayward, their influence potentially disturbing to cultural conventions. And the French investigated the fidelity of the birds. When the Emperor popped up onto the ice in early March did it seek its mate from last year or take just any partner? If it did not recognise its own egg, did it know its own baby? The scientists found that the Emperors were faithful lovers within each season and across seasons if they could be, and that they knew their infant by the sound of its voice. Living with the 'most primitive' of birds for a year enabled the French to discover their emotional and social sophistication. The birds' lovemaking could be exquisite. The four-month fast by the males offered a parable of parenting. The 'tortues' or huddles were an impressive collective adaptation to life on the edge. The strange inversion of the Emperor's reproductive cycle (nesting in winter) gave chicks born in July a chance of being independent in time for the break-up of the sea ice, and the best opportunity to harvest the abundant summer food.

The size of the Emperor is an adaptation to cold. The larger a body, the smaller the ratio of surface area to volume and the greater the capacity of a warm-blooded creature to conserve body heat. But the Emperor's larger size also means that greater time is required for incubation of the egg and raising of the chick, too great a time for the chick to become independent before the end of the short Antarctic summer. Thus size dictates breeding season, and breeding takes place in winter. The Emperor is not territorial and does not need to defend a nest, another adaptation to deep cold. The portability of its egg and the mobility of the brooding parent enables the Emperor penguins to huddle, which further reduces their heat loss.

The habits of the penguins determined the location of the new French base in Antarctica, founded in 1956 and named Dumont d'Urville. Replacing the burnt and abandoned station at Port Martin, it was established in this favoured corner of the continent amongst the nesting Adélie penguins on the Ile des Pétrels and beside the preferred sea ice of the wintering Emperors. It remains the only permanent Antarctic base near an Emperor colony.

In 2005, French scientists and photographers made the Pointe Géologie colony of Emperors famous with the release of a film called *La Marche de L'Empereur* (The March of the Emperor). Two photographers again crawled with their equipment over the sea ice through the long polar night in pursuit of nature and nurture. Their film gave voice to the birds' thoughts – expressed in soft and loving French – and the penguins' life on the sea ice was portrayed as a family drama of loyalty and self-conscious sacrifice. The American version of the film substituted a preaching narration by Morgan Freeman: 'This is the incredible true story of a family's journey to bring life into the world', explained the prize-winning and popular documentary. 'In the harshest place on Earth, love finds a way.'[14]

American religious conservatives promoted the film for its luminous family values. Anthropomorphism (an occupational hazard for cute bipeds) went wild. The film was an unexpected box-office hit; the second most successful documentary in American history behind *Fahrenheit 9/11* and said to be just as 'politically charged'. American churches block-booked cinemas and compared the passion of the penguins to Mel Gibson's *The Passion of the Christ*. In Antarctic terms, it was only slightly less religious to say that the penguins' odyssey on the ice rivalled Scott's; 'albeit with a happier ending'. The march of the Emperor became a Christian pilgrimage in the face of adversity. Emperor penguins are monogamous, claimed conservative and Christian reviewers. No, they are only serially monogamous, replied Marxist commentators, reminding their readers that Emperors often swap partners between seasons. But it is evolutionary pressures that explain seasonal fidelity, lectured scientists, because two parents are needed to rear the chick in such an environment. And Emperors often switch partners at the beginning of a new year on the ice because there is such a brief window for breeding and so little time to search for last year's mate. But they *do* at least search, others pointed out, and in winter they seek voice recognition with their partner and chick and thus maintain the family. This is just the selfish gene at work, replied biologists, because only scrupulous discrimination of your own kin will perpetuate your genes. The physical complexity and moral simplicity of the Emperor's life-cycle, declared religious enthusiasts, was proof of 'Intelligent Design'. These reviewers warned audiences that the film has a scene of penguins mating and 'includes a few lines that allude to evolution', and they criticised the narration for falsely alleging that the Emperors had been there 'for millions of years'. But they praised the film for its elucidation of love, fidelity and sacrifice. People debated the meaning of 'love' in a penguin's life and a human's: 'Is it really love if the characters have different sexual partners every year?' Human fathers worried about what the male Emperor's domestic commitment would mean for life in their own homes. Perhaps we were in danger of literally seeing family values in black and white? Conservatives wanted to agree with the English narration when it said that 'They're not that different from us, really', but radicals argued that this also meant that 'animal sex can be a lot like human sex' and then found gay penguins, promiscuous penguins and sexually playful penguins.[15]

Religious fundamentalism has rarely precipitated on the ice, although spiritual enlightenment often has. The evangelical embrace of the film disturbed people who valued Antarctica as a safely secular continent where Science was God as well as those who felt it was one huge cathedral of transcendental experience. The men who undertook the worst journey in the world, including the deeply religious Wilson, were indeed martyrs, but it was to earthly knowledge that they made their sacrifice. Those three small embryos were striven for 'in order that the world … may build on what it knows and not on what it thinks', wrote Cherry-Garrard. 'If you march your Winter Journeys you will have your reward, so long as all you want is a penguin's egg.'[16]

The French director of the film, Luc Jacquet, did not pursue arguments about climate change. 'It's obvious', he said, 'that global warming has an impact on the reproduction of the penguins. But much of public opinion appears insensitive to the dangers of global warming. We have to find other ways to communicate to people about it.'[17] The same conservatives who embraced the film and rejected evolution are also sceptical about the evidence for global warming. At the beginning of the twenty-first century, we know that the Emperors are indeed in decline at Pointe Géologie, but not because they are primitive and the last of their race. Diminishing numbers in this locality are probably due to warmer temperatures and a reduced extent of sea ice resulting in less adults surviving. From a 1952 population of over 6000 pairs, the numbers gradually slipped until the 1970s when they dramatically halved to under 3000 pairs. The population continued to decline gradually (about 1 per cent annually) but seems to have stabilised in the last decade.[18]

In their first studies of the Emperors, men tried to fit them into a global hierarchy of life. They looked past locality and individuality in their search for 'primitive' biological essence. Like indigenous peoples encountered by voyagers around the world, the Emperors were seen to be a doomed 'race', noble savages of a kind, destined to die out because they were a relic of past ages.[19] But as the humans came to know the birds better, they grew alert to variation and sensibility; they began to relate different colonies in time and place and to see the birds historically rather than as fossils, and so conceded them a future. The Emperors' year on the ice at Pointe Géologie in 1952 was different to that in 1956, as Jean

Prévost discovered when he returned. The life cycle of the Emperors was subject to change and contingency, not just over millions of years but from year to year. The birds now seemed to offer a parable of evolution- ary adaptation, or even a model of social behaviour. Soon the intruding people came to realise that their own presence on the ice and, indeed, on the planet – as scientists, as humans – might also change the life and prospects of the penguins in fundamental ways.

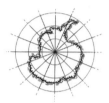

On the morning of Sunday, 12 October 1912, a sleek, adventurous Adélie penguin (*Pygoscelis adeliae*) shot up out of the water and landed perkily on the ice at Commonwealth Bay. He began to make his way to his traditional nesting site where he would begin collecting pebbles. Suddenly a large, agitated figure embraced him and carried him into a warm, dark, noisy place. His heart rate soared.

Mawson's men gathered around the traumatised Adélie in the hut and shouted with excitement over this harbinger of spring. It had been a long winter and this creature had just ended it. Archibald McLean had found him and brought him back in triumph, but it was Charles Laseron who would have to kill him. They wanted his skin and meat. Whereas leopard seals hungrily awaited the first Adélies into the water, men were ready to seize the first out of it.

Penguins helped people stay sane in Antarctica because they seemed so human. To anthropomorphise a penguin was a survival instinct. There on the ice, wrote the Scottish artist WG Burn Murdoch when observing Adélies in 1893, were:

> little fat men in black coats and white silk waistcoats … as
> respectable as can be … I am sure they discuss the new arrivals
> in their country; though 'quangk-quangk' is the only word I

distinguish, their attitudes are as expressive as a Shakespearian vocabulary. When they are not engaged making a living below water they come up and play games on the snow – have little debating societies, and King of the Castle and other games, and sometimes when they are in great numbers they have military manoeuvres. The men say they are the only things worth coming to see in the Antarctic, and no matter how melancholy a man may feel, if he sees one of these jolly little fellows he cheers up.[20]

The penguins provided a delightful mixture of comedy and formality, and offered both entertainment and audience. Human territorial behaviour on the ice took many forms, as we have seen, and was performed both for a distant, abstract international audience and an immediate, local, well-dressed one. Penguins were the nearest Antarctica could provide to indigenous inhabitants. Penguins constituted what looked like formal delegations and they nearly always approached visiting parties and exuded an air of confident ownership. From a distance, in the illusive Antarctic air, they looked like a troop of uniformed men. The French in 1840 carried them off as 'living trophies of our discovery'.[21] When the *Astrolabe* and *Zélée* anchored within the Pointe Géologie archipelago, men hustled penguins down from the rock on which they wished to plant their flag and described the birds as 'very surprised to find themselves so roughly dispossessed of the island of which they were the sole inhabitants'. Several kidnapped Adélies were paraded back on the ship: 'At the sight of our trophies there is general rejoicing, our discovery is confirmed and receives the name Adélie Land.' The land and its sole inhabitants were given the same name and both were appropriated. The British in the following year under James Clark Ross also abducted Adélies who defended their land, 'disputing possession' with their sharp beaks. The crew of the *Belgica*, trapped in the ice in 1898, thought that a posse of penguins on the winter horizon was a visiting mission and hurriedly got dressed. Roald Amundsen was sent out to investigate.[22] The Scottish Antarctic Expedition played the bagpipes to a resolutely unmoved penguin in 1903. Shackleton's

1908–09 expedition entertained penguins with a gramophone on the ice, and Scott's men regularly sang to them. A Nazi from the German Antarctic Expedition saluted an Emperor penguin with chants of 'Heil Hitler' in 1938.

Although the 'humanity' of the penguins enchanted the visitors, it also exposed them to the kind of bullying and violence often perpetrated by voyagers on indigenous inhabitants. The very passivity of a penguin was provocative. It literally stood up to you. During the Dundee Antarctic Expedition of 1892–93, the Emperors, in spite of their size, were found to be hard to catch. The artist on board the *Balaena*, Burn Murdoch, put it simply and brutally: 'We crept up to them, partly surrounded them, and let drive with our picks …' He continued in traumatic detail: 'I made a drawing of one of these extraordinary birds as it stood calmly on our poop after many vain attempts on the part of the crew to kill it. Driving a hole through its brain only saddened it, and all the most killing treatment usually applied to other animals only seemed to add to its expression of calm, eternal resignation.'[23] The crew became angry at its indifference so they all sat on it. They strapped the bo'sun's belt about it so they could haul it around the ship; they lashed it up with fathoms of whale-line; they harassed it with the ship's dog. It took them four hours to kill it, 'and it wasn't dead then'. They drove holes through its skull and beat it with clubs. The doctor sat on its back and worked at its brain 'till it lay on the deck apparently lifeless'. But two hours later 'it was waddling about with its head in the air as if it had neuralgia'.[24]

The Adélie that Archibald McLean kidnapped and ate in October 1912 had been heading for his regular nesting site on windswept rock. Before McLean surprised him, he may have had a pebble in his beak, carefully procured from the landing beach. Pebbles are the currency of the Adélie's spring. As Charles Laseron put it, 'the wealth of a householder lay in the number of stones he possessed … I am sorry to say, also, that the birds were inveterate thieves, and the invariable object of their thefts were – stones.'[25] Pebbles lined the nest, enabling the bird to sit or squat during incubation. Unlike the Emperor, the Adélie is territorial, fiercely so. They fight constantly with their neighbours over pebbles and territory. When the chicks are born, parents only feed their own. 'No longer, when it was calm, did the ears throb with the silence', wrote Laseron, 'the raucous cries of the penguins filled the air to the exclusion of everything else'.

The Adélie penguin, another warm-blooded creature of the sea ice, can dive for up to six minutes to depths of more than 150 metres (about 500 feet). Leopard seals stalk them, hence their famously tentative behaviour before entering water. Even on ice – especially new, semi-transparent ice – they may be snatched from below by a leopard seal that has spied them. Early human observers (except the Swedish ones) assumed that it was the female Adélies who arrived first, established territory, did most of the nest building and began the incubation of the egg, whereas it is the male.[26] The Adélie is smaller than the Emperor and hence more vulnerable to cold, but its smaller size also means that the chick can reach independence more quickly in the brief summer. So the Adélie escapes the winter to the outer pack ice and returns in summer to breed on the few areas of rock that Antarctica makes available. Each year they almost always return to the same colonies and nest sites.

There's a simple reason why those early voyagers who jumped ashore with picks and hammers in search of continental rock also found themselves catching and taking home Adélies. Penguins and people were attracted to the same sites in the same season. Voyagers found it hard to find earth and rock, and so do the Adélies; their available nesting places are few and precious. The penguins choose open, windswept, stony ground for their colonies because the wind keeps the snow clear of the rock and enables them to gather their pebbles. They need a special configuration of open sea, sea ice and rock. So do humans. People approach the coast where the pack ice is discontinuous and seek the few areas of exposed rock in a continent of ice. They colonise those rare sites where rock and the open sea are in reasonable proximity because it is there that their ships may find harbour and their shelters can be secured. They are also attracted to the friendly summer food sources.

David Ainley, a biologist who has written beautiful books about the Adélie penguin, describes a remarkable example of co-adaptation involving Emperors, Adélies and pebbles. Because Adélies need pebbles they do not breed on the sea ice like the Emperors. Thus the only two truly Antarctic penguins do not share either space or season in their breeding. But Emperor penguins occasionally swallow small stones as they feed on the ocean floor, and when they return to their winter colony on the sea ice, they sometimes regurgitate the stones. If the ice is permanent for many years and the colony is large and old, the pebbles may accumulate. On fast ice in the Weddell Sea region, there are enough stones deposited by Emperors for Adélie penguins to make nests.[27] Huddles and pebbles take their turns at breeding on the same tract of ice.

Adélies make do with those pebbles, even though they are on ice, because there is no other place in that region where land is easily accessible from the sea. Fast ice (held 'fast' to the land) is an impediment to the migrations of Adélie penguins because they can swim more swiftly than they can walk or toboggan. And extensive fast ice or giant floes prevent the Adélie from feeding in much of the water underneath because it cannot hold its breath long enough. So the Adélies frequent fragmented sea ice for most of the year and, in the breeding season, they seek land (or pebbles) accessible to the sea. They live in rhythm with the explosion and collapse of the pack. They work the ice edge, moving in and out with its seasonal breathing, retreating to the outer pack in winter and returning to that rare combination of rock and open sea in the summer. Because Adélies now live so closely with humans, congregating in the least Antarctic of Antarctic landscapes, they are among the best-studied birds in the world. Half a century of detailed scientific observation has brought humans to a tentative and preliminary understanding of the complexity of this bird's ecosystem. Small changes in their breeding biology might provide clues to large transformations in the environment. The Antarctic sea ice which is the Adélie's habitat is one of the most extensive and dynamic surface zones on Earth, and also now the most vulnerable.

Global warming is affecting polar regions more acutely than lower latitudes, and sea ice is a dramatic register of that change. As it recedes, it further accelerates warming because less sunlight is reflected and more heat is absorbed by the ocean. Over the last century, surface air temperature averaged across Antarctica has risen between 0.5 and 1 degree Celsius (0.9 and 1.8 degrees Fahrenheit). But temperature change varies; little change has occurred in the interior of the continent whereas average temperatures have increased by 5 degrees Celsius (7.2 degrees Fahrenheit) or more over the Antarctic Peninsula. Ocean temperatures have also increased around the Antarctic Peninsula, leading to the rapid disintegration of ice shelves. One ice shelf of 1300 square kilometres (over 500 square miles) was lost in 50 days in the 1990s. Adélies are already taking advantage of the newly-exposed coastal rock.[28] But their populations are decreasing near the Antarctic Peninsula, this region of rapid temperature change, while they are growing along the coast of Eastern Antarctica (which includes the Australian sector). In the third region of Adélie populations, the Ross Sea, no clear trend has emerged. David Ainley calls the Adélie a 'bellwether of climate change' because it is so dependent on the lifecycle of the sea ice.

Sea ice is a determinant of population variation, but not the only one. Sea ice is retreating, although this, too, is variable. Modern satellite photographs of the extent of sea ice (dating from the 1970s) can be supplemented by historical research amongst the documents of voyaging. From the late 1920s, whalers worked the edge of the pack ice and recorded the latitude of their hunting. Although the species of whale also determined the scope of their voyages, these ships' logs incidentally provide scientists with glimpses of earlier seasons of fluctuating sea ice.

Eric Woehler, who accompanied us to the Shirley Island colony, has studied the conservation status of Antarctic seabirds in the Australian Antarctic Territory. On our voyage south, he also kept a round-the-clock register of birds, whales and other sightings from the bridge of the *Polar Bird*. At Shirley Island, human presence and visitation seem to be influencing the long-term behaviour of the penguins. Although there has been a slight increase in the island population of Adélies since the 1950s (from 7500 pairs to 9000 pairs), there has been a marked shift westwards due to the abandonment of colonies closer to the station and the establishment of new ones farther away from people.[29] Breeding success at Shirley Island is significantly lower than at nearby Whitney Point which is an Antarctic Specially Protected Area and therefore much less visited by humans. The island I was allowed to clamber over (under expert guidance) lags behind all other breeding localities in the Windmill Islands in terms of population increases in the last 40 years. The numbers of Adélie penguins at Shirley are increasing but not as rapidly as elsewhere in this region. Perhaps the young birds that prospect for their future nesting site are rejecting this area due to the proximity of the station and its visitors? But at Pointe Géologie, where the building of an airstrip for Dumont d'Urville station between 1985 and 1993 obliterated ten per cent of the archipelago's nesting sites, Adélie populations have continued to increase dramatically. And on the Ile des Pétrels where the French base is located and Adélies live with the constant summer sound of engines and helicopters, their numbers increased from 6000 pairs to 15 000 pairs between 1985 and 2000. Scientists speculate that the background noise of machinery is less disturbing to Adélies than human visits. Global warming and human visitation, both emanating from the north, are the new factors in the lives of the penguins, and they are equally varied in their impacts.[30]

Thus people and birds continue their long wrestle over the pebbles, each craving the same few scraps of coastal rock. An Adélie penguin once courted Dr Edward Wilson by dropping a pebble at his feet. It was probably a misguided expression of love. But perhaps we can see it as a shrewd act of diplomacy.

## Tuesday, 31 December
**66 degrees south,**
**110 degrees east**

We let off an orange flare and weighed anchor two hours
ago (1330), thus leaving Casey a day ahead of schedule.
The weather is fine but we sail on a sea of emotion. Even us
roundtrippers are caught in the swell. Aboard are most of
the winterers and they don't know whether to look back at
the horizon of Casey or forward towards life in Australia.
Most had to leave their jobs in order to win Antarctica.
A week after our return they will be unemployed. This
community of 17 – including one woman – has just
signed off from their duties and have handed on the task
of managing the station to the new team of 19 (all men).
They are sad, uncertain, bereft, proud, exhilarated, excited,
apprehensive. They are considered to have been a 'very good
year'. They each speak of their winter as something very
special and of their fellow expeditioners as 'family'.

We roundtrippers – in so many ways the lowliest
of Antarctic expeditioners (but at least above tourists)
– are nevertheless the best-qualified observers of 'the
changeover'. We get to know the incoming winterers
well, we hear of their pre-Antarctic lives and of what
they expect and hope of their anticipated year away. We
witness their departure and share their excitement, and
we see their intense baptism in the urgency and pressure
of the five-day re-supply. And then, over the next ten
days, we effectively (and casually) de-brief the outgoing
winterers. The changeover ceremony, which we witnessed
this morning, constitutes the passing-on of the baton.
For everyone else it is a beginning or an end, but for us

roundtrippers it is the central symbolic moment of our journey, the hinge upon which our voyage hangs.

It was a most moving event where men cried and voices tremored, good speeches were made and gifts presented. The keys were handed over, and this itself was a joke, because one of the lovely things about an Antarctic station is that it doesn't need keys. Nothing is locked except for Fort Knox. Fort Knox is where the alcohol is kept. Then there are the keys to the safe, but nothing of value is kept there either. And finally, there is the key to the key cabinet! To make up for this deficiency in real keys, a symbolic key has been made and mounted, so that it may be inscribed and passed on. The changeover worked very well as ceremony: it allowed for the public expression of a whole range of emotions and it swept all present into their currents. So we voyage now in the wake of this event, and our winterers are exhausted and numb, and also (for the moment) excited.

By the end of our week at Casey, a small number of us were making regular pilgrimages to the station library. I had been telling my companions some of the stories I had found in the logbooks, and so a few joined me so that they could see the old documents. Soon we were finding a little time each day to sit down and pore over the logbooks, reading the best stories out loud to one another. One or two of my companions had not been in a library since school and were surprised to find themselves in one, and they shared my delight at how the past seemed to spill out of the pages.

Last night I slept on the continent, lay my ear to it. I threw a sleeping bag on the floor of the station leader's office at Casey and so was able to attend the farewell party in the Red Shed. It was a happy and convivial occasion (the station's 'home brew' was good). But what I will always remember is walking outside into the bright light of past-midnight and standing, without hat or gloves or parka, gazing for a long time at the view down to the harbour, the ship and beyond. The sky was opalescent – full of soft pinks and mauves and blues – and the air was completely still. The ice was luminous. I beheld a marble edifice.

# THE CHANGEOVER
## *Time, history and generations*

Antarctica first entered European consciousness as a consequence of the clock. Several hundred years ago, 'finding the longitude' was a metaphor for the impossible and the crazy. It was a quest that might drive you mad. Finding a reliable way to plot one's position on the east-west axis at sea was the great conundrum of navigation, and kings and queens offered rewards for its practical solution. Sailing in the open ocean was a dangerous lottery. On the foggy night of 22 October 1770, almost 2000 British sailors lost their lives due to a miscalculation of longitude. Their four warships, returning home from a successful campaign against the French, foundered on the rocks of the Scilly Isles, off Land's End.

Scientists knew – and had known for 2000 years – that time was the solution to longitude. The rotating Earth is a clock and the lines of longitude are its minutes. Twenty-four hours describe one revolution of the globe through 360 degrees, so each degree of longitude equals four minutes of time. So the secret to determining longitude was to compare the local time in two distant places. The time difference would translate into degrees. One's own local time could be determined by astronomical means, but how could the time of the home port be monitored? The challenge was to make an accurate clock, but it had to be a clock that would travel well, a clock that would keep reliable time on a stormy sea. Pendulum clocks could be upset by the motion of ships, and lubricating oils ran thick and metal parts contracted in different temperatures. Minutes lost were miles lost. The time travel had to be precise.

In a saga made even more famous by the success of Dava Sobel's book *Longitude*, the self-taught London clockmaker, John Harrison, solved the problem of longitude with his design of a 'watch-machine', a reliable sea-going chronometer. A beautiful copy of Harrison's original was made (over 30 painstaking months) by the London watchmaker Larcum Kendall and sent with James Cook on his circumnavigation of Antarctica, and the same timepiece (known as K1) later guided Arthur Phillip's First Fleet to Australia.[1] Proving the latest chronometers was one of the urgent scientific purposes of Cook's voyages. Indeed, to the chagrin and frustration of Johann Forster, naturalist on the Antarctic voyage, it seemed a higher purpose than natural history. Cook adored 'our trusty friend the Watch' and was forever sailing past lands blooming with undescribed plants and searching high latitude seas for continents that didn't seem to exist. Forster exclaimed: 'But instead of meeting with any object worthy of our attention, after having circumnavigated very near half the globe, we saw nothing, but water, Ice, and sky'. They even spent three days on a barren rock just to determine its longitude.[2]

The clock was revolutionary. 'Until the invention of the computer in the mid twentieth century, never was knowledge of the natural world so thoroughly transformed by a machine', wrote historian Graeme Davison in his book *The Unforgiving Minute*.[3] The clock – and the Earth as clock – were symbols of Isaac Newton's ordered, cyclical universe.[4] The clock was an instrument of such supreme design that God might be imagined as a watchmaker. Lilliputians wondered if Gulliver's pocket-watch was 'the God that he worships' because 'he assured us he seldom did anything without consulting it'.[5] One of the lesser purposes of the new colony at Port Jackson was to keep British standard time in the South Seas.[6] In April 1788, Lieutenant William Dawes, whose task it was (on behalf of the British Board of Longitude) to establish an astronomical observatory in New South Wales, anchored one of his Shelton regulator clocks to the rock of the promontory now known as Dawes Point on the western side of Sydney Cove. With hindsight, writes Graeme Davison, this action was 'as important in the transplantation of European civilisation as the raising of the British flag and the ceremonial volley of muskets that had marked the official foundation of New South Wales just a few weeks earlier. For what they symbolised was a process more profound and more enduring than the advent of British rule

– the permeation of the entire world by the European time-spirit.'[7]

Davison's study of time is ultimately haunted and inspired not by an inventor or a theorist but by a preacher – John Wesley – for his central concern is the morality of time. 'Australia was a child not only of the scientific and industrial revolutions of the eighteenth century', writes Davison, 'but of the religious and moral revolution known as the Evangelical Revival.'[8] Industrial time brought a distinctive morality too: of punctuality, discipline and efficiency, the 'competitive, guilt-inducing, masculine morality' of *the unforgiving minute*. In a classic essay on time, work-discipline, and industrial capitalism, the English historian EP Thompson analysed how, from the late eighteenth century, a new morality of time was imposed through the supervision and division of labour, the introduction of fines, bells, money incentives, preachings and schoolings, the suppression of fairs and sports and the rationalisation even of leisure.[9] Punctuality meant civility and time equalled money. A general diffusion of clocks and watches occurred at just the moment when the industrial revolution demanded a greater synchronisation of labour. Time became a moral and economic measure as well as a natural and geographic one. It was a currency to be saved, husbanded, redeemed, and spent wisely. One needed to be thrifty with time and to improve each shining hour. Mean time could indeed be mean.

Nations disputed which meridian should be the prime one. And longitude might determine which language you spoke. In the fifteenth century, Portugal and Spain warred over where the line should be drawn between their empires. In 1493 Pope Alexander VI settled their territorial disputes by dividing the world in half along a mid-Atlantic meridian of longitude. All new lands to the south and east were to be Portuguese, and to the south and west Spanish. But in the Treaty of Tordesillas of 1494, Portugal managed to push the line westward, thus ensuring that Brazilians speak Portuguese.[10] This line remained important in South American claims to Antarctica in the twentieth century.

The original prime meridian (used by Claudius Ptolemy) had been the western-most point of the known world, the 'Fortunate Isles' (Canary

Islands). But it later moved to the home capital of each mapmaker: to Paris, Lisbon, Copenhagen, Rome. In 1879, the astronomer royal for Scotland suggested that if there had to be a prime meridian, why not the Great Pyramid in Egypt?[11] Others suggested Bethlehem, or the frontier between Russia and the USA. The site had to be symbolic, politically resonant, and an established place of astronomical observation. By the late nineteenth century, most sailors were using the British *Nautical Almanac*, an annual publication with data from the observatory at Greenwich. In 1884, an International Meridian Conference held in Washington determined that Greenwich would be the prime meridian and that the mean solar day would 'begin there for all the world at the moment of mean midnight'. (The North American railroads had adopted a standard time system based on the Greenwich meridian only days before the invitations to the conference were sent out.)[12] The establishment of the Greenwich prime meridian (and Greenwich Mean Time) was a triumph of Victorian scientific imperialism, and a product of Britannia ruling the waves not to mention precision horology.[13]

So, although the placement of the prime parallel – the Equator – was dictated by environment and geometry, the choice of the prime meridian was political. If latitude was natural, then longitude was historical. And longitude was to cut the cake of Antarctica into national slices.

When John King Davis first contemplated a map of Antarctica, he was standing with his new friend, Ernest Shackleton, in an office in London considering an offer to become the mate of the *Nimrod*. Davis gazed intently at the representation of the southern continent before him. There it was, a vast, white empty space devoid of features, some hypothetical coastlines pecked in with dotted lines, and 'its parallels of latitude crossed by the meridians of longitude that met at the untrodden Pole, looking like some gigantic, empty spider's web'.[14]

Although Cook's chronometer had circumnavigated Antarctica and imperial time was to throw a cartographic net over the end of the

Earth and even partition it, the continent of ice ultimately eluded the regimen of the time spirit. Space and time took on new dimensions in Antarctica. The clear polar air was famously illusive, and there were no shadows to provide perspective. Light and looming could reveal features beyond the horizon. A man sledging on the ice could lift his gaze and see the party ahead of him projected as an inverted mirage some distance above their heads. A small part of one's clothing, glimpsed out of the corner of the eye, could appear to be a distant landscape feature. A matchbox could assume the size of a barn. The geologist Laurence McKinley Gould, always seeking rock amidst the endless ice, once identified a huge exposure of boulders ahead of him on the homeward trail, only to find out it was just dog droppings. Raymond Priestley and his companions found that a dog team travelling towards them on the horizon turned out to be a little scrap of black film paper fluttering near the skirting of their tent. A mite might be mistaken for a man.

Time, too, was warped. In Antarctica, it felt like time had not only skipped a beat, but had lost beat altogether. Frederick Cook of the *Belgica* found that it was so cold their watches 'refused to tell the time' on their winter sledging expedition.[15] The North American aviator, Lincoln Ellsworth, 'somehow' lost 12 hours in Antarctica in 1935 and missed Christmas by a day: 'We didn't know it until we saw a total eclipse and found out what day it was.'[16] At Mawson station in 1967 there were no reliable clocks or watches on the base and they relied on the wireless and a stopwatch. When in late summer they could get the right time, they turned the stopwatch to 'Mawson bastard time' which gave them two hours of 'daylight saving'.[17] At Little America, Gould used to 'long for the soothing velvety feel of the darkness on my eyes. I really believe that this unending light was far more tiresome to me than had been the long dark.'[18] In high summer the sun just circles the horizon, and the only indication of night in a large station like McMurdo is a slowing of the pace and a decrease in machine noise.[19] The extremes of climate and geography, and the distortions of high latitudes and compressed longitudes, made a weird nonsense of the passage of a day.

The US scientist, Bill Green, working in the lunar landscape of the McMurdo dry valleys in the 1980s and '90s, considered that 'time had become a useless convention, an artifact of monasteries and corporations, a trick to divide the seamless day. Here, with constant daylight, nothing rang the hour. What time was it really? What eon?'[20] His colleague, Mike Angle, down on his knees looking for life in a streambed and finding only blue-green algae, decided: 'We're in the Archaen … We're back a long, long way.'[21] Antarctica, wrote Green, is 'a land of frightening antiquity'. 'Time is everywhere and everywhere visible.'[22] Antarctica, which some counted as the seventh continent, was also, perhaps, the seventh day of creation, a place out of time. It has, in Barry Lopez's words, 'an ethereal aloofness time won't stick to'.[23] Jennie Darlington found that, in Antarctica, 'The relationship of space with time was altered.'[24] 'The concepts of past and future seem to vanish', observed Alexa Thomson from her tent on a glacier in 2001. Helen Garner, visiting on a tourist ship, found that her memory started to pack up, and severe existential anxiety set in.[25] One's biological clock also 'refuses to tell the time' at the poles, sleep becomes disordered, and 'Big Eye' – the summer affliction of insomnia – sets in. Time assumes different rhythms. There is the deeper pulse of the ice ages, the seamless months of eternal light or night, the fourth dimension of a blizzard, the breaking up of the sea ice, the return of the Adélies, the schedule of the ships, the race to resupply, the changeover.

Antarctica has been described as 'a frozen time capsule' because of its role as a global archive of historical pollution and climates. On its surface, it preserves the vestiges of the past almost cruelly. The wind freeze-dries the flesh of dead, stranded seals so that they become hard and polished like monuments, but from a distance they seem alive. When you approach them, they look as if they died last week, but some have been carbon-dated at thousands of years old.[26] You can actually trip over a seal carcase that Douglas Mawson prepared for his dogs in the last days of his stay in 1914, but which was never cut up and simply abandoned. There is still a Harrods bottle and a trilby hat and much else to be wondered at in Scott's hut at Cape Evans, and if like philosopher RG Collingwood you believe that history is the recollection of past thought and experience, then you may pursue that quest by lying your head on Scott's pillow. And the ice preserves the food rations of previous expeditioners so that, when discovered decades later, they remain tantalisingly delicious and sustaining.

For many men, a trip to Antarctica had the attractions of a return to boyishness, 'a second childhood', as Belgrave Ninnis put it. For women as well as men, the polar winter – in its sheer generosity of time – recalled an early innocence. John Béchervaise wrote of his year at Mawson station in 1959: 'I have always said that time achieves a childhood dimension in the Antarctic; days are long, stretching on and on, as attenuated as the polar sunset that ends the dawn ... This year will have the value, in terms of certain kinds of experience, of ten spent in a settled continent.' Writing in the dark of midwinter, he tried to evoke the fabric of polar time, 'whose stuff is with me now: great, slow halyards of it passing through an Antarctic night where ordinary time does not exist. Here is always the time of watching, the time of waiting, the time of contemplation, the time of the little child, time that has not been impoverished by being thought valuable.'[27] In Antarctica, as well as hearing the silence, you can feel the 'slope of time'.[28] Time had contours in Antarctica; it had seasons and a geography that resisted rationalisation. The task-orientation of peasant societies endured there in surprising company with technology and modernity. The time-spirit of the industrial world struggled throughout the twentieth century to tighten its net around the continent.

In Antarctica, even in an age of instant communication, one looks back on Earth with a sense of being estranged. I felt this transcendent distance, too, even as a roundtripper. Time and space shifted and one's perspective was altered. Bill Green said: 'I don't know what happens here, but after a short time everything above the Antarctic Circle seems chimerical, an invention of one's isolation ... News ... has about it the quality of implausible rumour. You realise just how distorted time and distance have become, how, in mere calendar months, whole decades, whole centuries, have sped away.'[29] Richard Byrd, during his lonely vigil at the Bolling Advance Weather Base in 1934, had occasional world news items relayed to him and 'they seemed almost as meaningless and blurred as they might to a Martian'. The changeover is a brutal collision not only of seasonal generations but of senses of time. Cultural transmission has to be accomplished in days; history has to be honed in hours. And weaving the two worlds back together again is part of the challenge of homecoming. Brian Murphy, an electrician at Casey who took us out in the Zodiacs among the Windmill Islands, wrote to me later, saying: 'A lot of people talk of me taking a year out of my life to work in an amazing place. They were right, it was amazing, but it was not a year out of my life. It was a year *in* my life ...'

There is a stunning moment in the winter of 1912 when the remaining men of Scott's party debate which way they will sledge in the spring. They are haunted by the fate of the two lost parties of the previous summer. The one led by Scott was last seen 241 kilometres (150 miles) from the Pole (by the support party) but never returned. The other, led by Victor Campbell, was stranded on the coast when the ship could not rendezvous with them because of the ice. The choice before the group in the winter hut at Cape Evans is between sledging north to rescue the possibly live men or sledging south to find the definitely dead. They choose almost unanimously to go south.

Partly it was because they thought the members of the coastal party, if they were alive, could rescue themselves, as indeed they did. But it was also because their most urgent psychological need was to know the end of the polar party's story. They researched history rather than conducting a rescue. If they had not chosen to sledge south, the tent 11 miles from One Ton Depot with its cache of legendary words would have been swiftly embraced by the snow and ice and never found. The dying Scott had trusted that his men would make that journey and they were loyal to the end.

Edward Atkinson, the expedition's replacement leader, was Scott's first historian. Having found the tent, the men waited for Atkinson to catch up with the dogs. Men stepped back from the pyramid of snow, dumbfounded, recoiling from its 'awful secrets'. It was Atkinson's duty to enter the tent, contemplate the bodies and retrieve the archive. He sat down and read Scott's journal, beginning with the last entry and working backwards, reversing time and the desperate march as if resisting the stark conclusion. And then he called the men together and told them the story, from the beginning.[30]

The saga instantly took hold of them, immobilising and haunting them. It so threatened their dreams that Apsley Cherry-Garrard feared sleep that night.[31] The tragic story that was dug out of the tent was carried back to Cape Evans and there it was endured day and night in the hut, yet Scott's men still had no clear sense of the wider potency of their personal torment. In London in the same month, Roald Amundsen was grudgingly honoured for his clinical attainment of the pole while Scott, it was announced at a public meeting, remained working in Antarctica 'with unostentatious persistence, and in the true spirit of scientific devotion'.[32] The polar party had already been dead for most of a year, but Kathleen Scott's marriage remained alive. She wrote to her husband: 'I'm so glad, so very glad that you are staying another year.' Edward Wilson's wife, Oriana, had been travelling in New Zealand while awaiting her husband's return on the *Terra Nova*. When taking the train to Christchurch she heard a newspaper vendor parading along the platform shouting, 'Scott's dead! Scott's dead!'[33] When the *Terra Nova* returned south to pick up the men in late January 1913, Campbell shouted the dreadful news from the shore. Teddy Evans, commander of the ship, remembered 'a moment of hush and overwhelming sorrow – a great stillness' that, in later years, he would recall in the annual silence of Armistice Day.[34] When the news finally reached Earth on the expedition's return to New Zealand on 10 February 1913, even the tortured Cherry-Garrard was unprepared for the extent of mourning and history-making that broke out. They had been too long away and this story was so personal, yet the world owned it.

On frontiers every action seems self-consciously historical and theatrical, and yet frontiers also threaten the survival of stories. It is intriguing to see the birth of a legend, beginning with the lone, desperate writing of Scott's last words, then the first reading of them on the ice, and their eventual enshrinement in the newspapers of the world. There was no moment when they were not History. Fred Middleton had that experience with Shackleton's story of the *Endurance*; he heard it from the man himself on the *Aurora* and knew its future. The Norwegian whalers at South Georgia recognised the legendary status of the *James Caird* and its story instantly, with spontaneous reverence. In such moments, past, present and future become seamless and time refuses linearity. It is not only the isolation of Antarctica that creates the space for dreams and ironies and not only the experience of voyaging that creates cycles. Nor is it simply the desperate need of those on the ice to evoke a distant audience, and thus posterity. It is also that the present is only ever a conciliation of past and future. The present is always caught in the act, always in the process of becoming, sometimes consciously and precipitately so.

This book is about the enduring power of these stories. At first I thought I might navigate around the heroic era of Antarctica in order to privilege later personal and institutional experience. But Mawson, Scott, Shackleton and Byrd established the metaphors of language and experience on the ice. One must voyage with them and through them to new territory, not around them. When Bill Green was camped in the McMurdo dry valleys at the end of the twentieth century, he dreamt of Edwardian figures sledging and the wind that he heard was 'the wind of Scott's death'. Michael Parfit, visiting Antarctica in the 1970s, observed that 'As a rare form of life here – human beings in Antarctica – we seemed haunted by our own history. The voices of people long gone were part of the present wind.'[35]

What does history mean on a continent where nature is deadly and time is deformed? Barry Lopez wondered about how the forces of life must be construed by people in the Arctic who live in a world where swift and fatal violence is inherent in the land, where suddenly in the middle of winter and without warning, a huge piece of sea ice could surge hundreds of feet inland, like something alive.[36] How does one write about the relations between people and nature in such a place? The founding American generation of environmental historians – Donald Worster, Alfred Crosby, William Cronon, Richard White, Stephen Pyne, Carolyn Merchant – have all been strongly influenced by ecological concepts. Environmental historians have followed the fashions of ecology with special interest – from climaxes and superorganisms to energy flows and ecosystems to patch dynamics and landscape mosaics. In one sense, ecology humanises the biota, bringing concepts from the social sciences to bear on our understanding of the natural sciences: applying notions of community, neighbourhood, interdependence, sense of place. Ecology and history, we might argue, are both evolutionary sciences. And particularly in its more recent guises, ecology has developed a strong historical sense by embracing 'disturbances' as a necessary part of any living system. So ecology seems a science with which historians can feel at home, and it has shaped the style and predisposition of environmental history. We look for relationships across the human and natural worlds (even if they are only to be breached), we seek evidence of interdependence and attachment between humans and other biota and between people and the land, we suggest mutual acknowledgment between species, and in these ways we document a broader community than historians have before.

So how do these intellectual strategies apply to a place such as Antarctica? Our first response might be to ask: What history of Antarctica could not be 'environmental'? Since the signing of the Madrid Protocol on Environmental Protection to the Antarctic Treaty in 1991, environmental issues have become paramount in the administration of the Antarctic Treaty. Science and state are intimately entwined on the ice. Everything you do in Antarctica is now governed by strict environmental protocols. Furthermore, this is a place where 'the environment' is so completely dominant and overwhelming that it can never be tamed or taken for granted. The adjective 'environmental' almost becomes redundant on a continent where nature is so belittling of humanity.

Yet this 'continent of science', as it has been aptly named, might also be called the least ecological of lands. This is a place where nature is lethal, humans are always just visitors, and the land is covered by ice kilometres deep. This is a landscape in which the laws of chemistry and physics – and indeed the power of metaphysics – predominate, and terrestrial biology looks very marginal indeed. The ocean is where the life is: the largest land animal is a mite. The ice is massive, deadly and – in spite of its own variety – reductionist. It simplifies and universalises. There are no ecological niches on the ice, no biological differentiation on the ice shelf and polar plateau. In Antarctica there is no dialogue between humanity and ecology except at the marine edges. The environment of the ice is so simple, elemental and alien that the idea of an 'ecology' that might acknowledge humanity – and thereby create environmental *history* – seems absurd. Where is the opportunity in such a place to study culture and nature in community?

In Antarctica there were no indigenous peoples resisting or guiding the expeditions; nature was unmediated and often unrecognisable. It was surely no accident that, at the beginning of the twentieth century as Modernism swept through art, literature and physics, the abstract quests for the substance of the atom and the structure of the elements coincided with the quest for the Pole. Both were journeys in spacetime. Both were a probing of limits, products of a 'desire to take the measure of the world'.[37] Stephen Pyne calls Antarctica 'the most intellectual landscape on Earth'.[38]

Barry Lopez, who wrote a superbly lyrical book about the Arctic called *Arctic Dreams*, was shocked by Antarctica. In Antarctica he found himself unable to enter into any affectionate conversations with nature or history. He wrote:

> You look in vain for any conventional sign of human history
> – the vestige of a protective wall, a bit of charcoal, a discarded
> arrowhead. Nothing. There is no history, until you bore into
> the layers of rock or until the balls of your fingertips run the
> rim of a partially exposed fossil.[39]

It was, he said, 'a terrifyingly abiotic' landscape, one where humans feel not so much insignificant as superfluous, and the elements are worse than hostile; they are indifferent. This was 'environment' without 'nature', at least no familiar nature.

So environmental history looks a little different in Antarctica. The confronting strangeness of the place helps us to think in new and useful ways about the relations between people and nature in inhabited lands. On a continent of ice, nature is both more dominant and less complex. It is both impossible to ignore and impossible to engage with. Antarctica is the Earth's only true wilderness, and yet one that is inimical to life. It makes us realise how much of ourselves we unconsciously invest in the concept of wilderness. Biologist David Campbell describes terrestrial Antarctica as 'the antithesis of the Amazon ... It is like the silence between movements of a symphony'.[40] On the ice, it is physics and chemistry, not biology or ecology, that are the environmental sciences with which historians must engage. The environment is reduced to the elements: the weather and the periodic table.

At this end of the Earth, all students of life (historians included) will be more conscious of the sea than the land, and of the wind than the soil. And if terrestrial biology is your interest, then your most intriguing subject becomes human biology, under stress. Medicine is an environmental science in Antarctica, and the doctor is an ecologist. The physical, mental and social dimensions of living become dangerously seamless in small, isolated communities of people outside nature.

The storm in one's soul might well be more destabilising to an expedition than the blizzard outside. When humans visit Antarctica, they are both extremely vulnerable to nature and curiously outside it. On a windless day, they can hear the sound of their own blood coursing through their arteries. They can easily become stranded in their own little ebbing pool of life.

So humans in Antarctica are well beyond their biome, they are tiny ecologies of their own, waning heat sources in a vast terrain of ice, condemned always to retreat. Nature is different there and so is history. Down south, each year begins anew with the break-up of the winter ice. Can history and culture resist that devastating rhythm?

Mad rush all day. Changeover business
No time even to turn around
Changeover, thank God.[41]

These last three logbook entries by the *oic*, John Erskine, at Mawson station in 1967–68 constitute an Antarctic summer poetry, a hectic haiku. John Béchervaise wrote in 1960 that 'When a relief ship enters an Antarctic harbour, a new kind of time arrives. Men are all wildly busy, yet they don't really achieve any more for their hours; the days are suddenly brief and fleeting, as in cities.'[42] Robert Easther, writing in his annual report of Davis station in 1986, recorded that 'The arrival of the ship was everything people said it would be; the end of a phase, overrun by new people, a change-over that felt like a take-over, fresh food, final packing, losing a job, but going home…'[43]

When we arrived at Casey station on the *Polar Bird* in late 2002, scientific staff and roundtrippers were corralled offshore for a while and then allowed to visit in small groups to minimise the sense of invasion. We were warned that winterers might behave strangely, and that they were experiencing 'a change-over that felt like a take-over'. All of us, to varying extents, were catapulted into the maelstrom. 'We only work during daylight', remarked one of the barge operators ruefully.[44] The refuelling that year was genuinely hazardous. Both station leaders, outgoing and incoming, wrote about it later in their annual reports. During the evening of 26 December, the wind picked up to 15 to 20 knots, bringing more ice into the bay. At the same time, the *Polar Bird* swung around into the wind and tidal current and started to move away from the shore, dragging its anchor. The refuelling line became hooked around the stern of the ship. With fuel still being pumped, the fuel hose became all that was stopping our ship from continuing to swing and move. The line grew taut, which made it harder to lift over the drifting ice, now streaming out of the bay. John Rich feared that the line was 'going to be stretched to breaking point'. Incoming and outgoing expeditioners worked together in the crisis. Bloo had no sooner arrived in Antarctica than he was working through the 'night', hands frozen, back aching, anxiously monitoring the fuel line and the pump, ever fearing a leak. His cheerful face looked strained and tired: he was where he most wanted to be in all the world and he was exhausted. Ivor Harris was filled with admiration not only for the uncomplaining crews, but for 'whatever material the fuel line is constructed of'. 'It was not a nice introduction to Antarctica!' he wrote later, adding that it was 'a potential nightmare situation' from a health and safety point of view as well as environmentally, and that 'the fuel spill scenario doesn't bear thinking about'.[45]

To observe the changeover is to see the most crucial operational exchange in Antarctica, a period of urgent refuelling in every sense, a passing on of learning and wisdom in a matter of days. It is the frantic turnover of generations in Antarctica. What other societies may do in years, Antarctica has to achieve in hours. Anticipation, experience, memory and history are telescoped into one frenetic moment and become indistinguishable. Bloo remembers standing on the wharf as the last ship sailed away for the winter and thinking: 'What did that bloke say about that generator?'

When Stephen Murray-Smith voyaged to Casey and Davis in 1985–86, he was a keen and learned observer of this Antarctic ritual. Murray-Smith, as we've seen, had an academic and practical interest in small, isolated communities. He was also an expert on technical education and deeply admired practical intelligence. At Casey, he watched with wonder as 2300 tons of supplies were unloaded efficiently and safely in near-freezing conditions and with good humour by his shipmates: 'We were in danger of not realising that the greatest marvel of all was right under our noses; that we had the privilege of observing, and for many of us taking part in, the major annual resupply of an Antarctic base.' Everyone on board, he noted, suddenly had a place to go and a job to do. Murray-Smith was well qualified to assess the operation. Every summer, when re-establishing his camp at Erith Island in Bass Strait, he masterminded the stores and their stowage, always bringing more than anyone else thought was needed and expecting their help to unload it. So, here at anchorage in Antarctica, in the calm between blizzards, he knew that he was witness to the climax of years of work. The dropping of the anchor in Newcombe Bay, he reflected, 'was a moment for which hundreds of people unknown to us had been working for years'. Orders had been placed, designs made, manufacturers cajoled, committees convened and politicians briefed so that crates and pallets and drums could be lined up, loaded and now landed. It was hair-raising watching it all, and he supposed that every safety regulation in the book was broken a hundred times over, but the job was done and done with speed and grace. In spite of himself, Murray-Smith felt a surge of embarrassing patriotism: 'I was moved, very moved. I began to feel well of my own countrymen. All this immense labour, carried out without complaint, without shouting, in good humour, without congratulation. Could anyone else do it as well?'[46]

Murray-Smith was not usually an admirer of bureaucracy. When he was a commando in New Guinea during World War 2, he had been contemptuous of the arrogance and stupidity of the members of the officer class he encountered, and the way they interfered with the decisions of the men on the spot. Recoiling from the horrors of the Depression and world war, he returned to Melbourne in 1945 and joined the Liberal Party, the Australian Labor Party and the Communist Party all in a space of 12 months, but dropped the first two and devoted himself to the Communist Party. Thus, he wryly reflected, although he had already learned from boarding school and the army 'the idiocy of authority', he once again set himself 'on a path towards becoming a rebellious subordinate within a rigid and demented authoritarian system'.[47] He left the Communist Party soon after the Soviet invasion of Hungary in 1956 and the revelations of Stalin's atrocities, disillusioned with the whole formal process of politics and deeply suspicious of ideology and bureaucracy. But Murray-Smith remained grateful for those years in the Communist Party, for they cultivated his international political conscience and helped him strive for the integration of ideas and action. He was thankful that membership of the party, together with his marriage into a Jewish family, prevented him 'from being just another middle-aged, middle-class ex-public schoolboy'.[48] Through his association with the Jewish community, he came to feel personally the horrors of fascism, and so a keen sensitivity to evil and injustice permeated his complacent and comfortable Australia.[49] His political energies became directed towards education in the broadest sense. Through his editorship from 1954 of the literary magazine, *Overland*, Murray-Smith invested in intelligent popular culture, aiming to develop Australian writing and thinking as a continuation of a local democratic tradition and to counteract powerful, imported mass culture.[50] *Overland's* motto was an adaptation of a proud phrase of the Australian writer, Joseph Furphy: 'Temper democratic, bias Australian.'

Therefore, Murray-Smith took to Antarctica social democratic ideals that were also pragmatic, a distrust of authority and bureaucracy, a sympathy for the ordinary worker, a commitment to international cooperation and the politics of peace and disarmament, and a belief that good policy must be embedded in a knowledge of local history and culture. His democratic temper left room for educated wisdom and his Australian bias demanded stringent national self-criticism.

*Sitting on Penguins* was the confidently opinionated book of an older man, and much of its appeal is the intelligent and honest mixture of news and views. In the judgment of his friend, John McLaren, Murray-Smith, 'while he may have become grumpier, is one of the few to whom age brought wisdom'.[51] Murray-Smith's book had its origins as a report commissioned by his friend, Barry Jones, then minister for science (and Antarctica) in the Hawke Labor government. The book (like his private report to the minister) was not meant to be 'a study in repose' or 'a disinterested history'. Murray-Smith placed a warning at the start of *Sitting on Penguins*: 'Readers may note inconsistencies or contradictions in my responses. My answer is "Of course." This is the way we all work towards an understanding of what is happening to us.'[52] So, the book is an argument with himself as much as others about what Australia was doing in Antarctica. Murray-Smith's criticisms of the Australian Antarctic Division, some of which were first published as feature articles in *The Australian*, stirred official responses, even while he was still voyaging. 'Don't they see that, behind it all, I respect and believe in our commitments here?' he mused. 'That is the real message of what I've been writing. But I'm buggered if I'll write publicity handouts for a government department.'[53] He believed that all true societies should be involved in a perpetual *apologia*.[54]

Since his return from World War 2, Murray-Smith had hankered after Antarctica. In 1945, recently back from New Guinea, he wrote from Melbourne to Douglas Mawson inquiring about a place on an expedition to the Antarctic. Mawson acknowledged the letter but explained that the plan to establish a permanent Australian research station on the continent was far from realisation.[55] When, 40 years later, Murray-Smith boarded the *Icebird* in late December 1985, he was fulfilling an early dream and had become an official assessor of Mawson's legacy.

He had already made himself an experienced voyager of the roaring forties, for he had been an explorer of sorts in his own backyard of Bass Strait. All his favourite holidays – at Port Fairy, Wilsons Promontory, and at Erith Island – played at the edges of the vast southern realm. Once, when bent double in a storm high on Erith, he had looked across the Kent Group of islands and glimpsed 'a scene of such wildness that for the first time, we were forced to visualise New Zealand to the East and South America to the West as the only windbreaks at 40° South'.[56] Between 1966 and 1971, in the company of scientific friends, he made several short voyages of discovery amongst these peaks of submerged mountains on the vast plain that once connected the Australian mainland to Tasmania. They were nunataks in an ocean of meltwater. Landing on them, describing plants and animals, scaling them and investigating their intriguing human history, enabled him to make his own contribution to the literature and science of exploration. Murray-Smith greatly admired all seafarers and lighthouse-keepers, and a stormy ocean made him think 'of our ancestors committing their souls to God as they stepped on board the sailing ships for the long, uncertain passage to the new country'.[57] *Sitting on Penguins* was dedicated to his neighbours in the Kent Group, 'Shirley and Stan Grey of Deal Island, Keepers of the Light'. And when approaching Tasmania on his return voyage from Antarctica, he was awake at 5 am peering out his cabin window into a welcome darkness and looking for 'a spot of light which grew and exploded and then diminished, with another spot following close behind it'. It was the light of Maatsuyker Island and he knew that Julie and Graham Heynes were back of it, and that Julie was the daughter of Shirley and Stan Grey. Murray-Smith may have been a roundtripper, a boffin and – even worse – a 'Ministerial Observer', but his Southern Ocean was already storied and peopled, and he settled into the *Icebird* with the relish of a man who loved messing about in boats.

Unpacking his typewriter in his cabin, he was about to renew a favourite form of writing. During the war in New Guinea, he had started a diary 'as an exercise in contemporary history', and afterwards he wrote it out as a way of preparing himself for scholarly study in history.[58] In most issues of *Overland*, he wrote a thoughtful commentary on intellectual and literary life called 'Swag' because he wanted a personal, human presence in his journal rather than distant, authoritative editorialising. 'Swag' allowed him to be funny, complex and contradictory, to showcase evolving thought, rather than to be definitive. After his experience with the Communist Party, he wanted *Overland* to 'avoid the dreadful humorlessness and dogmatism of the fully convinced'.[59] It is this organic concept of intellect that attracted him to the diary format and which, as we've seen, he championed as the rationale of *Sitting on Penguins*. As he headed south, he felt that '[t]he story-books of generations were coming alive', and that he was adding to them.

The voyage of the *Icebird*, Voyage 6 in the summer of 1985–86, offers us a privileged porthole onto Australian Antarctica of the period. The expedition took six weeks and visited Mawson's Hut at Commonwealth Bay and Casey and Davis stations, but could not anchor at Mawson station because of ice and time. The *Icebird* also briefly visited the French station at Dumont d'Urville where, for the first time in 20 years, Australia exercised its rights of inspection, under the Antarctic Treaty, of another nation's base. They anchored near the black rock of the Ile des Pétrels and watched helicopters ferrying construction materials to the controversial airstrip being built across the Adélie penguin colonies. As is the norm, the French base was given only 24 hours direct notice of the inspection, and the Australian contingent carried a gift of a dozen selected bottles of Australian wine, and a presentation plaque. Murray-Smith wondered if the French had any inkling of the heavy cargo of Australian resentment the *Icebird* carried. Just a few months earlier, French secret service agents had sunk the Greenpeace ship *Rainbow Warrior* in Auckland harbour, because of its role in protesting French nuclear testing in the Pacific, 'something that was in all our minds'.

The *Icebird* also set out to locate the wandering South Magnetic Pole, now over water, a task they achieved with memories of Mawson's era. And they were to have a reminder of Shackleton too, and of the vulnerability of ships in Antarctic waters. While on the ice at Wilkes, they learned by radio that a private expedition ship had been sunk: the *Southern Quest*, crushed by ice, without loss of life, in the Ross Sea near Beaufort Island.

But Murray-Smith generally discovered too few resonances with the era of exploration and adventure, and found too much evidence of the influence of 'the new managers' based not on the ice, but in Hobart and Canberra. He was shocked by the ugliness of the human impact on Antarctica. He was deeply moved by the transcendent beauty of the place and thus appalled by the recent, sordid mark of humanity. He found visiting Antarctica to be an emotional roller-coaster: 'if you were constructing a psychological drama in which people were dumped from elation into depression in a very short space of time, you couldn't have done better'. There was the ironic contrast that you were not allowed to pick up a stone because of the precious environment, yet you could quarry the rock in big lumps and destroy ten tons at a time. Joining Murray-Smith on the homeward voyage was Ron Lewis-Smith, a Scottish botanist and senior member of the British Antarctic Survey who was completing a highly critical report on Australian environmental behaviour. 'I must say' said Ron to Stephen, 'Casey station is the ugliest I've ever seen, and that's saying a lot.' He added: 'In fact all your Australian stations stand out.'

Murray-Smith arrived at Casey in the middle of a building boom. Old Casey station – which he described as 'an environmental disaster of the first order' – was in the course of being replaced by the new station, a kilometre away. 'Yes,' he sighed, 'we've gone a long way towards buggering up one continent. There's no reason why we shouldn't start on another.' Walking from old Casey to the site of the new was 'more than a brief walk,' recorded Murray-Smith: 'It was the crossing of a divide between the old way of doing things in Antarctica and the new.'

Walking into the new living quarters, the Red Shed,

> caused our mouths to drop open … Not all the talk we heard
> on our way down about our building program in Antarctica
> prepared us for this … It was all so bewildering in its way,
> such a treacherous attack on what had been my mental image
> of Antarctica … Our foreign observers were as taken aback
> as I was: what we saw could only be interpreted one way, as a
> massive statement by Australia that it was in Antarctica in a big
> way, and there to stay.

Ron Lewis-Smith commented to Murray-Smith: 'You do seem to go a
bit hard at making your mark. Wander around on shore and try to find
the science all this is supposed to serve.' Murray-Smith decided that
'these fantasy buildings we are putting up are for aggression-display,
not accommodation for scientists'. He was bemused by the imperial
ambitions of his nation: 'A country unable to organise a decent taxi
service in its national capital now has an empire.'[60]

As Murray-Smith was surveying Casey, the minister for science,
Barry Jones, was admitting at home that Australia's research effort in
Antarctica was 'falling behind' that of other nations. Australia was on
'thin ice' because it had devoted so much of its energy and resources to
the building program at the expense of scientific research and logistical
support.[61] Since the negotiation of the Antarctic Treaty, science of the
highest quality had become the currency of influence in Antarctica.
Australia had begun well: Mawson's 1911–14 expedition was regarded
as the most scientific of the heroic era expeditions, and the establishment of Mawson station as the earliest permanent base on the continent
had led to some fine, pioneering science. In 1959, the British scientist
Joseph MacDowall judged that Australians had a much broader public
interest in science and Antarctica than the British and that 'Their permanent base at Mawson had an enviable record for scientific work.'[62] In
the late 1970s, with buildings at all three continental stations getting
shabby, Australia committed to a ten year rebuilding program. This
decision was meant to open the way also to solving the question of

transport and investigating the level of the scientific program. But the building program engulfed everything else and, as we have seen, even corroded the communities on the ice. By the mid-1980s, it was commonly accepted that Australia had slipped in the scientific prestige stakes. Late in 1985, the Australian Senate Standing Committee on National Resources found that Australia's research effort in Antarctica was 'grossly inadequate' and totally insufficient to support its claim to the continent. It pointed out that in recent years the total number of wintering scientists at the three mainland stations was fewer than the Japanese at a single base (and Japan had no territorial claim).[63] 'Once a pre-eminent presence in Antarctica, Australia is now heading for the second division', wrote Jeffery Rubin in *Time* in 1988.[64]

In 1984, a year after he took over ministerial responsibility for Antarctica, Barry Jones wrote himself a memorandum acknowledging that 'We do face a massive credibility gap in Antarctica, claiming so much (42% of the whole) and performing so little.'[65] As a new minister with a strong commitment to research, his hands were tied by the re-building in progress. In the period of his ministry, 1983–87, only ten per cent of the Australian Antarctic budget was left for research. In 1989, journalist Keith Scott wrote that the new Red Shed at Mawson epitomised Australia's policy: 'presence first, science and environment later – depending on how the money goes'.[66] There was a strong view among scientists that the bureaucracy had run off the rails. Barry Jones believed that Australia had made a poor policy choice in the 1970s between funding aeroplanes and airstrips or putting money into stations. The result was that 'like penguins, we cannot fly'. He noted that 'building is not an end in itself in Antarctica, in the way that science *is*'.

The problem was not just that the buildings exhausted the budget; they also represented a different philosophy, as Murray-Smith had observed. Glaciologist Ian Allison considered the buildings to be 'almost an aggressive statement that we can conquer the environment'.[67] The new stations used energy inefficiently and were not well designed for waste management, qualities that were hard to reconcile with the conservation ethic that strengthened with unforeseen speed as they were built. The buildings made scientists a minority on stations and seemed to be monuments to distant managers. Some expeditioners felt they were *too* comfortable, cutting people off from the environment and the ideals that had attracted them to Antarctica, severing any last continuities with the heroic era. They discerned an Antarctic culture that was becoming more superficial and less elemental, one that was insulated from responsibilities as well as dangers, where station conversations were less about science and more about maintenance and logistics. Graham Robertson, a biologist who in 1988 spent a winter studying the feeding ecology of the Emperor penguin at colonies at Auster and Taylor Glacier near Mawson, believed that the buildings added 'a degree of superficiality to our lives as expeditioners and a detachment from the deeper themes that Antarctica has to offer (presumably the reason most of us go to Antarctica in the first place)'. 'One of the problems with modern ANARE', wrote Robertson in 1989, 'is that it stifles an expeditioner's contact with the land ... [There is] a bureaucratic stranglehold that ensures they do nothing truly adventurous. This problem arises partly because of the size and complexity of the station.' The station demanded constant attention. 'With the present system expeditioners can spend the year cleaning, fixing, repairing, painting, burning and rearranging things simply to sustain the station.'[68] The re-building program solidified the grip of modernity on the continent. The 'pre-industrial' rhythms of Antarctic life were threatened, and casual, 'rural', seasonal exposure to the elements was curtailed. Meanwhile, outside, the silence was calling.

Murray-Smith felt he was observing a changeover of a larger kind, the loss of vital continuities of knowledge and tradition. He was moved to ponder the nature of historical consciousness on the ice. Murray-Smith depicted Australians in Antarctica as 'the ultimate existentialists'. He was shocked in the 1980s by the poverty of the Australian Antarctic Division's historical imagination, and by the severity of the annual discontinuity between past and present. He found that working data was lacking for every aspect of Antarctic operations. There were no records of the extent of fast ice in the bays, no easy access to information about ground covered by field parties in earlier years, no way even that a plumber could find out the age of a building, no history books or videos available at the stations, little popular knowledge of even the most famous of Antarctic heroes, and the officer-in-charge's daily logs of activities and achievements were, for a time, officially discontinued. And the logs, even where they did exist, remained unopened, disorganised and therefore practically inaccessible. At the start of every year, at the breaking up of the ice, the accumulation of knowledge began anew, but then only for a few months.

The journalist Jean-Paul Kauffmann also commented on the short generations of the south. Of Kerguelen Island in the 1980s, he observed: 'There is no continuity. Whalers and scientific mission have nearly always followed one another in ignorance of what went before ... One mission leaves, another takes its place. Each one knows nothing about the one before. The Kerguelens have no memory.'[69] Kauffmann found the only sense of history on paper. He wrote of his visit to the library at the French base on the island as a sensual experience. Situated in a wooden building that was the former weather station, Kauffmann described the library as 'the only thing of any aesthetic appeal in the southern ocean village'. It was regarded as the centre of the base.[70] Outside, in 'Charles de Gaulle Square', is a flagpole flying a tricolour that is so regularly shredded by the west winds that the base requires 'a plentiful stock of flags'. Inside, Kauffmann found himself heady with the scent of old paper, dust and the glue from worn spines of books. The library had 'that musty iodine smell of seaside villas shut up in winter'. Opening an old book (that is, one dating from the foundation of the French mission there in the early 1950s) was 'like taking the stopper out of an old perfume bottle'. The writer of the French Mission Report for Kerguelen for 1959–60 advised his successor: 'Keep a log: it is tiresome but irreplaceable. It is like Dangeau's memoirs: there is nothing more boring to read, but when you are looking for a detail of such and such a day at the court of Louis XIV, that is the only place you will find it.'[71]

These are remarkable portraits of societies without history or memory, frozen not just by temperature and energy gradients but also by a challenging information gradient, by a severe disconnection between the past and the present, a disjunction of time in places already severed in space. While Murray-Smith marvelled at the technical competence of the changeover, he also wondered about the meaning of it all. In the 1960s, he had written a PhD on the history of technical education in Australia, an academic inquiry that built on his radical political activism through an interest in social equity and practical knowledge.[72] In Australian Antarctica of the 1980s he discerned a profound failure of technical education, because he felt that technology had outstripped the politics that had given it birth and had become disconnected from its environmental and social consequences. It was a prime example of 'a rogue technology', technology that had got out of hand. He found that people were trained to perform their tasks, but not educated to understand the context in which they were working.[73] Australia was building a vast edifice down south. Where were those 2300 tons going and why? During a heated debate in the ship's bar on his return voyage, Murray-Smith challenged his companions: 'Don't you think ideas and ideals are important? I tell you this, if this country has a future in Antarctica it will be because people have *ideas* about it.'[74]

Almost two decades after Murray-Smith's voyage, I found a different Antarctica. Murray-Smith sailed at a time of severe rupture in Australian Antarctic life. There was, in the 1980s, accelerating global uncertainty about the future of Antarctica, as the next chapter explores. There was the stress of a building program in progress. There was a haemorrhage of morale and sense of purpose. In some ways, the 1980s recall the disabling tensions of the BANZARE era 50 years earlier, the open conflict of science and sovereignty that corroded even the old friendship between Mawson and Davis, and the farce that was flag-planting for its own sake. With *Sitting on Penguins* at my side, I looked for echoes of Murray-Smith's world in the early twenty-first century.

Some of the serious divisions he had observed remained as undercurrents on my own voyage: those between scientists and tradespeople, between managers and fieldworkers, between petty nationalism and global consciousness. There is still a surprising scarcity of wintering scientists and an overwhelming devotion to keeping the buildings humming. I heard some grumbling, but there were no coups. I might have been lucky to be a roundtripper sandwiched between two 'good years', but the commitment and goodwill of the expeditioners and leaders I sailed with, there and back, were truly impressive. Some of these things have got better because people like Murray-Smith stirred, coups were embarrassing, and systems of selection and training improved. The talk on the ship and at the station during my voyage was certainly as much about ideas as logistics, and many of the logistics were intellectually exciting too. After reading *Sitting on Penguins*, I did not expect to find so much historical literacy down there. History is much more a part of the conversation now than it was in the 1980s, and I mean not just the telling of Mawson stories but the everyday practical continuities that Murray-Smith feared were being lost: the changing extent of the sea ice, the historical background for any policy initiative, the care and attention to the keeping of logs and yearbooks. I think he would have been gratified at this change, but, of course, still grumpy, because he said himself that he 'hated dances and dinner-jackets and team sports' and (as a schoolteacher once told him) had 'an overdeveloped sense of injustice'.[75] And he would have found the increased bureaucracy of the era of the Madrid Protocol infuriating. But that would have given him a lot of joy, too, for environmental causes became increasingly important to him throughout his life. This is what changed most between Murray-Smith's voyage and mine. The white continent became green. Stephen never knew this, for he died suddenly of a heart attack in 1988, two years after his voyage, sitting on the steps of his library at home. *Sitting on Penguins* was in press.

## Wednesday, 1 January

**63 degrees south,**
**111 degrees east**

Last night we saw in the new year on the bridge. The ice
is relentlessly beautiful and captivating, luring you back
on deck, a hypothermic bait, holding your eyes as a shift
in light or configuration of berg or floe brings new gasps
of wonderment. The sun set just before midnight but not
enough to make the day dark. It set just behind a great
iceberg. We all shook hands or kissed and wished one
another a happy new year up there at the top of the ship
with ice all around, and the horizon burning orange.

When we awoke today we found ourselves in the open
sea and the swell gathered us quickly in. A four metre (13
foot) swell with wind-waves of an additional two metres
(6.6 feet), and the ship – now high in the water – has been
rolling as well as tossing. (As I am writing this in the B
Deck lounge, the ship has suddenly started rolling more
and newspapers and pens and remote controls are sliding
across tables and dropping to the floor, people are trying
to catch them, and every inanimate object seems to have
sprung to mischievous life.) Tonight, after a lovely dinner
and talk around the table, I went outside and was delighted
to sight a few lone, outlying icebergs near the ship, ghostly
and haunting presences, thrown out in farewell from
Antarctica. I greeted them like old friends.

I am enjoying the company and conversation of
Peter Cook, a Labor senator and former minister in the
Hawke and Keating governments (1983–96) who is on
this voyage to educate himself further about Australian

Antarctica. As a former resources minister, Peter played a role in Australia's successful campaign in the late 1980s to introduce a ban on mining in Antarctica. Unlike some politicians who have taken trips south, Peter has won many friends on board because he is a good listener and genuinely interested in others. He's a keen sailor and relishes being a part of this floating community of myriad expertise, where he can learn from barge operators and biologists about what they do. I've been telling him how his colleague, Barry Jones, sent Stephen Murray-Smith south at a time of escalating resource politics about Antarctica. I have loaned Peter my copy of *Sitting on Penguins* and look forward to discussing it with such a thoughtful primary source on Australian politics in the 1980s.

# GREEN CRUSADERS
*Greenpeace and greenhouse*

In 1964, still flushed with the success and excitement of the International Geophysical Year (IGY), Phillip Law, director of the Australian Antarctic Division, tried his hand at prophecy. He imagined what the continent would be like in 1984, that famous year of twentieth-century futuristic visions. He looked into the crystal ball and this is how he saw the crystal desert:

> The change-over of mine workers taking their summer furlough will be in progress. Mining, deep inside the rock of the peripheral mountains of Antarctica, will have proceeded unceasingly through the long winter months. With the advent of summer the mine work will have ceased and the workers and their families will be flying to their home continents for a short New Year vacation before returning for a further year underground. Hollowed out of the rock, irradiated with 'simulated' sunshine generated electrically from nuclear power, the mining townships will be independent of surface meteorological conditions and shift work will proceed around the clock through the months between March and December. House wives will work and cook and tend their infants, children will go to school, doctors and nurses will attend their patients in the small efficient hospitals and all the busy activity of a normal township will proceed in this human anthill.[1]

At the end of his lecture, Phillip Law revealed that this was more a mission statement than a prophecy. It was a future, he declared, that we must make possible. It was a Jetson-like dream of nuclear families in a nuclear city offered from the perspective of the early 1960s – the atomic age – when social complacency, technological optimism and economic growth reached their heights, and even penetrated Antarctica.

Discovering and using Antarctic resources – especially minerals and marine animals – had always been one of the spurs to exploration, and Douglas Mawson, like Phillip Law, found no contradiction between conservation and regulated industry. Economics rather than sensibility dominated the early twentieth-century conservation movement which aimed for 'wise use' of resources. 'Utilitarian' conservation and controlled exploitation were Mawson's aims. When he argued in the 1920s and '30s that Macquarie Island be declared a sanctuary for wildlife, he also advocated the culling of skuas, petrels and leopard seals because they preyed upon penguins. He suggested further that redundant bull elephant seals could be harvested for their blubber oil to pay for conservation science.[2] When, in the late 1940s, Australia (through ANARE) was planning long-term logistical and scientific initiatives down south, Mawson urged that they should be partly funded by associated commercial ventures such as whaling.[3] We have already seen the close relationship between industry and science exemplified in the career of the Norwegian whaler, Carl Anton Larsen. Politics was also always part of the mix. When, in 1925, the British launched a series of scientific investigations into the status of whale populations in Antarctica with the aim of moving towards sustainable yield harvesting, the project was to be funded by an increase in whaling fees in the Falkland Island Dependencies, which was itself an assertion of British sovereignty.

The Antarctic Treaty did not explicitly deal with resource politics, and from the mid-1970s this began to seem a fundamental and possibly fatal flaw. The twentieth-century world ran mostly on fossil fuels, especially coal and oil, and in 1973 the Organization of the Petroleum Exporting Countries (OPEC) restricted production. Oil and natural gas prices sky-rocketed, and exploration for alternative fields quickened. In 1975 Antarctic Treaty nations agreed to an informal moratorium on possible mineral development in preparation for a more formal approach to the question. Knowledge of possible offshore oil and natural gas reserves in Antarctica was growing and by 1980, experts believed that exploratory drilling for these resources was only a decade away.[4] Some geologists predicted that Eastern Antarctica would prove to be the world's largest coalfield.[5] The new understandings of ancient Gondwanan geological connections – the scientific realisation of the 'Great South Land' – underpinned speculations of links between the mineral resources of Australia and South Africa and their southern cousin, Antarctica.

From the late 1970s, there was a significant increase in the number of states acceding to the treaty. Krill and hydrocarbons were the chief riches expected of this icy 'treasure island'. The terms of the Antarctic Treaty dictated that a review conference could be called after 30 years, and this opportunity would soon be available. A new political regime was brewing in Antarctica, and nations began to jockey for a place in the kitchen. In Australian Antarctic Territory, the scale of Soviet operations exceeded that of Australia. Australia launched its massive re-building program. A number of Chilean families were sent south to settle 'permanently' in Antarctica. In 1982, treaty nations initiated negotiations over a Minerals Convention, the purpose of which was not to initiate mining in Antarctica but to set down rules should it ever happen. Many environmentalists believed it was better to have a regime than no regime. Political and economic considerations again began to eclipse scientific purposes in Antarctic affairs and some commentators expressed fears that the continent would become an international crisis zone in the 1980s. Phillip Law told Stephen Murray-Smith in the mid-1980s that they were witnessing the end of the Scientific Era in Antarctica and the beginning of the Resource Era. Science would continue, but would facilitate agreements on the exploration of Antarctic resources and the distribution of the profits.[6] Nineteen eighty-four had arrived and Dr Law's prophecy was coming true.

After six years of drafting and debate, all treaty nations agreed to adopt a Convention on the Regulation of Antarctic Mineral Resource Activities (CRAMRA) in Wellington, New Zealand, in June 1988.

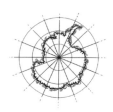

Meanwhile, an alternative vision of the future of the ice had been gathering political momentum. Antarctica, the last continent to be colonised, was envisaged by many not only as a realm of peace and science, as established by the treaty, but also as a 'nature reserve' or a 'world park'. In 1935, Frank Debenham imagined 'this last and least useful continent' as an 'International Park of vast proportions', and Robert Cushman Murphy suggested in 1962 that Antarctica might properly be regarded as an international park or sanctuary.[7] Following the signing of the Antarctic Treaty, several landmark conservation measures were negotiated: the Agreed Measures for the Conservation of Antarctic Flora and Fauna (AMCAFF, Brussels, 1964), the Convention for the Conservation of Antarctic Seals (CCAS, London, 1972) and the Convention for the Conservation of Antarctic Marine Living Resources (CCAMLR, Canberra, 1980).

The earliest international agreement in the Antarctic had concerned conservation. In 1935 the League of Nations brought into force a Convention for the Regulation of Whaling. Neither Japan nor Germany were parties to the agreement and it proved to be ineffective, but it was expressive of a conservation and management ideal that continued to be pursued throughout the century. Britain, the United States and Norway had dominated whaling in the Southern Ocean, but Japanese whalers extended their fishing into Antarctic waters in 1936–37 and the Soviet Union sent whaling fleets south from 1946. Conservation measures that were negotiated at conferences of whaling nations in 1937 and 1938 were mostly disregarded, and record captures continued to be posted.[8] Robert Cushman Murphy, writing in 1940, was shocked by the 'stupendous quantities of flesh and blood now being taken from the sea'.[9] In December 1946, 14 governments signed the International Convention for the Regulation of Whaling and established the International Whaling Commission (IWC). But this, too, proved of limited effectiveness. There was a serious need for protein in impoverished post-war Japan and whaling was important to the recovery of the Japanese economy. Illegal hunting was rife, and actual killings greatly exceeded reported killings. Catch data submitted by the Soviet Union to the IWC in the 1960s and 1970s was falsified by massive margins.[10]

In the late 1960s, Bob Murphy was still campaigning for the creatures he had seen slaughtered from his whaling brig in 1912. He lamented that the treaty nations' Agreed Measures for the Conservation of Antarctic Flora and Fauna (1964) excluded whaling operations 'which have long been the most shamefully short-sighted of overexploitations'.[11] In 1982 the IWC agreed to an indefinite moratorium on all whaling, commencing in 1986, and a Southern Ocean Whale Sanctuary was declared in 1994. Japan continued to kill at least 400 minke whales each year, often much more, for 'scientific research' (an allowable exception). Many of them ended up in supermarkets and restaurants. Conservationists claimed that this was commercial whaling by stealth, and Australian politicians condemned Japan as a 'barbarian nation'. The Japanese argued that whale meat was a traditional food that had been eaten by ordinary people since the mid-nineteenth century, although today only about 1 per cent of Japan's 127 million people eat whale meat regularly, a declining proportion of the population. Unwanted whale meat has ended up as pet food in Japan and a very large stockpile remains unsold. In 2006, a slim majority of the IWC voted to reconsider the whaling ban. Southern fishing remains a tense issue. Tens of thousands of tonnes of Patagonian Toothfish (*Dissostichus eleginoides*), a deep-sea species found around sub-Antarctic islands, have been illegally caught and traded since the 1990s.[12]

In January 1987, the international pressure group that united the ecology and disarmament movements, Greenpeace, established a station on Ross Island in Antarctica, called 'World Park Base'. It was the first long-term non-governmental base to be established in Antarctica, and its practical purpose was to document and expose the environmental effects of humans on the ice, and to provide a focus for the campaign to have Antarctica declared a World Park. With the negotiations towards a Minerals Convention reaching fruition,

Greenpeace activists regarded Antarctica as 'under siege'.[13] They feared that resource competition would undermine the cooperation and sharing of data that was essential to the treaty. The term 'World Park' had been used by the New Zealand government in the 1970s, and Greenpeace gave it new political momentum. In the 1980s there was growing criticism of the human injustices and dispossession associated with the creation of protected areas and national parks in the rest of the world. The 'wilderness' ideal lost the moral high ground when conservationists began to hear protesting voices coming from the forests and deserts. The purity of the ice beckoned. Although Antarctica was a 'park' without trees, it was also a park without people.

Wintering in their little green box at Cape Evans in 1987 were four volunteers: Kevin Conaglen (mechanic and team leader), Gudrun Gaudian (scientist), Justin Farrelly (radio operator and technician) and Cornelius van Dorp (doctor). They were about 30 kilometres (19 miles) from the US base, McMurdo, and New Zealand's Scott base. Quite close to them was the hut from Scott's last expedition, which they found enveloped by a 'strange atmosphere': 'The rough wooden skis still rest against the wall, the mummified body of a sledge-dog sits at the door, and bales of hay for the ponies are piled high nearby.' They wondered what it was like inside; for it was locked and they were not allowed entry. Visitors from McMurdo or Scott base or any passing tourist groups were readily given the key to the hut from the officer-in-charge at Scott base, but it was impossible for members of Greenpeace to obtain it. Officials at McMurdo also refused to pass on local weather information if the Greenpeace helicopter was in the air, even if it was on its way to McMurdo.

By 9 February, the Greenpeace members made their first uninvited inspection of Scott base and McMurdo, to see how New Zealand and the United States treated their rubbish and effluents. They found that the giant McMurdo rubbish dump included plastic and rubber, steel pipes, a battery, trucks and other materials that violated the 1975 Code of Conduct for Antarctic Expeditions and Station Activities.

Further down the road near the ice-wharf they found a similar scene: a pile of truck skeletons, wheels, oil drums, and a pipe discharging brightly coloured liquid straight into the water. 'A lot will have to change here', wrote Maj de Poorter (who left on the ship a few days later) in her diary.[14]

Gudrun Gaudian, the only woman staying for the winter, relished the 'vast, white silence' of this place. Sometimes the only sound she could hear in the autumn was the creaking of the thickening ice in the bay as the tide moved beneath it: 'It makes an eerie sound, just like an old door in a secret castle being opened very slowly.'[15] Gaudian, a marine biologist from West Germany, cut and maintained several holes in the ice so that she could monitor fish and krill populations. A winter task for the small group was sewing or attaching rainbows to all field equipment. In mid-June Cornelius van Dorp, the doctor, 'moved out for a few days' as part of his study on the pineal gland. Blizzards engulfed their little green box in snow. But they were also at the centre of another storm. 'New Zealand and the US have refused to carry our mail', recorded Justin Farrelly. 'As you can imagine we are very annoyed, not so much at the lack of mail, but at the insinuations that we are not a "fully independent" expedition.'[16] A week later the Greenpeace team leader, Kevin Conaglen, wrote: 'Sorry to say that not much of our mail managed to slip through the US Navy sort-out in New Zealand. It appears that "they" don't like some of our photographs of their McMurdo waste dump. It's a shame they don't spend the same amount of time cleaning up their dump as they did going through our mail.'[17] The presence of Greenpeace on the ice was officially resented. But, the following summer, when MV *Greenpeace* was moored off McMurdo for a few days, the ship was visited by over 100 US personnel who donated more than $1000 to Greenpeace.[18]

The first expeditioners in Antarctica had dumped their rubbish onto the sea ice where it 'disappeared' with the summer melt. Or, like the Soviets at Leningradskaya until 1986, they threw glass, tins and plastic into a crevasse 30 metres (100 feet) wide. In January 1959, the

Australian expedition ship, the *Thala Dan*, was holed by a reef when approaching Davis station, releasing oil into the ocean. Volumes of it welled and gurgled to the surface, and penguins started leaping straight out of the water 'as though onto invisible ice floes'. John Béchervaise, who was the incoming *oic* at Mawson station, recalls that when reassured that the ship could be repaired, they had a 'Shipwreck Party' on this beautifully calm, sunny day, and 'The oily water, almost motionless around the ship, carried scores of floating beer cans.'[19] Even in the 1980s, as Australia managed its building program, there was little regard for environmental impact. Casey station was sited near areas that had been valued since the 1960s for their diverse colonies of moss, lichen and algae; John Rich introduced them to me as 'The Daintree of Antarctica'. The new station was built in the 1980s right next to a Special Site of Scientific Interest (SSSI no 16) and the moss beds were affected by plumes of alkaline cement dust that emanated from the building works. Australian scientists quickly declared that 'Considering how recently the new station was planned its location can openly be described as environmentally irresponsible.'[20] The Thala Valley tip, which our voyage was helping to repatriate, contained about 4000 tonnes of waste dumped during station operations between 1969 and 1985. Leaching of pollutants from the tip, especially heavy metals, decreased the diversity of marine species to a level similar to that around Sydney sewage outfalls.[21] As Stephen Murray-Smith and Ron Lewis-Smith found, Casey in the mid-1980s was a massive building site and an environmental disaster. The 'beer dominated social atmosphere' was uncongenial, too. When Lewis-Smith put up notices about caring for the environment, they were defaced.[22] In the year following Lewis-Smith's damning report on environmental practices at Casey (1986), the Australian Antarctic Division began the return to Australia of long-standing rubbish at Casey and Wilkes.

In targeting the McMurdo dump, Greenpeace was scrutinising the largest permanent human community on the ice, where the urban history of Antarctica finds its focus. The main US base, known as

Mac Town, is now over 50 years old and has a population that varies from 1200 in summer to 200 in winter. It has more than 100 buildings including laboratories, observatories, dormitories (named, for example, Hotel California or the Mammoth Mountain Inn), canteens, a clinic, church, firehouse, bowling alley, barber shop, video store, gym, three bars, a coffee house and an aquarium. The base is serviced by a helipad, three airstrips and a harbour.[23] You will even find automatic teller machines. There is a rumour that McDonalds has opened a franchise in Mac Town, but Americans are too offended by the question to confirm or deny it. A weekly newspaper, the *Antarctic Sun*, is published throughout the summer and Radio McMurdo is 104.5 FM. The station's oldest precinct is called 'Downtown McMurdo' or 'the historic district'.

From the moment of its foundation next to Scott's old 1901 *Discovery* expedition hut, first as a tent camp and then a cluster of 11 prefabricated buildings in 1955–56, this Antarctic 'city' looked ramshackle and untidy, with one or two makeshift streets defined by bulldozers. At McMurdo, vehicles drove vaguely on the right, and a mile away at Scott base they drove vaguely on the left, until (so the urban legend goes) a road connected the two stations and a US Navy caterpillar tractor weighing 35 tons met a New Zealand jeep, and the 'right' of way was asserted.[24] Jamesway huts salvaged from the Korean War provided some accommodation, showers were rationed to one a week, and huskies roamed the streets for food scraps and human attention. By 1970, McMurdo in summer was described as 'a dry, dusty, American frontier town, with saloons containing full-length nudes in the best Wild West tradition'.[25] The official language was English but the dialect, at least in the 1950s and '60s, was expletive. Byrd's younger colleague, Paul Siple observed that 'At McMurdo the use of profanity had been developed into a fine art.'[26] The sound of summer was the thunder of helicopters: 'you could feel the beat of their coming and going in your chest'.[27] Today, under the steaming white cone of Mount Erebus, numbered buildings sprawl along unsignposted streets, and people can get lost even after two months' residence, without any help from a white-out.[28] There are urban easements and backyards, fire alleys, truck turnarounds and rows of stores. The speed limit on 'Antarctic 1', as the residents call the main road, is 30 kilometres per hour. Another road – this one an ice route – is being marked from McMurdo to the South Pole. Antarctic bases are often compared to shabby mining towns: there is a brutal practicality about them and an abrupt disconnection with the landscape. No matter how substantial the installation, it remains a camp.

In 1961 the US installed a nuclear power station at McMurdo, a 1.8 megawatt experimental reactor nicknamed 'Nukey Poo' which was sited halfway up Observation Hill near the active volcano of Mount Erebus. The plant operated for ten years but also suffered shutdowns, fire damage and radiation leakages. In 1972 it was shut down permanently and over 10 000 tonnes of radioactive soil and rock were shipped back to the US, and the site was cleaned and decontaminated.

When Greenpeace established World Park base and photographed McMurdo, the garbage dump was a huge feature of the local landscape, extending towards Scott's *Discovery* expedition hut. The station's ecological imprint was steadily expanding. Each season the biologists had to go further afield for their ecological studies, and Bob Murphy reported that the bottom of Winter Quarters Bay was so polluted it was losing the semblance of a 'natural area'.[29] In the late 1960s, underwater dives near McMurdo found beer cans outnumbering sponge species. Following the exposure and embarrassment of the Greenpeace campaign, waste management at McMurdo changed rapidly. By 1994, waste was categorised in 18 different ways, and today you might sight a line-up of washing machines waiting to be fed into the metal baler for repatriation.[30] It is important to distinguish between waste and the station's indigenous aesthetic. Bill Fox, who studied the vernacular art of McMurdo, observed 'the careful recycling of waste materials into value-added items', a common human response to extreme environments, whether they be polar terrain or Depression-era poverty.[31] Art and sculpture at McMurdo rely on the creative use of surplus materials. When Alexa Thomson visited the Russian base of Novolazarevskaya in 2000, it took her a while to see the 'auto-wrecker ambience' of the station as just a different kind of recycling.[32] They had a resource for every possible repair. In earlier times, the Russians had introduced pigs to recycle their garbage. Out in the Southern Ocean, plastic rubbish is increasing as fishing intensity grows. Monthly rubbish collections along the west coast of Macquarie Island show an increase in net and plastic rope fragments since the mid-1990s. In the summer of 2000–01, when Stu Fitch walked the beaches of the remote Heard Island, he collected about one and a half tonnes of garbage, almost all of it plastic.[33]

Resource and environmental politics were two major challenges to Antarctic management in the 1980s, and there was a third, related political issue that also emerged in this decade. In December 1983, the United Nations (UN) became part of Antarctic politics for the first time, although there had been earlier attempts to make UN control of the continent possible. The UN General Assembly adopted a draft resolution on 'The Question of Antarctica' as a result of arguments made by an alliance of Asian and third world nations that the continent of ice should be managed as the common heritage of humanity rather than as the exclusive preserve of nations claiming sovereignty or exercising scientific muscle. Led by Malaysia, and supported by Antigua/Barbuda, Bangladesh, Malaysia, Pakistan, Philippines, Singapore, Sri Lanka and Thailand, the draft resolution aimed to build on the United Nations Law of the Sea Convention signed in 1982 and the Moon Treaty of 1979. The Malaysian prime minister, Dr Mahathir carried the campaign, arguing that the Antarctic Treaty was a relic of colonialism and 'not accountable internationally', and seeking greater democratisation of decision-making on the international scene. 'I have heard that the South Pole is made of gold and I want my share of it', declared Mahathir. If there was going to be a resource regime in Antarctica, then third-world nations wanted to be a part of it. The Antarctic Treaty powers, led by Australian ambassador to the United Nations, Richard Woolcott, responded with defensive unanimity, for this was an issue that brought together even the USA and the Soviet Union, and Britain and Argentina. Woolcott pointed out that Malaysia had only to accede to the treaty if it wished to participate more actively in Antarctic affairs.[34]

At the end of the 1980s, as all these pressures built up, something unexpected happened in Antarctic politics. Even environmental campaigners at Greenpeace and the Antarctic and Southern Ocean Coalition did not foresee it.

Perhaps it was one of the outer ripples from the establishment of World Park base by Greenpeace the year before. On southern shores of that other dry continent, the Australian Cabinet gathered, unusually, in Melbourne on 28 March 1988. The Antarctic meeting in Wellington was just over two months away, and a proposal to support CRAMRA, which was endorsed by the foreign minister, Gareth Evans, and the environment minister, Graham Richardson, was on the agenda. At the meeting, two eloquent people spoke against it. Senator Peter Cook, the resources minister, objected to CRAMRA because he felt that Australia's interests as an exporter of minerals would be disadvantaged by making mining possible in Antarctica. Paul Keating, the treasurer, made the argument on both economic and environmental grounds. But the two men did not win the support of the Cabinet, and Australia went on to adopt the convention, along with all other treaty nations, in Wellington in June.

In the period between agreement and ratification, the Australian government invited community debate on the issue. The Department of the Environment received thousands of letters and postcards against signing the convention, Greenpeace and other non-government environmental organisations intensified their public campaign against CRAMRA, and political support in the Australian Labor Party began to swing behind exploring alternatives. Cook and Keating elaborated their earlier objections to their Cabinet colleagues: Australia would be forfeiting its sovereignty and exposing its own mineral producers to competition from mining operations subsidised in Antarctica for strategic purposes. Cook felt that the lack of an anti-subsidies provision was a serious flaw in the convention. In September 1988, Keating met the French prime minister, Michel Rocard, and sowed the seed for an alliance with the French, and an alternative, joint critique of CRAMRA on environmental grounds. The member for Dunkley, Bob Chynoweth, a roundtrip voyager on the *Icebird* in early 1989, watched the moon rise over Antarctica and became an influential convert. The French government indicated that it, too, had misgivings about CRAMRA in its current form. In April the Australian Democrats called on the government not to sign CRAMRA. On 2 May 1989, the leader of the opposition in Australia, John Howard, announced that the Coalition parties would not support mining in Antarctica. A few weeks later, at a Cabinet meeting on 22 May, Prime Minister Bob Hawke committed his government to what he called 'Mission Impossible': to reject CRAMRA and argue for the protection of Antarctica as a nature reserve and province of science.[35]

This was a huge political gamble. Management of the treaty is by consensus, and so a single dissenting nation is enough to derail an agreement. By refusing to sign, Australia was committing itself to an international diplomatic mission. Other treaty nations responded to Australia's stance initially with disbelief and then with bitter opposition. Andrew Jackson, a policy manager in the Australian Antarctic Division, recalls that Australia was cast as a spoiler, a nation prepared to walk away from the consensus principle and threaten the stability of the treaty. It was branded 'a Southern Hemisphere radical that would tire and fall into line once the [1990] Australian elections were over'.[36]

Hawke's change of mind was firmly and publicly rejected by US President George Bush (senior) and Britain's Prime Minister Margaret Thatcher, and it was still subject to attack in his own Cabinet. The president of the Scientific Committee for Antarctic Research (SCAR), Richard Laws, later blamed 'a number of vociferous, well-financed environmentalist groups' for 'wrecking' the convention. Phillip Law condemned the Hawke government's about-turn as naïve, ignorant and obstructive, and he tried his hand again at prophecy: Australia, he said, was pushing 'a lost cause'.[37]

But, over a period of 18 months, the Australian government mobilised its best diplomats and won support first from France, and then Italy and Belgium, and gradually and tenaciously built a new consensus against mining and in support of a new environmental regime. The dramatic oil spills of the *Exxon Valdez* in the Arctic and the *Bahia Paraiso* in the Antarctic in 1989, with their dramatic images of slicked polar seas and suffering wildlife, strengthened the hand of the environmental campaigners. In mid-1991, President Bush announced that the United States, the last government to hold out against the Australian and French campaign, would finally support the no-mining position. And on 4 October 1991 a Protocol on Environmental Protection to the Antarctic Treaty was signed by the treaty nations in Madrid. It included a ban on mining in Antarctica and put into place comprehensive and legally binding measures to protect the Antarctic environment.

Richard Woolcott was secretary of the Australian Department of Foreign Affairs and Trade at the time of the campaign for environmental protection of Antarctica, and he was also Australian ambassador to the United Nations in the mid-1980s when Malaysia challenged the treaty system. Both challenges, he believes, strengthened the treaty. Woolcott first became fascinated by Antarctica when he read Douglas Stewart's radio play, *The Fire on the Snow*, and he was able to visit the South Pole and the Beardmore South field camp in January 1985, where he participated in a conference and a very cold cricket match.

His personal experience at the United Nations and with Antarctic politics 'demonstrated to me that one system – the UN – works less effectively than I had hoped, probably because of its sheer size, while the other system – the Antarctic Treaty – works more effectively than I had expected'. 'So far', he proclaims, 'the continent's only export has been knowledge.'

Andrew Jackson describes the campaign as a watershed in Antarctic politics: as a dramatic shift from a resources view to an environmental view of the continent, as a demonstration of Australia's influence in the management of Antarctica, and as a successful test of the robustness of the Antarctic Treaty.[38] The negotiations over the Minerals Convention and the drafting of the Madrid Protocol on Environmental Protection to the Antarctic Treaty effectively constituted the 30-year review of the treaty that had been due in 1991. The new agreement took the sting out of Malaysia's criticism of the treaty for no longer could select nations be seen to be hoarding mineral wealth. Phillip Law's prophecy of mining towns under the ice was, for the moment, proven wrong, but even more importantly, it became clear that the mission statement for human endeavour down south had changed radically.

The huskies concluded their final run at Mawson in 1993 with a snarling brawl, as if they knew it was their last performance on Antarctic ice.[39] Britain, Argentina and Australia were the only countries that still had huskies in Antarctica at the time of the signing of the Madrid Protocol, and the dogs had to be removed because of the new, strict environmental guidelines on introduced species (apart from humans, of course). The Australian government, knowing the immense popular sentiment towards huskies, had tried to win an exception, but the Americans – who were still irritated by Australia's rejection of CRAMRA – were not about to compromise on this issue. There were scientific concerns about the dogs infecting seals with canine distemper, although this was not known to have occurred in Antarctica. The sadness and outrage felt by Antarctic expeditioners at the removal of their faithful friends, and at the repudiation of history implied in the action, made the fate of the huskies the most public and intense issue faced by the Australian minister for the environment in the early 1990s. The dogs had not only represented transport and friendship; they were full members of expeditions and lifted Antarctic morale. The huskies at the British base, Rothera, were the last to go – on 22 February 1994 – and they quickly died of a virus in Canada. Their demise was reported in the *Times* newspaper as being 'due to pressure from Australia and the environmental campaign group Greenpeace', and Sir Vivian Fuchs, the veteran Antarctic explorer, declared that the huskies were victims of misplaced environmentalist zeal and that their expulsion was absurd. Douglas Mawson, who had been kept alive by dogs, would have welcomed their removal. In 1956 he wrote to Bob Dovers, author of *Huskies*: 'For some time past I have come to the conclusion that henceforth scientific expeditions in the Antarctic regions should not take dogs. They are too great a risk for the maintenance of native life, and not a patch on modern mechanical means of transport. I am all for helicopters, aeroplanes and tractors.'[40]

The removal of the huskies was a symbol of the new era introduced by the Madrid Protocol. The environmental consequences of every action in Antarctica would now be carefully assessed, and not even old friends were exempt. But the triumph of green politics in Antarctica had greater implications, for it was the expression of a shift in the way humans understood the usefulness of Antarctica. A constant challenge for post-war endeavours down south, and one articulated constantly by Phillip Law, was to build the significance of science in Antarctic culture. Law recalled that ANARE was established by the Australian government in 1947 'for purely political reasons'; to secure Australia's territorial claim and its right to commercial opportunities in the south. Research was 'a side issue', explained Law. He resigned as director of the Antarctic Division in 1966 because of continuing problems concerning the status of science in the organisation. Science again felt marginal during the resource politicking and sovereignty anxiety of the 1980s. The Madrid Protocol greatly empowered science, not just as a currency of prestige but as an urgent and practical requirement. It ensured that research was no longer 'a side issue', although it certainly remained subject to political, strategic and bureaucratic pressures. Some scientists now worry that, under pressure from the protocol, science in Antarctica may become mostly a form of environmental monitoring, and that issues of human impact threaten to overwhelm the Scientific Committee on Antarctic Research (SCAR).

With the end of the Cold War and the signing of the Madrid Protocol, issues of sovereignty became less urgent, and the treaty system was confirmed in its importance and robustness. But had sovereignty really taken a back seat? Australia's 42 per cent claim includes 5000 kilometres (3107 miles) of coastline, and in the 1990s its government declared an Australian Fisheries Zone and an Exclusive Economic Zone off its Antarctic coast in response to the United Nations Convention on the Law of the Sea. The declaration of continental baselines and adjacent maritime zones is the act of a sovereign coastal state. In 1999, the Australian government committed to spending $40 million to survey the seabed in order to define the limits of about 1.8 million square kilometres (695 000 square miles) of the extended continental shelf off its claimed sectors. Because Australia was the first of the Antarctic claimant nations to sign the Convention on the Law of the Sea, it also became the first to lodge the necessary documentation with the Commission on the Limits to the Continental Shelf. This was required within ten years of signing the convention. Some treaty nations have argued that it is contrary to the Antarctic Treaty for claimant states to extend their pursuit of rights under the Law of the Sea into Antarctic waters because the treaty disallows new claims or enlargements of existing ones. Australia argues that it is merely documenting an existing claim which pre-dates the Antarctic Treaty. The compromise solution has been for Australia to lodge its claim with the UN Commission, but to ask the Commission not to act on it for the time being. This is a classic Antarctic strategy, the freezing of a claim, the retention of a right without asserting it; protecting the precious *status quo* avoids tensions. For years, Australia issued the Russians in Antarctica with annual radio licences, free of charge.[41]

In the second half of the twentieth century, a geographically marginal continent has become intellectually and environmentally central to the world. As British polar scientist and diplomat John Heap observed, the challenge of the post-war period for Antarctic research was no longer simply to find out what was down there, but rather, to integrate sustained Antarctic science into the mainstream of global research. And this has indeed happened, more dramatically than might have been expected. Antarctic science made itself fundamental to world concerns about climate change, ocean processes, marine biodiversity and human environmental behaviour. 'In the old era of Antarctic research', reflected scientists, 'people used their knowledge of the rest of the world to find out what was in Antarctica. In the new technological era, we use our knowledge of Antarctica to learn about the rest of the world.'[42] It had become a global laboratory as much as a world park. If the 1980s were a decade of pragmatism and nationalist resource politics in Antarctica, they were also the years when Antarctic science emerged to become central and urgent in world affairs.

In the 1990s, the senior Australian Antarctic scientist and geologist (and shipmate of Murray-Smith's), Patrick Quilty, believed that two events had brought Antarctica to the centre of global concerns: one was the view of 'Spaceship Earth' given us by the Apollo missions of the late 1960s – a finite planet floating lonely and blue in space, doubly capped in white – and the other was stratospheric ozone depletion which shocked people into seeing Earth's atmosphere as a single system.[43] Each spring since the mid-1970s, a vast rupture has opened up in the ozone layer over Antarctica. The ozone hole, discovered by British scientists in 1985 using long-term Antarctic data, is created by chlorine from chlorofluorocarbons (CFCs), and allows damaging ultraviolet radiation from the sun to infiltrate and threaten life on Earth. Antarctic science gave the world the warning, and nations acted cooperatively to reduce the output of harmful gases.

The continent of ice is playing a similar role in helping us to monitor and understand human-induced climate change and its likely effect on temperatures and sea-levels. By the beginning of the twenty-first century, Antarctica was becoming green in a different and disturbing sense. In the southern summer of 2004, 'great green swards' of Antarctic hair-grass began to appear on the peninsula, forming meadows in 'the home of the blizzard'.[44] The burning of fossil fuel resources without reducing or capturing carbon emissions is transforming life on Earth. Isotherms (lines marking average temperatures) are racing towards the poles with the effect that polar animals will be pushed off the planet. The greatest threat for human beings lies in the potential destabilisation of the vast ice sheets of Greenland and Antarctica.[45]

The long-term view of the twentieth century may be that environmental and demographic changes were the most significant events of that period. The world's population quadrupled in the twentieth century, energy use increased 16 times, carbon dioxide emissions went up 13-fold, water use rose nine times. In his environmental history of the twentieth century, *Something New Under the Sun*, JR McNeill argued that 'humankind has begun to play dice with the planet, without knowing all the rules of the game':

> The human race, without intending anything of the sort, has undertaken a gigantic uncontrolled experiment on the earth. In time, I think, this will appear as the most important aspect of twentieth-century history, more so than World War II, the communist enterprise, the rise of mass literacy, the spread of democracy, or the growing emancipation of women.[46]

Environmental change became so intense and pervasive that concerns which at first seemed local or remote became global in significance. Politically and intellectually, the southern ice cap is a symbol of that revolution. And in Antarctica at the end of that century, a highly specialised fossil-fuel based civilisation turned its back – for how long? – on minerals.

## Friday, 3 January

**56 degrees south,
126 degrees east**

It feels warm on the deck (relatively!) and now it's getting
dark (a little). It's approaching 10 pm and I've been out on
deck watching a beautiful sky of pink and purple clouds
above an ocean of deep blue-black flecked with white crests.
There's an albatross soaring and wheeling behind the ship,
never flapping its wings, just sporting with the wind and
waves. We sail towards the north-east horizon and marvel
at the coming darkness. Ice and light haunt our dreams.

Deep in the midst of our voyaging and swamped by
this endless void, we seek solace in thinking, reading and
eating. Fred Middleton in 1916, once his stomach settled
down, followed a great Antarctic tradition by writing
a lot in his diary about food. Sailing south with Ernest
Shackleton on the *Aurora*, Fred began each day at 7.30
am with that cup of tea and biscuits in bed; then at 8.15 a
breakfast of steak and onions or fried eggs and bacon, or
grilled steak; 'dinner' at 12.30 of roast beef, boiled mutton,
roast mutton, pork, or *haricot* (any two of these every day),
plus potatoes, beans, and asparagus on some days, followed
by two lots of pudding and fruit (apples and oranges). At 4
pm it was afternoon tea time, and then at 5.30, tea, which
consisted of boiled eggs and some other meat and bread
with jam and cheese. Sitting on the thrumming engine
of the ship, Fred wrote in his diary: 'Now fancy that fare
and we are supposed to be roughing it? Ye Gods, may we
always thus rough it. T'would be a pleasant death to be
thus starved.'

Soon Fred would meet men who had suffered true starvation. The voyage of the *Aurora* in 1916–17 was shadowed by concern about the fate of Shackleton's other party. Fred was in the first party ashore at Cape Royds in McMurdo Sound in January 1917. He was therefore witness to one of the most poignant Antarctic reunions. For two years, the men of Shackleton's Ross Sea party had laid stores and awaited Sir Ernest's arrival by sledge from the south. When Shackleton appeared by ship from the north, his men were, in one moment, rescued and redundant. All their toils and hardship suddenly seemed in vain. Their story became a minor footnote to the drama of survival on the other side of the continent. When Sir Ernest, Fred Middleton and Morton Moyes approached the wild, smelly creatures they took to be men, excitement at first made communication difficult. Shackleton found himself asked the same question he had posed eight months earlier when he had stumbled into Stromness on South Georgia: 'Tell me, when was the war over?' And the answer was still the same: the war, Shackleton told his men, was 'worse than ever'. And the *Endurance* had been crushed. So the war had not ended and the trans-Antarctic sledging expedition had never even started. What kind of limbo had they been inhabiting?

They made heroic journeys south to lay depots for Shackleton. Their ship, the *Aurora*, broke free of its moorings in a blizzard at Cape Evans and was driven northward in pack ice, leaving them marooned. They fought scurvy with meals of seals. They were ravenous for fresh meat and searched the tide cracks for their prey. At their first sight of seals on returning from a long depot-laying journey, it was all they could do not to devour them raw. Three of their party had died, they told Shackleton. Arnold Patrick Spencer-Smith had suffered scurvy on one of the southern depot journeys and had been hauled back on a sledge only to die within a day or two's journey of Hut Point. Their leader, Aeneas Mackintosh, and Victor Hayward had gone missing eight months earlier during a reckless scramble from Hut Point to Cape Evans in the face of an oncoming blizzard. On hearing news of these fatalities, Shackleton, Middleton and Moyes walked solemnly away from the rescued men and lay down on the ice. It was a pre-arranged signal to their anxious shipmates indicating how many men of the Ross Sea party had perished.

'It is a severe blow to EHS,' wrote Fred in his diary, 'particularly after he saved those other 22 men after almost miraculous escapes.' 'He of course is very much disturbed by the death of his three comrades.' The last sledging journeys of the heroic era were made in search of any trace of Mackintosh and Hayward. Shackleton found it hard to let the continent go and to accept that those men were lost. He kept the *Aurora* in the Ross Sea many extra days searching for the two who were missing, in the vain hope that he might rescue them, too, or at least their bodies.

Having Fred's diary with me on this voyage, and reading of his close encounter with human extremity, has helped me to think about survival. As the expedition doctor, Fred examined the seven ravaged men who became his shipmates. He wept at the signatures of their suffering. Generally, he found their physical condition surprisingly good. But what of their mental privations? Hunger has other dimensions, and sustenance comes in many forms.

# FEEDING BODY and SOUL
## Hunger and wonder

When the United States expedition led by Finn Ronne arrived at Stonington Island on the Antarctic Peninsula in 1946 they found – to their surprise – a secret British expedition already ensconced just a few hundred yards away. The British were there to keep an eye on the Argentines, and the Americans were very put out to find them there, so near their old base. The American flag was raised, a British complaint was made, relations became cold and terse, and Commander Ronne forbade fraternisation. But the British afternoon teas and scones proved irresistible to the Americans. Men would announce that they were just going out to check the ice – they may even have said at the door 'I am just going outside. I may be some time' – and they would head down towards the harbour and when out of sight, double back over the hill for a bit of old world civilisation and crumpets.

Antarctic expeditioners, past and present, talk a great deal about food. But early Antarctica was full of men who hadn't cooked much before, working with very limited ingredients. It was not a promising recipe. Remember that Admiral Byrd even forgot his cookbook on his lonely vigil. Even today, the Lonely Planet guide to *Antarctica* warns that 'Gourmet cooking is simply not the emphasis in Antarctic tourism. If five-star dining is critical to you, you'd be better off taking QEII across the Atlantic.'[1] The privations of exploration and starvation, however, meant that rarely have people celebrated food with such an edge of desperation. Cookbooks for temperate lands do not talk about hunger. Unsatisfied hunger brings gravity to the handling of food.

Shackleton's and Scott's men, when out sledging, designed a dinner ritual to ensure that there were no grievances about smaller and larger portions. The cook would divide the meal scrupulously, taking every care to make sure the pannikins of hoosh were as equal as possible, and the heaps of biscuits too. But, even after such attention, perhaps one serving would look bigger, and crumbs could not be divided. When everyone had agreed that the food had been divided fairly, one man would turn his back on the hungry crew. Another would then point to one pannikin or group of biscuits and say 'Whose?', and the man with his back turned would say someone's name. In this way the food was assigned without rancour, nevertheless ensuring that each person somehow felt that it was they who had drawn the smallest share. (On Scott's last expedition, where Naval divisions between officers and men were enforced, the men could not stop themselves from going through the names in order of seniority, which rather undermined the ritual.) But even with the food fairly divided, disaster could strike. In the confines of the tent, eating elbow to elbow, a precious morsel could fall outside the mouth. Apsley Cherry-Garrard wrote that 'We have forgotten – or nearly forgotten – how the loss of a biscuit crumb left a sense of injury which lasted for a week …'[2] But if the crumb could be seen, recalled Shackleton, 'the others would point it out, and the owner would wet his finger in his mouth and pick up the morsel. Not the smallest fragment was allowed to escape.'[3] In a culture of such fastidiousness with food, to drop the serving pot was (in the words of Priestley) 'a catastrophe which ranks in the minds of a sledging party with the fall of empires'.[4]

'We thought of food most of the time', wrote Shackleton of his southern sledging journey in 1908–09.[5] Admiral Richard Byrd considered that 'Nowhere in literature has food been so rapturously beatified as in Shackleton's praise of the ever-thinning "hoosh" that sustained him and his party on their starvation-dash to within 97 miles [156 kilometres] of the Pole.'[6] Byrd was right: Shackleton's pages of his published journal about food, and dreams of food, are exuberant indeed. The men discovered how it feels to be 'intensely, fiercely hungry'. During the last weeks of their journey towards the pole and throughout the long march back, when their allowance of food per man had been reduced from 35 ounces (1 kilogram) a day to 20 (567 grams), they thought of little but food. Hunger propelled them: 'Our food lies ahead, and death stalks us from behind.'[7] The proximity of the pole, the glory of the mountains were as nothing to the next meal, or a vividly imagined meal. Shackleton was forced to reflect upon the danger extreme hunger had for civilisation for he discovered that 'no barrier of law and order would have been allowed to stand between us and any food that had been available'. And they could not joke about food. They could see nothing funny about it, not even their most fanciful dinner dreams, or their arguments about culinary delicacies. It was a subject that demanded due reverence and complete seriousness. There was no smiling to yourself or one another about food.

Once they got down the Beardmore Glacier and on to 'the Barrier', they could more easily march together, and so would take turns in planning and describing the enormous meals that they proposed to have back on the ship or when they returned to civilisation. 'No French chef ever devoted more thought to the invention of new dishes than we did', boasted Shackleton. And what they invented were meals of stupendous fattiness and sweetness. 'Have you ever had a craving for sugar which never leaves you, even when asleep?' asked Cherry-Garrard on Scott's expedition. 'It is unpleasant.'[8] The acknowledged height of gastronomic luxury amongst Shackleton's returning party was dreamed up by Frank Wild and became known at over 80 degrees south as the 'Wild roll'. *Take a supply of well-seasoned minced meat, wrap it in rashers of fat bacon, and place around the whole an outer covering of rich pastry so that it takes the form of a big sausage roll. Now fry it in plenty of fat.*[9]

And the perfect day, imagined after a breakfast of half a pannikin of semi-raw horse-meat, one biscuit, and half a pannikin of tea, was as follows:

> Now we are on board ship. We wake up in a bunk, and the first thing we do is stretch out our hands to the side of the bunk and get some chocolate, some Garibaldi biscuits and some apples. We eat those in the bunk, and then we get up for breakfast. Breakfast will be at eight o'clock, and we will have porridge, fish, bacon and eggs, cold ham, plum pudding, sweets, fresh

roll and butter, marmalade and coffee. At eleven o'clock we will have hot cocoa, open jam tarts, fried cods' roe and slices of heavy plum cake. That will be all until lunch at one o'clock. For lunch we will have Wild roll, shepherd's pie, fresh soda-bread, hot milk, treacle pudding, nuts, raisins, and cake. After that we will turn in for a sleep, and we will be called at 3.45, when we will reach out again from the bunks and have dough-nuts and sweets. We will get up then and have big cups of hot tea and fresh cake and chocolate creams. Dinner will be at six, and we will have thick soup, roast beef and Yorkshire pudding, cauliflower, peas, asparagus, plum pudding, fruit, apple-pie with thick cream, scones and butter, port wine, nuts, and almonds and raisins. Then at midnight we will have a really big meal, just before we go to bed. There will be melon, grilled trout and butter-sauce, roast chicken with plenty of livers, a proper salad with eggs and very thick dressing, green peas and new potatoes, a saddle of mutton, fried suet pudding, peaches *á la Melba*, egg curry, plum pudding and sauce, Welsh rarebit, Queen's pudding, angels on horseback, cream cheese and celery, fruit, nuts, port wine, milk, and cocoa. Then we will go to bed and sleep till breakfast time. We will have chocolate and biscuits under our pillows, and if we want anything to eat in the night we will just have to get it.[10]

One point on which they were all agreed was that they did not want any jellies or 'elusive stuff' like that. And that nothing about food was funny.

It is time we talked about *pemmican*. It was a food of Arctic indige-nous peoples whose hunting cultures were based on an almost pure meat and fat diet. Large pieces of fresh meat were cut into long, very thin strips and laid across frames above smoky fires. When the strips were completely dried and smoked, this concentrated protein was spread and eaten with fat or lard. The pemmican that Antarctic explorers came to know was finely ground dried beef with 60 per cent added beef fats and a little seasoning, and it was packaged in cans or square cakes. It was greasy and rich, and valued for its compressed nourishment. Its water content was less than three per cent, and one pound was supposed to be the equivalent in food value of six or seven pounds of raw beef.[11] The standard sledging ration in Antarctica for the first half of the twentieth century consisted of pemmican, biscuits, butter, cocoa, sugar, tea and powdered milk, with various optional additions such as oatmeal, chocolate, cheese and raisins. There could be unexpected bonuses. Raymond Priestley remembered preparing his team's sledging provisions by opening hermetically sealed tins of pemmican, and by the time he had finished the job his hands were so badly cut that his own blood had thoroughly mixed with the pemmican, making the mixture 'more nutritious'.[12] *Hoosh* was the hot, thick soup that constituted the standard sledging meal. It was made of crushed pemmican boiled up with water, thickened with biscuit or oatmeal, and perhaps garnished with curry powder or other flavourings. As they ate with disciplined slowness in their tents, explorers watched one another closely and felt a distinct grievance if one managed to make his hoosh last longer than the rest. When Shackleton pressed an extra biscuit on a sick and starving Frank Wild during their Farthest South journey, he won his love. 'I do not suppose that anyone else in the world can thoroughly realise how much generosity and sympathy was shown by this; I DO by GOD I shall never forget it', wrote Wild in his sledging diary. 'Thousands of pounds would not have bought that one biscuit.'[13] Scott's northern party of six marooned men cast lots for the crumbs at the bottom of a biscuit tin, 'a mixture which consisted of about equal parts of biscuit, granite, paraffin, snow, ice, and frozen meat'.[14]

The biscuits were an art form, and eating them even more so. They were no ordinary biscuit. Consistency was more important than taste. Portability was more important than appearance. Durability was more important than your teeth. They had to be made for rough handling, for broken biscuits were unwelcome and – as we have seen – unattributed crumbs might be a cause of dissension. They were *plasmon biscuits*, a trade name for the soluble milk protein with which they were baked, ensuring their added food value and also their extreme hardness. Sometimes they had to be broken with a geological hammer (just as knives were sometimes sharpened with geological specimens). Cherry-Garrard swore by the 'Huntley & Palmers' biscuits of Scott's expedition: 'their composition', he confides to us, 'was worked out by Wilson and that firm's chemist, and is a secret'.[15] In some circumstances, an overbaked biscuit might be 'a luxury we all looked forward to' because they were crisp and easy to break. But in conditions of extreme privation, a box of overbaked biscuits could cast gloom over a whole party for weeks, for they were too thin and without substance. An underdone biscuit lingered longer in the mouth. There were great debates about texture and consistency. Was it better to dunk a thin, overbaked biscuit to give it more bulk? But, if you chose that strategy, at what point did the biscuit's growing sogginess allow it to be too easily and quickly eaten? It was not unlike that other dinner-time argument about whether one should stand one's pannikin of hoosh in the snow so that cooling after cooking would make it thicker and thus more filling, or whether the heat itself was more critical? As Shackleton's southern party stumbled home in early 1909, they came across the recent tracks of another party. They knew the ship was in because they found discarded cans bearing different brands from the original stores, and they scoured the ice at the other party's noon camp and triumphantly found three small bits of chocolate and a little bit of biscuit. They 'turned backs' for them, and Shackleton was unlucky enough to get the biscuit. A 'curious unreasoning anger took possession of me', he wrote. 'It shows how primitive we have become, and how much the question of even a morsel of food affects our judgment.'[16]

When Cherry-Garrard moved tents during one sledging journey, he found himself in a different kitchen. It was the same ration, the same ingredients, the same stove, but a different culinary culture. It was Scott's and Wilson's tent that he had joined and he immediately felt embraced by the comfort of a careful routine. 'I was hungry', remembered Cherry-Garrard, 'and said so. "Bad cooking" said Wilson shortly; and so it was. For in two or three days the sharpest edge was off my hunger.'[17] Wilson and Scott had learned to be inventive within the constraints of the weekly ration and ensured that the meals varied in small and satisfying ways. One evening you might have the pemmican plain, and the next a thicker pemmican with arrowroot mixed with it. Perhaps you might persuade your sleeping companions to surrender a biscuit and a half apiece for the public good, and thereby concoct a *dry hoosh*, which was biscuit fried in pemmican with only a little water added, followed by a big cup of cocoa. Or you could follow it instead with *teaco*, a combination of tea and cocoa, of stimulant and substance. And it was amazing what you could do with raisins! A dessert-spoonful was your daily ration and they were good soaked in tea and a delight in a dry hoosh.

All these dried and preserved foods were essential for travelling light on journeys inland, but fresh food was crucial for long-term health, especially in avoiding that dread disease of voyagers on sea and ice, scurvy. By the coast in summer there were seals and penguins galore, providentially attracted to the same sites as humans, but one had to stock up for winter, unless you happened to live near an Emperor colony. Polar explorers had to overcome some prejudices against eating seals and penguins. De Gerlache, captain of the *Belgica*, regarded the serving of freshly hunted meat as an insult to his provisioning, and severe scurvy broke out. Frederick Cook had to urge his companions on the *Belgica* to eat penguin, and it was not until past midwinter that they learned to like and crave it.[18] They certainly detested the endless canned food: 'How we longed to use our teeth!' They would have welcomed even pebbles or sand with their food, just for some roughage. Food was one of the sources of unhappiness on the *Belgica*: 'Food now a cause of general dislike', noted Cook, 'the entire food store is used as a subject for bitter sarcasm'.[19] Other expeditions had no such inhibitions about eating the locals, especially when little else was available: Carl Larsen and his stranded crew on Paulet Island survived the winter of 1902 by eating 1100 Adélie penguins.[20] The men of John Rymill's poverty stricken but highly effective British Graham Land Expedition of the Depression era (1934–37) were seal-eaters for financial reasons, and killed 550 Weddell and crabeater seals for food.[21] Scott's northern party, wintering in an igloo, were forced to develop a taste for seal blubber as well as meat. Their staple was seal hoosh, and on Sundays it was made with seal liver, heart and kidneys. But the consummate culinary delicacy of their winter was seal's brain. *Walk down to the icefoot. With your ice-axe, cut the top off the head of a butchered seal (which you have already laid down for winter). Chip out the frozen brain with a chisel. Collect the fragments in a tin and add to a hoosh.*

Shackleton's Ross Sea party also gorged themselves on seals. 'Richie' Richards recalled that 'The effect of seal meat is simply miraculous. That is all we had at Hut Point because there was virtually nothing else there.'[22] At Hut Point today, you can still find an unfinished meal of seal meat in a pan. Roald Amundsen, haunted by the experience of the *Belgica* and remembering Frederick Cook's successful treatment of the disease, was determined to feed seal and penguin to his own expeditioners twice a day, fresh and undercooked, as well as regular doses of cloudberries, a known antiscorbutic.

What was the role of diet in the race to the pole? Scott's men ate white bread; Amundsen's brown. Scott had regulation pemmican; Amundsen made his own with added vegetables and oatmeal. Scott ran out of food; Amundsen brought back provisions from the polar journey as souvenirs for his suppliers. There has been a long controversy over whether Scott's polar party died because they were weakened by scurvy, and whether that was due to reluctance to eat the available wildlife. Amundsen, when he heard of Scott's death, commented that it was 'heartbreaking to picture these tempest-tossed, disease-eaten, frozen, starving men encamping in that final blizzard to die', and Shackleton and several other polar explorers agreed that the 'disease' was scurvy. Scott's men had certainly suffered from scurvy during the *Discovery* expedition, and in the years immediately following Scott's death, scientific breakthroughs were made in understanding the role of vitamins in diet and the importance of Vitamin C in combating scurvy, and these new insights may have strengthened the retrospective diagnosis of the polar party's ills. But there is no conclusive evidence from the diaries of the polar party that they suffered from scurvy, and Dr Edward Atkinson's examination of the bodies in the tent in November 1912 found no signs of it. The British Navy since the days of James Cook had fought scurvy with fresh food and citrus fruit, and Scott observed these traditions. But it is clear that Scott and Atkinson were also persuaded by new medical theories that scurvy primarily had other causes: 'I understand that scurvy is now believed to be ptomaine poisoning', wrote Scott in 1904. According to this view, the disease was caused by eating tainted food or decaying meat and was exacerbated by a general lack of hygiene. Atkinson lectured on these new theories of scurvy in the Cape Evans hut in the winter of 1911. Such understandings may have diminished the ardour of Scott's men as hunters and made them more fastidious and less adventurous about the unfamiliar, undercooked food their bodies craved.[23]

Mid-winter dinners were – and still are – grand affairs, and generally featured a mixture of local and imported delicacies. Fricassee of penguin might be one of the highlights, as an entrée before a roast sirloin or chicken. Champagne and fine wines made an appearance at the table, as they did on other special occasions. French hospitality at Kerguelen proved so attractive to Australians returning from Mawson in 1954 that Phillip Law had to muster them back on board the ship. But he and a few others were able to return the following day for a two hour lunch accompanied by a Montrachet 1948, a Red Burgundy 1954 and a Chateauneuf-du-Pape 1943.[24]

If you are caught in extreme privation, completely out of tobacco and yearning for a smoke, try Frank Wild's famous Hut Point Mixture: *Mix equal parts of tea, coffee and sawdust. Flavour with herbs.* Shackleton's men on Elephant Island, awaiting rescue in 1916, smoked the feathers of chinstrap penguins, and one man smoked the wood of one pipe in the bowl of another. When Shackleton returned to rescue his men in the Chilean Navy cutter, *Yelcho*, he threw bags of tobacco ashore before he landed.[25]

The director of the Australian Antarctic Division in the post-war years, Phillip Law, acknowledged the importance of food for morale. 'Good eating is certainly one of the greatest single pleasures that an expedition can afford', he declared. He argued that, whereas 'a man in the field' expects it to be tough and will tolerate and even enjoy a basic 'hard' ration, 'a man at base' is 'living practically a normal routine life for twelve months' and 'is entitled to the best food that we can give him'. He found that, at first, many men at Heard Island and Macquarie Island would not eat the local game. 'The reason', he reported, 'mostly was sheer prejudice.' So he 'indoctrinated' them before their departure, 'describing how luscious were such items as roast penguin breast or fried seal's liver or crumbed brains'. He could have mentioned that Douglas Mawson was known for doing a delicious crumbed seal in its own blood.[26] Law hinted that any man who bridled at eating such things 'was "a bit of a sissy" and not a worthy expedition member'. But the men were perhaps more adventurous (and 'manly') than he knew: the Heard Island station log recorded meals of braised penguin, penguin steaks 'better than rump steak', elephant seal brain fried in bread crumbs and described as 'very tasty', and elephant seal liver and kidney.[27] Law reported that men in Antarctica tended to put on weight rather than lose it. And if there were any emotional or social problems, they registered them physically in the gut: 'when personal relations between the members of a party are not harmonious the digestive organs are generally the first to register the fact', wrote Law in the *Medical Journal of Australia*. 'Indigestion, gall bladder trouble, appendicitis, neurotic stomach troubles – these are the principal ills of an antarctic party.'[28] Certainly, Frederick Cook noticed a craving for fat and much noisy indigestion among his shipmates during the *Belgica*'s unhappy winter.[29]

Robert Cushman Murphy surprised himself by developing a taste for humpback whale meat even though sickened by the slaughter of his whaling brig. While at South Georgia, Captain Cleveland of the *Daisy* ordered his steward to prepare a supply of *muctuc* or pickled whale, an Arctic delicacy which the skipper said would be prized by the gods on Mount Olympus:[30] *Take one whale. Remove a chunk of blubber from it.*

*Slice it neatly into black squares of one inch thickness. Each of these cubes should be half ebony and half ivory, because the black skin of the whale is about a half inch thick and the white blubber is beneath it. Boil the cubes in a large pot of sea water until a fork will easily penetrate them. Ensure that they are tender but still firm. Drain the cubes, pack them in jars filled with vinegar and cap them. Store for a month and put them on a shelf where you can admire them while you await a gastronomic delight.*

On an Antarctic base, laboratory and kitchen were sometimes indistinguishable. On Commander Ronne's expedition to Stonington Island in 1946–48, the chief commissary steward kept chickens to supply the expedition with fresh eggs. But he also had an experiment underway. He had set aside 24 eggs in the hope of hatching them artificially. But one evening he appeared at dinner and announced with disgust: 'Somebody ate my scientific program!' The eggs had been enjoyed by the men of the twelve-to-four watch, henceforth known as the Egg Watch.[31] On Australian Antarctic bases today, eggs are stored at about 4 degrees Celsius (39 degrees Fahrenheit), which keeps them edible for eight months. The shells are oiled to prevent the air penetrating them and to slow the aging process, and they are turned weekly to stop the yolk coming in contact with the membrane or shell.[32] If the eggs were not oiled, it was worthy of comment in the station leader's annual report.[33]

Penguin and skua eggs, sacrosanct in modern Antarctica, were once welcome alternatives. Xavier Mertz had made himself a bit of a reputation in the hut at Commonwealth Bay for the quality of his penguin-egg omelettes, and when Mawson was trying to nurse him home along their tragic journey, they would remember Mertz's omelettes fondly and look forward to the day when he could again wield the frypan. Mertz promised to cook his friend a special omelette on their return, and explained the unusual flourish that would make it unique: 'I will name it Omelette Mawson', he said in his sleeping bag on the glacier. When Mawson dragged himself home without Mertz and began to settle into another year of isolation, he cooked a solemn meal for his companions in the hut. *Break a dozen penguin eggs in a bowl. Set three pans on the stove and melt butter in them. Beat the eggs into a foam. Now, in honour of Mertz, add a quarter of a bottle of Scotch whisky. Cook with reverence.*[34]

The first *ANARE Operations Manual* (1947) recommended choosing a station site close to a penguin rookery 'from the point of view of fresh meat'. The *Leaders' Manual* continued as late as 1974 to recommend supplementing food supplies with native fauna, and we know that the *Field Manual* given to Stephen Murray-Smith in 1985 explained how to sit on a penguin if, in an emergency, you needed a 'palatable' stew.[35] Fresh food is still a joyous treat in Antarctica, but it can no longer include the local wildlife. At Novolazarevskaya, every bedroom has tomatoes and cucumbers growing in boxes of soil against any window.[36] At 'Blue 1', an airport camp in remote Antarctica, boxes of eggs are welcomed with excitement and freshly delivered milk is 'held aloft like a world heavyweight trophy'. At McMurdo station, the resident greenhouse technician is also the unofficial morale officer. Using hydroponic techniques, this Antarctic gardener grows about 1600 kilograms (3527 pounds) each year of cucumbers, peppers, tomatoes, limes, and lots of lettuce under thousands of watts of artificial light. The 198 square metres (236 square yards) of growing space also attracts expeditioners who are starved of the colour green: several hammocks occupy a corner of the lettuce room.[37]

Frozen food can also occasionally deliver delight. When Australian traversers plotting the thickness of the ice sheet arrived at the uninhabited Russian Vostok station in late 1962, after covering 1500 kilometres (932 miles) and climbing 3500 metres (11 483 feet) in tractors, they pushed aside polar bear skins hanging in the doorway of the main living quarters and found a big pan of frozen steak and onions on the stove and table places laid out for three people. They surmised that the previous occupants had suddenly been summoned to a plane about seven months earlier. So the Australians heated up the stove and finished the meal.[38]

Supplying a station with water remains a constant demand in this ocean of ice. A plumber who worked at Davis station in 1987 spent four hours of every day melting ice for water, and sometimes a double shift of eight hours. In summer, he had to drive the tractor a kilometre or more beyond the station to reach clean ice (sometimes doing three trips a day) and then brought it back and loaded it into the melt tank, which was equipped with a heating coil. The next day he pumped the melt to the water supply tank. Showers were rationed at twice a week, a maximum of seven minutes or so each.[39]

Tea and cocoa were the staple sledging drinks, and home brew is a creative and essential tradition at Australian and other bases. At Mawson in the 1950s, it was said that the Ginger Beer Plant stalked the station at night, undoing the nightwatchman's work. Anything that went wrong could be blamed on it. Members of Mawson's 1911–14 expedition blessed their leader for providing absolute alcohol for the primus instead of methylated spirit, for it undoubtedly enhanced the tea and cocoa.[40] At Stonington Island during the 1940–41 US expedition, little alcohol had been brought south, so men worked their magic with dried fruit fermented with baker's yeast, and on special occasions, the biologist would ration out the alcohol from his bottles of preserved specimens. Thus drinking became known as 'draining a fish'.[41] It is hard to beat sipping an Antarctic G and T while the sun fails to set. *Put a liberal splash of gin in a glass or mug. Add tonic water delivered on the latest plane or ship. Finally, slip into it a piece of historic ice quarried from the glacier.* It will pop and spit with the release of ancient air bubbles from the ice, the very bubbles that your scientific colleagues might have studied to understand past climates.

And what of the giant tub of vanilla ice cream I was asked to carry, with some urgency, down Main Street, Casey station? It had just arrived on our ship, and the container in which it sat had been unloaded and driven to the Green Shed where we had unpacked it. Now it was needed instantly in the station kitchen to feed the summer multitudes. It was a luxury item and the station must have run out of it. I was carrying the long-awaited first ice cream of the new Antarctic year. There is a rich human history of ice harvesting, of gathering ice from distant mountains for valley communities to preserve food, cool drinks and treat the sick. In sixteenth-century Florence, elegant living required a private snow store in the grounds of your villa, and even peasants wanted snow as much as bread and wine. Snow was so important to the Sicilians that soldiers sent up Mount Etna to seize it were invested with the powers of life and death. In the seventeenth century, Italy entered an Ice Age – not the related Little Ice Age of lower temperatures which still prevailed in Europe – but an era of snow wells and ice cellars, where ice was an accoutrement of civilisation. The ice trade was a lucrative business and carrying the goods from mountain to Mediterranean was perilous. It was not done by donkey, you can be sure, but by fast horses. I had now joined a cavalcade of ice couriers across time, people performing 'the trick of calling back winter in midsummer'. In the nineteenth century, London confectioners had sent ships to collect from icebergs in Greenland's seas. In the twentieth century, Australians had dreamed of wrapping and towing southern icebergs to the edge of their wide, brown land and using them as a precious water source. The Antarctic mineral that might become most sought after in the future is ice itself, and Antarctica's greatest gift to the world may yet prove to be fresh water; it is also its greatest threat. But here I was bringing ice cream to the ice. There is a well documented taste for ice cream in cold countries; the Russians and Scandinavians had good reason to develop sophisticated tastes in cold foods because natural cold storage was at hand. Ice cream is paradoxically a product not of the manipulation of cold, but of heat. And the modern world's hunger for manufactured ice and the use of chlorofluorocarbons in refrigeration have helped produce the ozone hole over Antarctica. Eating ice cream in Antarctica is both reassuringly normal and an extraordinary luxury. Taming the ice, tubbing it, making it sweet: these are triumphs of technology and ingenuity and are the mark of civilisation amidst the roaring elements. And eating ice cream offers a curious intimacy with the substance of the place. Perhaps it is the oral equivalent of the Russians rolling in it.[42]

Frederick Cook noticed that people see in icebergs what they are looking for. The captain of his ship saw a beautiful woman; the naturalist saw a polar bear; the cook a pot of boiling soup.[43] (Stephen Murray-Smith saw 'small islands' and 'rococo delights for an eighteenth-century gentle-man'; I saw meringue.) A pot of soup, already at the boil, may well be the dream of a busy Antarctic cook. A cook is famously the most important member of any remote expedition, and can make or break the experience. At one Australian station, it was said that the plumber volunteered to do the bread baking because it gave him a chance to get his hands clean. He attributed the special flavour of his bread to his custom of wearing his 'lucky' sewerage overalls as he baked.[44] At the end of one year, Law received a requisitioning cable that read: 'Send one good cook, pack well in cotton wool.'[45] The cook at the Russian base of Novolazarevskaya in 2000–01 was a criminal from Moscow, sent south as an alternative to prison back in Russia. Visitors were warned: 'You should be a bit careful of him. He is no good.'[46] Alexa Thomson from Sydney was employed in 2000–01 by a private company, 'Ends of the Earth', to cook at their isolated summer camp of tents beside a blue ice runway in Dronning Maud Land. For most of the year there was just a bamboo pole marking the buried stores, but in summer it sprouted into an international airport mostly used by government scientists of various nationalities. With an English literature degree, some professional cooking experience and an assortment of pots and pans, Thomson flew from Cape Town and arrived in Antarctica in time to cook dinner. She stepped off the aging Russian aircraft into scentless air and a silent void. Her kitchen was a tent with no electricity, no running water and a gas oven with a broken door. Her freezer was 'the approximate size of the United States', and more specifi-cally, a 3-metre (10-foot) pit outside the dinner tent. The camp reservoir, located well away from the yellow ice, was an area marked off by Tibetan prayer flags from which ice was dug and daily melted.

But Thomson wanted 'to cook not add water to dried food'. There was meat of indeterminate age in the natural freezer and she had to manage its defrosting with exact timing; she preferred not to have to ask 'would you like your pasta with or without botulism?' Fortu-nately, living at an ephemeral airport had its advantages, and fresh food arrived occasionally by plane. She conducted a bake-off between bases. The 'neighbouring' Russians (160 kilometres (99 miles) away) baked bread for them and Thomson responded with the mother of all chocolate cakes. The experience of a full blizzard in a camp of tents was so unsettling and frightening that cooking a roast dinner in the white-out was the only way she could hold onto her sanity. Thomson relished the routines, sociability and challenges of cooking on the ice, although she sometimes tired of 'running around after the boys, fretting about their stomachs'. She found that gazing from her kitchen towards distant ice peaks was a wonderful inspiration to the beating of egg whites. *Chop two fresh mangoes just flown in from South Africa, and marinate in Cointreau for an hour. Make a roux of butter, almond meal and sugar, then fold into the mixture some of the beaten egg whites which you have whisked into soft, snowy peaks while contemplating the horizon. Add the egg yolks to the marinated mango, and fold the rest of the egg whites into the mixture. Line a soufflé dish (which you have carefully brought south) with the roux. Run your thumb around the side of the dish and put it into the oven. Slam the partly-broken door shut with your knee. Feel smug when you serve soufflé in Antarctica.*

Bernadette Hince, an Australian lexicologist, high-latitude historian and renowned cook, produced *The Antarctic Dictionary* as a homage to the ice (where she once worked as a seal researcher) and has now written a history of the sub-Antarctic islands.[47] For a cook keen on cold words from a nation that boasts pavlova as a national dish, 'The Big Pav' is an irresistible term for Antarctica, but a lexicologist needs it in print to put it in a dictionary. So here it is. And here also is her mother's recipe for an 'economical pavlova' in honour of the extravagant one at the base of the world. *Mix one and a half cups of sugar and one teaspoon each of cornflour, vinegar and vanilla essence, four tablespoons of boiling water and two egg whites. Using an electric mixer (unless you are in your tent on blue ice), beat until stiff (the mixture, not you). Shape into a round on an oven tray lined with baking paper. Cook at 180 degrees Celsius for 30 minutes and 120 degrees Celsius for a further 30. Carve two embayments from the cooked meringue and piece them together into an Antarctic*

*Peninsula of a tail. Mould the fragments into one ice cap with the help of whipped cream. Plant a little flag in your chosen sector.*

When Thomson lifted her eyes to the Antarctic horizon as she beat egg whites, she reflected on the metaphysics of the place. With her days tethered to the material and corporeal dimensions of life in Antarctica, she discovered that she inhabited a landscape of the mind. The savage, elemental supremacy of the continent was such that she found herself both powerless and free. She was giddy with a sense of infinity. 'This is an unfettered environment with little boundary between the mental and the physical', she wrote.

Just how porous is that boundary between mind and body is made clear by the history of dreams in Antarctica. The British explorer, Sir Ranulph Fiennes, completing the first unassisted crossing of the Antarctic continent with Mike Stroud in 1992, reported that he never thought of sex during polar journeys. He believed that this was due to the temporary withdrawal of testosterone in the absence of any prospect of activity. 'Food took the place of sex', he decided, 'anticipated with salivating eagerness and savoured to the last lick.' However the great German mountaineer and lone conqueror of Everest, Reinhold Messner, found that his own Antarctic manhaul in 1989–90 exposed him to constant erotic day-dreaming. 'Time and again', he confessed, 'I encountered figures of women, living, obstinate, wanton women. I pictured them naked and sometimes as my partner.' Fiennes dismissed Messner's response as the result of too many years of high altitude climbing having mucked up his testosterone.[48]

We have heard how Shackleton and his men, wracked with hunger on their southern journey of 1909, daydreamed constantly of sumptuous meals. They were certainly lascivious. And at night, the mind elaborated its tricks: 'We all have tragic dreams of getting food to eat, but rarely have the satisfaction of dreaming that we are actually eating.' If they did get to eat the dream food, it was worth a diary entry: 'Last night I did taste bread and butter.'[49] The Swedish men stranded on Paulet Island in 1903 experienced the same dreams: 'Why, we could dream through a whole dinner, from the soup to the dessert, and wake to be cruelly disappointed.'[50] Mawson, when struggling homewards with Mertz after the loss of Belgrave Ninnis, remembered lying in his sleeping bag and dozing off 'into the land of food once more'. This time he found himself in a confectioner's shop, grandiose and opulent. None other than the proprietor courteously led him up a winding stair-case to the roof where Mawson was amazed to find two long rows of gigantic cakes, each about four feet in diameter. Blissful, he ordered one, as that was all he could hope to carry, and hurried downstairs to pay for it. Some time later, as he walked down the street, he realised that he had forgotten to pick up the cake. When he returned in haste, the shop was shut and a placard hung on the door announcing 'Early Closing'. This was the pattern of the dreams. Each man sat down at night to a wondrous meal in a seductive restaurant only to find that something happened to prevent the eating of it. On very rare and wonderful occasions, the dream culminated in actual eating and even, Mawson reports, in a feeling of repletion, but such experiences were dangerous to share. The telling of such a dream to one's companion 'invariably raised some feeling of disappointment in the other fellow who had not been so fortunate'.[51]

The most extraordinary case of food dreaming was reported by Raymond Priestley from his ice cave during the winter of 1912. Scott's marooned northern party of six men wintered for six months on less than half-rations of monotonous food. They had a constant craving for food that was physically painful. Their dreams were about three things: food, relief and disaster to their companions. They experienced the same food dream, the one about a long walk to the shop behind the snowdrift only to find it was a day of early closing. All of Priestley's dreams were about *almost* getting food. Three of his companions experienced this same nightly temptation and frustration. But two of their number had quite different midnight experiences. They actually got to eat their dream meals! They sat down to imaginary feasts every night and polished them off. These two men, it was observed, were much more lethargic during the day than the other four. Their physical state, remarkably, seemed linked to their midnight satisfactions. 'This fact', reported Priestley, 'was in danger of becoming a grievance with the rest of us'. There was an irresistible feeling that these two, by partaking of these dream meals, somehow had an unfair advantage. The remaining four seriously wondered about reducing their rations accordingly.

Rationing food was not the only challenge. What book do you take on a sledging journey? You had to make a judgment about its weight, both physical and spiritual. Is it light enough to carry but heavy enough to sustain you? Will it offer you 'compressed nourishment' like pemmican? In the words of Francis Bacon, 'Some books are to be tasted, others to be swallowed, and some few to be chewed and digested.'[52]

When Laurence McKinley Gould and his companions set off from Little America on their geological sledging journey in 1929, each man was allowed one book. Ordinarily Gould would have chosen Robert Browning, but he had enjoyed a solid diet of this poet and playwright through the winter. So he packed a thin paper edition of William Shakespeare, complete in one volume. The tragedy of King Lear seemed all the more powerful, reading it out there in the midst of the Ross Ice Shelf. He read the whole volume, regaling himself with a daily play. On this dedicated scientific expedition, almost half their waking hours were spent reading, for they had five to seven hours each day with their chosen books 'without any kind of interruption'. With Shakespeare finished, every last word savoured and digested, it was time for Gould to work his way through his companions' literary choices. Mike Thorne had brought WH Hudson's *Purple Land*, and although Gould had read it just a few weeks earlier, he was delighted to feast upon it again. Eddie Goodale had brought a volume of English poetry which Gould also 'devoured'. He was especially thrilled to discover that Browning had come with him on the sledging journey after all, for the volume contained Browning's 'The Bishop Orders His Tomb', one of Gould's favourites. He loved to roll the poet's words on his tongue. John O'Brien had brought a thick volume of short stories by HG Wells. Gould read this, too, because literary provisions were low and there wasn't much choice, but he would gladly have dropped it down the deepest crevasse he could find. But O'Brien wouldn't let him. He had borrowed the book from Russell Owen and had promised to return it.[53]

Shackleton was also a fan of Browning, loved reading and could wax lyrical about books as he did about food. On his sledging journey south with Scott and Wilson in 1902–03, he read Charles Darwin's *The Origin of Species* to the others in the tent.[54] On his own expedition's southern journey six years later, it was a sad day when, approaching his farthest south, he reluctantly shed weight and buried books in the snow. 'We have nothing to read now, having depoted our little books to save weight, and it is dreary work lying in the tent with nothing to read, and too cold to write much in the diary', he wrote on 8 January 1909.[55] Readings of Charles Dickens were by far the most popular pastime in Shackleton's winter hut of 1908. Professor Edgeworth David would read aloud for hours after supper. Sometimes these midnight readings were only stopped by a firm reminder from Shackleton, called out from his cabin in the corner, 'that it was after one o'clock and time all "good" explorers were in bed'.[56]

On Mawson's voyage south in 1911, John King Davis recorded in the log of the *Aurora* that:

> We got up some cases out of the hold today ... The mate who was with me was delighted at the discovery of a case of books which he remarked he would rather have found than the case of bacon he was supposed to be looking for. I asked him if he did not consider it a reflection on him that as chief officer he should not have some idea of where the stores were stowed and

told him had I been chief officer I should have taken very little time for reading until my job was running properly.[57]

But the very next day, Davis recorded in the same log: 'I have been reading all day Mahon's *Life of Nelson* a delightful book which makes one understand what a great man Nelson was and appreciate him as an example.' Davis, resentful of the half-hearted support that Mawson's expedition had received from government, drew strength from the way that Nelson dealt with the blundering of the Admiralty.[58]

The history of reading is an intriguing subject and deserves a polar chapter.[59] In 1957, John Béchervaise believed that 'The quality and quantity of a man's reading is not a bad criterion, amongst others, of his suitability' for an Antarctic expedition.[60] In Antarctica, books got blubbery with beloved use. We must wonder how much greater is the power of the word, the written text, at the end of the Earth, in an abiotic land with the longest night, a world of black figures on a white background? Books, libraries, letters, notes and emails assume disproportionate influence in Antarctic history. Words stood out starkly against the snow. Polar expeditions took vast libraries, not only as a source of vital information, but also as a kind of insulation against the elements. The English were renowned for their ability to take their home with them wherever they went. Their imported literariness may have aided their imperviousness to local influences. Sir John Franklin, for example, did not know that the Arctic was rich in food and would not learn from 'savages'. But, as one historian reflects, 'there's something endearing, as well as exasperating, about a commander who set sail with 1700 books aboard one ship, and 1200 on the other, but with no furs for the men, only regulation wool, which would freeze to their backs once they began to sweat'.[61] Did Scott spend too much time with written words, or too little? Again, his experience and fate provide the archetypal grounds for debate. In some ways, he knew too little about Antarctica and its literature before leading expeditions there; in other ways, he sought too much refuge in his polar study, where the expedition library was kept. And, when he was in trouble, he would write. He was sometimes more sensitive to the distant, delayed audience for his words than he was to the breathing men next door.

Some very fine libraries were taken south. Richard Byrd boasted that his first expedition to Little America possessed 'one of the best polar libraries in the world'. It contained some 3000 volumes and provided 'the most important single source of recreation'. The books had been collected according to Commander Byrd's prescription: 'Dickens, detective stories, and philosophy.'[62] Detective stories were the most popular nourishment, but accounts of polar expeditions came a close second. The most widely read single book was WH Hudson's *Green Mansions*, a story set in the lush shadows of the Venezuelan rainforest. Green is the colour people most miss in Antarctica and 'is the colour of dreams'. By the end of the winter, the books were worn, shabby and discoloured, although the complete set of Rudyard Kipling's works was scarcely touched by the Americans. Scott's men would have been amazed, for Kipling was a great favourite amongst them. 'I can say quite truthfully', Cherry-Garrard later wrote to Kipling, 'that there were no books which we had which were so much used, gave so much food for conversation or more enjoyment'.[63]

'Probably the wisest thing we did, when we went South', reflected Richard Byrd of his first expedition, 'was to bring a set of the *Encyclopaedia Britannica*, the *World's Almanac* and *Who's Who*. These repositories of essential information were a godsend.' They were essential for settling arguments. One man at Little America determined to spend his winter learning aerial surveying and navigation, and reading the *Encyclopaedia Britannica* right through. But he only got as far as Ammonium tetrachloride before throwing Volume One aside.[64]

The stranded Ross Sea party of Shackleton's *Endurance* expedition also became dependent on the *Britannica*. They might not have taken all the food and clothes they needed ashore before the ship blew away, but at least they had a good book. When the stranded men were rescued by the *Aurora* in January 1917, Alexander Stevens wrote the following letter to Captain Davis:

When I begged from you – quite shamelessly – a set of the Encyclopaedia Britannica for the Ross Sea Party of Sir Ernest Shackleton's Expedition, I knew it would be invaluable in filling every gap in our library, but never dreamed it would be almost our sole literary possession for more than two years. It was fortunate that when the shore party was established at Cape Evans we made sure of having the Encyclopaedia ashore.

There was not a man of the party who did not make frequent and eager use of the volumes. Information was sought on every conceivable subject from costume and cookery to philosophy and finance. We appealed to it for everyday information, and for settlement of our interminable discussions. Indeed, it was our only text-book on the Antarctic itself.

Now nothing pleases us more than that the book has returned to you with all the blubbery traces of Antarctic study, for we know that its value to you is measured by its interest and value to us for twenty six months.[65]

Mawson's expedition of 1911–14 was well supplied with books, and his own favourite writer was Robert Service, known as 'the Canadian Kipling' for his stories and poems about the Yukon and Klondike gold rushes. (John Rich, himself originally Canadian, perpetuated the tradition of reading Service at the Casey mid-winter celebration of 2002.) Brigid Hains has described the reading of Mawson's men and their liking for romantic literature of the masculine, vitalist frontier, which helped them to place Antarctica and themselves in a frontier tradition. Mawson read Service aloud to his men after dinner. Books were their 'food for conversation', as Cherry-Garrard would have put it. Morton Moyes, a voracious reader at Mawson's western base, recorded his reading in his diary and wrote that 'Literature is absolutely at zero, if any man mentions a name all the others can fill in the story in detail even to the sentences.' One day at the main base, John Close was nominated to return to the hut to prepare lunch for the others. But when the rest of the party returned, instead of the anticipated lunch they found Close lying in his bunk reading *The Strenuous Life*, Teddy Roosevelt's book venerating 'the life of toil and effort' over 'the life of slothful ease'.[66]

Almost 20 years later, on the BANZARE voyages, Mawson set off in the *Discovery* with only 100 books because 'space was at a premium'. Although the *Discovery* was not wintering, it was symbolic of the difference between the expeditions, one romantic and idealistic, the other dogged by pragmatism. The Melbourne *Herald* was certainly doubtful that such a small collection would 'supply the intellectual cravings of 26 men'. But what were those books? First on the list, of course, was the *Encyclopaedia Britannica*. And the *Concise Oxford Dictionary* was also there to settle disputes. There were several books of poetry, drama and philosophy, including a Shakespeare, Bernard Shaw, John Masefield, and Harrap's selection of one-act plays. There were 24 volumes of history, travel, biography, popular science and polar exploration, including Scott's *Voyage of the Discovery*, Shackleton's *South* and Mawson's own *Home of the Blizzard*. Then there were 53 books of fiction: a George Eliot, a Thomas Hardy and a Joseph Conrad, Leo Tolstoy's *War and Peace*, Fyodor Dostoevsky's *The Brothers Karamazov* and all the novels of Jane Austen, as well as Rolf Boldrewood's *Robbery Under Arms*.[67]

When the list of books was made public, the Dickens Fellowship protested that their author was not included. How, they wondered, could a ship sail south in 1929 with a hundred selected books − a compressed capsule of the world's best literature − and not include Charles Dickens? The Melbourne Dickens Fellowship held an urgent meeting and decided to send a cabled protest to Mawson. But, interjected a member, the *Discovery* sailed this very day! The president was momentarily stumped. Then another member piped up: 'Yes, but they will stop at South Africa, and I suppose there are booksellers there.'[68] The protest cable was sent, but I doubt that a Dickens made it onto the *Discovery*.

When the North American scientist, Bill Green, met Russian scientists at Lake Bonney in the 1980s, they 'talked of Tolstoy and Turgenev, the Petersburg of the Czars'.[69] Literature provided them with a common language, too, and so did their science. Ideas and idealism keep drawing people south and sustaining them there. 'Science' lured Cherry-Garrard, Edward Wilson and 'Birdie' Bowers out in the dark of the polar night on the worst journey in the world. A hunger for knowledge – a sheer sense of wonder and curiosity about an extraordinary place – remains the currency of community in Antarctica. 'Science' is the word that we use to describe disciplined inquiry, a way of turning wonder into shared insight and understanding. 'History' is a good word for that, too.

Peter Medawar was a famous scientist and commentator on the processes of scientific thought. 'Science', he was keen to point out, 'is no more a classified inventory of factual information than history a chronology of dates'. He deeply resented the equation of science with facts and of the humane arts with ideas. A scientist, he argued in 1965, must be 'freely imaginative and yet sceptical, creative and yet a critic. There is a sense in which he must be free, but another in which his thought must be very precisely regimented; there is poetry in science, but also a lot of bookkeeping.' 'What exactly are the terms of a scientist's contract with the truth?' mused Medawar. A scientist, he decided, far from being someone who never knowingly departs from the truth, 'is always *telling stories* in a sense not so very far removed from that of the nursery euphemism – stories which might be about real life but which must be tested very scrupulously to find out if indeed they are so'.[70] Medawar wrote a famous little book called *Advice to a Young Scientist* which celebrated this narrative, processual experience of the scientific researcher.

When Bill Green studied the Antarctic lakes, this was his experience, too. It surprises people that there are lakes in Antarctica. It has to do with the peculiarity of water. 'How absolutely strange it is that ice should float', writes Green. Solids normally sink in their own liquids. 'So', he reflects, 'if water behaved like a normal liquid, there would be no lakes in the Antarctic Dry Valleys, only blocks of ice.'[71] The three-dimensional structure of hydrogen bonds makes water an unusual liquid, one that expands rather than shrinks as it freezes. The buoyancy of ice means that the lake freezes from the top down. But the freezing of the top layers caps the lake, insulates it, and creates a fascinating and warmer underworld.

Green first went to Antarctica in 1968 to study the microbiology of a cluster of permanently ice-covered lakes in the dry valleys near the Ross Sea: Lakes Vanda, Bonney, Miers, Hoare and Fryxell. The extraordinary existence of dry valleys in Antarctica was discovered by Scott on his first expedition. Scott wrote of Taylor Valley in *The Voyage of the Discovery*:

> I cannot but think that this valley is a very wonderful place. We have seen today all the indications of colossal ice action and considerable water action, and yet neither of these agents is now at work. It is worthy of record, too, that we have seen no living thing ... It is certainly a valley of the dead; even the great glacier which once pushed through it has withered away.[72]

One of Green's research team, a gifted chemical engineer named Larry Varner, found himself unable to imagine or believe in Antarctica, even when he was there, especially when he was in the dry valleys. 'There is no language for this', he exclaimed. 'I'll tell you what's out there. Nothing ... We're not meant to be here at all.'[73] Varner was disturbed by the autism of Antarctica and longed to abandon it, but his intellectual passion kept him there. He had to construct a flume to capture the stream flow from melting glaciers so that they could measure the rate at which invisible chemical elements entered the lakes. In a landscape otherwise devoid of meaning, his science was his sanity, his sustenance.

In a typical temperate lake, the warm, light water near the surface floats on cooler, denser strata below. It was a shock, therefore, to discover that at the bottom of Lake Vanda the water temperature is a very warm 25 degrees Celsius (77 degrees Fahrenheit). Like lakes elsewhere, it is a stratified lake, but in Vanda the stratification is upside-down. The first science of the lakes, begun in the 1960s, was dismissed as unbelievable and was excluded from the refereed journals. Green's own mission in the dry valleys was to study the behaviour of nutrients and heavy metals in Lake Vanda and its inflow, the Onyx River, Antarctica's longest river. What was the history of this lake, and of the other lakes in these dry valleys? Why were they so different in their chemical compositions? How were chemical elements brought to them, and how were they then removed?

One day, while camped amongst the ancient rubble and diatom skins beside Lake Miers, a young scientist named Walt who was working with Green came running out of his tent with a book in his hand. The book he was flourishing was Peter Medawar's *Advice to a Young Scientist*. 'Look at this', he said. 'This is just the way it is.' He flipped the pages into the wind and then read aloud to his companions the passage where Medawar talks about the way that science works, how it is experienced by the scientist not as boring, tedious puzzle-solving, but as a continuing revolution of thought. Bursts of passion feed you and carry you along. During any study, everything appears to be in flux. You never know what you're going to find next. Writing history is like this, too.[74]

Walt was studying how much phosphorus and nitrogen were brought to the lakes each year in the melt streams.[75] He closed Medawar's book and said: 'I know now why I love this stuff. It's not like the textbooks say at all. I can't wait till the nutrient data come rolling in. We don't understand anything yet.' Research, he was discovering, is about mystery and passion. It is about intuitive leaps as well as the gathering and identification of evidence. Hypothesis is story by another name.

Bill Green is a scientist who writes history: the history of the elements, the metals especially, manganese, calcium, cobalt, sodium, copper. They were 'voyagers' too. 'What would it be like to be an atom of manganese?' he wondered. Green observed that 'Because they seemed to cleanse themselves of the metals brought into their waters, might [these lakes] not tell us something about the way the Earth regenerates itself?' He was interested not only in what the lakes contained but also in what they lacked. He was attracted to the Antarctic lakes as geochemical microcosms that pose clear questions about more complex lakes in the temperate world. As for so many in Antarctica, absence and simplicity were his research edges. Stephen Pyne observed that 'This is not merely the simplest environment on Earth but one that aggressively simplifies. The quintessential Antarctic experience is of something taken away, not something added.'[76] Sometimes, though, what is added is also of scientific concern. The lead content of Antarctic snows is four times what it was before the Industrial Revolution.[77]

Green and his team discovered that these lakes – all ice-covered, all located in similar glacial terrain and fed by melt-streams, and all with few living things in them – nevertheless harbour remarkably different chemistries.[78] They have quite different qualities of brine, different concentrations of metals, different flushes of fresh water. They found themselves explaining these variations through an enquiry into landscape history. As well as imagining the histories of water in the dry valleys, Green also analysed the lakes' processes of self-management, the means by which they systematically and fastidiously cleansed themselves of the metals that flowed into them. The purifying cycles of the Earth are exposed in Antarctica. The scientists of the lakes found 'a benevolent conspiracy' at work; metals pour into the lakes and the lakes remove them. Is the cleansing fast enough, they wanted to know? Is it being overwhelmed?[79] And what does the increased flow of the rivers and the current swelling of the lakes mean? Are we pushing the maintenance systems of the Earth too far, in Antarctica and elsewhere? Vanda station, operated by New Zealand since 1968, once stood on high rugged ground overlooking the lake. But the lake has been rising about a metre a year for 30 years, and the station was destined to be flooded. It was dismantled in the early summer of 1992.[80] When water flows in Antarctica, the world now watches with concern.

Green wants to write 'the biography' of manganese, 'the history' of the lake: these are the words he uses. He feels he has to fight a stereotype of science as being wholly rational and without imagination, creativity or passion. 'We think of science, at best, as discovery. It is the artist who is a creator.'[81] But his own experience of science is of imaginatively constructing and testing stories. 'I began to construct a little story about the manganese', he writes. 'I didn't know whether it was true or not, but I knew eventually we could test it and find out. That was the way science worked. You wrote a story. It was pure imagination bounded by a few, usually weak, constraints. Then you tested it, saw whether the world out there could really abide your notions of what was so. Usually it could not. So you tried again and again until you got it. Until you had something that might be so.' You learned to trust your gut. Divining meaning is a visceral and intuitive exercise. Researchers are driven by the adrenalin of ideas and by a hunger for understanding. Their appetite for intellectual nourishment is insatiable and they are omnivorous. Science doesn't have a monopoly on wonder, and history doesn't have a monopoly on stories. 'There is too much suspense in this work', exclaims Green.[82] In his quest, he cannot tell the difference between science and poetry, or where one starts and the other ends.[83] Nor can he separate his awe at the landscape of Antarctica from the exhilarating logic of his elements.

## Monday, 6 January

**47 degrees south, 141 degrees east**

We just had a fascinating presentation and discussion about
Mawson's Hut. The speaker was Ian Godfrey from the
Western Australian Museum and his title was 'Mawson's
Hut – Just another tip site?', and some wit promptly erased
the question mark on the noticeboard.

What constitutes heritage and what rubbish in the new,
environmentally responsible Antarctica? And, even if you
accept that Mawson's Hut is an exceptional site – an iconic
heritage precinct of outstanding significance – what do you
do about a fragile wooden building that has filled up with
ice blown through gaps in the timbers and which occupies
the windiest place on Earth and therefore has been ice and
wind-blasted to paper-thin cladding?

There is a straw poll being conducted now on ship.
The options are to leave the hut there and do nothing, to
overclad the original timbers with new timbers, to build
some sort of overstructure that would protect the building,
to repatriate the whole hut to Australia and install a replica
on site, to 'stuff around' (in the value-laden words of the
noticeboard) which means to make ad hoc and sensitive
repairs to the original structure, or finally (to quote the
list on the board), 'to put in a chair lift'. I am shocked that
about half the people on board want to see Mawson's Hut
brought back to Australia, 'so that more people can see it'.

My main contribution to the debate has been to try
to explain the special value of original buildings *in situ*,
and further, to broaden the heritage argument beyond

sentiment and culture by articulating a political and strategic case for keeping the original hut there. In other words, that in maintaining Australia's territorial claim or indeed, in strengthening our influence in the treaty system, Mawson's Hut is worth ten Red Sheds as a way of demonstrating to the world the longevity and seriousness of our investment in the continent.

# CAPTAIN SCOTT'S BISCUIT
## *The archaeology of return*

In 1968, the writer Thomas Keneally took one of Captain Scott's biscuits from a tin in the 1901 *Discovery* expedition hut at McMurdo Sound. It seemed forgivable at the time. There, beyond the outer edges of the sprawling garbage dump at McMurdo, the old hut stood unlocked and unattended. No-one was officially responsible for its preservation, some ice was accumulating inside, and there was lots of stuff spread about the hut, including boxes of Fry's cocoa on the shelves, tins of preserved fruit, old harnesses, newspapers and magazines. Even when first erected, the hut at Hut Point had never felt snug to its occupants; over 60 years later, it was easy to feel that everything in it was destined to neglect and decay. Keneally approached an open tin of Huntley & Palmers digestives, the 'hard-tack' biscuits that sledgers lovingly mixed with pemmican and whose every crumb was jealously mustered into the mouth. Bolstered by the rationale that removing was preserving, he reached out possessively. He didn't even take a whole biscuit; it was just two-thirds of one. The biscuits were famously hard when fresh, but the souvenir that Keneally extracted from the tin was now 'near ossified by Antarctica's perpetual freeze'. It was not even a biscuit anymore. He took it home 'as if it were more a fossil than a food, and displayed it, in a glass case' in his home in Australia. So the biscuit was indeed preserved and revered, but out of context on another continent.[1]

Part of what allowed Keneally to take the biscuit, I suspect, was a disbelief that he was a tourist. And strictly, he wasn't. Keneally might have visited Antarctica for less than a fortnight, but he was an invited member of an official party in the company of Bill Crook, the US ambassador to Australia. One enduring definition of a tourist is someone who pays for their visit, although by this criterion Captain

Oates was a tourist, for he paid £1000 for a place on Scott's last expedition. Keneally was a full expedition member, and in 1968 he could not easily imagine tourism in Antarctica. He did not realise that it had already begun. He did not recognise what he was. He did not understand that he was a forerunner, a humble pioneer of a different kind of explorer. Behind him, stretching into the future, was a long line of visitors, and the forlorn hut could not afford each one a biscuit.

By the end of the twentieth century, hundreds of tourists could visit one of Scott's huts in a single day.[2] And they had no doubt they were tourists. They had paid their way onto a cruise ship, carried cameras in their pockets or around their necks, and generally proceeded – like the penguins – in single file, wearing matching jackets. From a distance they had the same dutiful, plodding stoop. Their single file was stiffened with environmental consciousness, for they were literally minimising their footprint on the continent.

When did Antarctic tourism begin? Was it when Carsten Borchgrevink paid for his berth on the whaling ship, *Antarctica*, and bustled ashore first at Cape Adare in 1895, ensuring that the first person to set foot on Eastern Antarctica was a tourist?[3] Was it when the men of Scott's and Shackleton's expeditions played tourists to one another's huts and souvenired one another's things? Was it when Fred Middleton got a cruise to the Antarctic with none other than Sir Ernest Shackleton as his guide? Fred had a chance to go ashore in Antarctica in early 1917 and he recorded that he 'got a few curios from the hut [Shackleton's hut at Cape Royds] ... which I shall bring home and exhibit'. He had hoped also to 'collect some memoirs' from the Cape Evans hut but was thwarted: 'As I've not been ashore much I find it hard to get mementos of the trip.' Certainly, mementos had monetary value at home. When Shackleton's expedition ship, the *Nimrod*, berthed in Sydney on its return late in 1909, tens of thousands of people swarmed over it in search of souvenirs. The sailors sold every fragment of spare wood and rope available, every scrap of penguin or seal skin, every penguin or skua egg, every tiny piece of Antarctic rock. Captain John King Davis noticed people surreptitiously cutting the standing rigging of his ship so that they could take a polar souvenir home. When visitors asked about Professor Edgeworth David's geological collection, one sailor had a bright idea. He quietly stepped onto the quay and filled his pockets with specimens of 'Kiama' basalt, the local roadmaking stone from the coast of New South Wales, and sold them at half a crown a piece as specimens of Mount Erebus lava. While at port in Sydney, a watchman on the *Nimrod* could make £76 a week from gratuities and the sale of souvenirs.[4] By 1988, the United States commissary at South Pole station was doing a US$40 000 annual business in polar souvenirs.[5]

It is generally accepted that Antarctic tourism officially began in 1966 when the Swedish entrepreneur, Lars Eric Lindblad, took fare-paying passengers on a ship to the Antarctic Peninsula. One of the early cruises was led by Peter Scott, son of Robert and a greatly admired naturalist, making this scion an Antarctic pioneer, too.[6] Lindblad established the pattern that has prevailed ever since. By 1980, about 17 000 people had visited Antarctica by cruise ship, overwhelmingly to the Antarctic Peninsula. In the decade of the 1980s, almost that number again of ship-based tourists visited. By the early twenty-first century, that many people were visiting Antarctica by ship in a single summer. In 2004–05 some 22 000 visitors made over 200 000 individual landings in the Antarctic Peninsula alone. Overall, tourist numbers have increased 68 per cent over the last three years.[7] As well as this exponential increase in human feet on scarce Antarctic ground, huge cruise ships (not ice-strengthened and carrying one or two thousand passengers) are now sailing in Antarctic waters, causing concerns for the environment and travellers' safety should they sink or run aground.[8] An unexpected boost to tourist traffic was the collapse of the Soviet Union in 1989 which released Russian ice-breakers into southern polar service. Still more tourists flew over Antarctica or visited by plane. As early as 1956, a chartered Chilean airline flew 66 people over the Antarctic Peninsula. During the late 1970s, day-trip flights over Antarctica became popular until an Air New Zealand DC10 aeroplane slammed into Mount Erebus in 1979, killing all 259 passengers and crew. It was a shocking event. Any survivors of the crash would not have survived the elements.

During the 1980s, as the numbers of tourists grew steadily and governments invested further in installations on the ice, tensions broke out over the responsibilities of private expeditions to Antarctica. Most of the heroic-era expeditions had been privately funded, but the increasing role of government after World War 2 and the growing Antarctic bureaucracy that followed the signing of the treaty began to leave private expeditions out in the cold, literally. Two Norwegians who trekked to the Pole in 2000 were officially refused entry to South Pole station and pitched their tent in the snow (while sympathetic staff snuck out with food). The rise of tourism began to blur the boundaries between work and pleasure, and between heroism and irresponsibility. Were all private expeditions now just a form of tourism?

In 1985–86, a private British expedition with government endorsement set out to re-enact Scott's journey to the South Pole. The 'Footsteps of Scott' expedition caused a dramatic collision between the values of exploration and bureaucracy. It was a private expedition like so many of the heroic era, and it was motivated by a respect for that history, especially Scott's story. Britons Robert Swan and Roger Mear, together with Canadian Gareth Wood left from Cape Evans on their commemorative polar journey along Scott's route in October 1985. On the morning that they set off for the South Pole, Robert Swan lay down on Scott's bunk in the Cape Evans hut, looked at the ceiling and thought: 'Well, that's what you looked at the morning you set off.' Then he left the hut and turned south with tears streaming down his face. Seventy days later they manhauled their fibreglass sledges into South Pole station. At virtually the same moment, they heard the news that their ship, *Southern Quest*, had been crushed in the ice of the Ross Sea (this was the sinking reported to Stephen Murray-Smith while he was at Wilkes). US helicopters recovered the ship's crew from the ice and flew them back to McMurdo.

The 'Footsteps of Scott' trio, once they arrived at the pole, had arranged to be flown back to their ship by their own aircraft, which was on standby at Ross Island. But the loss of their ship now gave the United States, which wished to discourage private expeditions to Antarctica, the opportunity to humiliate them. 'There is no place in the Antarctic for adventures such as yours any more, and we are going to make that point', explained Peter Wilkniss, the new director of the National Science Foundation Polar Program. US personnel at the South Pole were informed that fraternisation with members of the expedition could be detrimental to their future employment. When officials visited by plane for a few hours from McMurdo, the expeditioners were advised not to declare their presence and to keep their sledges and equipment out of sight. At the end of their energy-sapping journey, the three men could not find the strength to wrestle bureaucracy. Back at McMurdo, the support team was getting the same frosty treatment. Captain Shrite, the officer-in-charge at McMurdo station, 'greeted the survivors with stern and humourless formality'. 'It is not our policy to assist private expeditions', he insisted. Not even a tube of toothpaste was to be sold from McMurdo stores to the 'Footsteps' people. The expedition pilot was not allowed to fly to the pole to retrieve the three manhaulers. The expedition members reported 'a rigid clamp-down on freedom of speech and association in this barren territory'. Fifty-six members of McMurdo and Scott base stations later wrote to the expedition apologising for the 'shamefully hostile treatment' meted out by their own masters. Murray-Smith, following the drama from another part of the ice, wrote later that 'If any evidence is required that the old Antarctic spirit is dead, and that that unfortunate continent is now firmly in the hands of soulless bureaucracies, then that evidence will be found in Mear and Swan's book, *In the Footsteps of Scott.*'[9]

The 'Footsteps of Scott' expedition members felt trapped and disempowered by ungenerous officiousness. They wanted to complete their expedition independently and felt that they could have done so. They had a plane and pilot ready for the South Pole flight. They were also poised to rescue their people from the ice when the ship sank. They felt that the United States, which had refused to give their pilot or ship any weather information, had taken over the rescue operations so that the expedition could be embarrassed. The 'Footsteps' people wanted to maintain control over their expedition and claimed to be equipped to deal with their own emergencies. They were in search of what they called 'this basic freedom – to live by one's own judgment, wit and skill'.[10] They wanted to rescue themselves. They had just done a Scott and now they wanted to do a Shackleton.

A red and white striped pole circled by the flags of treaty nations now marks 90 degrees south. Adventurers trekking to South Pole station continue to be shocked by the barber's pole bureaucracy and commercialism of this hallowed tract of ice, and by the fact that the flag of the United States greets them at the Amundsen-Scott station. When 27-year-old Tom Avery walked 1100 kilometres (684 miles) to the South Pole in 45 days in 2002, he declined to buy from the gift shop a T-shirt declaring 'I reached the South Pole'. He was the youngest person to get there on foot – for five days anyway – for his record was broken later that week by 23-year-old Andrew Cooney.[11]

Tourism, which now seemed to mean any private expedition, was perceived as a threat to science. People trooped through government bases, making a spectacle of the 'real expeditioners' and a nuisance of themselves. A United States delegate to the Scientific Committee for Antarctic Research (SCAR) alleged that in the 1986–87 season, 1600 visitors to Palmer station on the peninsula caused a loss of 40 working days. Chinese delegates reported that about 2000 people had visited their Great Wall station over the two years of 1985–87. In 2002, John Rich's team enjoyed welcoming tourists from the *Kapitan Khlebnikov*, and treated them in style, but acknowledged that this was partly because tourists at Casey were very rare. By the early 1990s, with a Minerals Convention rejected, some people wondered if tourists posed a greater environmental threat in the near future than miners.[12]

In 2004, the leader of *Aurora Expeditions*, the renowned moun-taineer Greg Mortimer, reflected on these issues. He could say that he had travelled to Antarctica 'about 100 times'. He had carefully observed the most visited beaches in Antarctica every summer for the preceding 13 years and he could report that there is 'no sign of visita-tion, no diminution of penguin populations, nothing'. He pointed out that guidelines for the conduct of Antarctic visitors and tour operators had been voluntarily drafted as early as 1989 by the tour operators themselves. The critical issues for tourist management, he felt, were more to do with wilderness values than with environmental distur-bance. People go there for purity, solitude, otherworldliness; they go there for the silence. Andrew Dodds recalled his tourist trip to the Antarctic Peninsula in the summer of 1998–99. He was sitting on a rock on Cuverville Island watching a Gentoo penguin land and preen when a rhythmic booming sound grew, and suddenly, from between two icebergs, a giant cruise ship, the *Clipper Adventurer*, chugged into sight. Words were exchanged between the two tourist operators; bookings and timetables had been altered at the last minute. There were three cruise ships seeking solitude and wilderness in the very same area.[13] At the Cape Town Antarctic Treaty Consultative meeting in June 2004, a whole week of discussion was dedicated to tourism.

On the cusp of the millennium, the writer Helen Garner voyaged to the 'regions of thick-ribbed ice' on one of Greg Mortimer's cruises. She was self-consciously a tourist and had read the *Lonely Planet Guide*. She joined a Russian ship, the *Professor Molchanov*, leaving from Ushuaia, an Argentine port in Tierra del Fuego. It was called a 'voyage' not a tour or trip. Garner soon found herself on deck refusing to wear a hat or to believe in the cold and 'hanging out for a short black'. From the moment she saw ice, even humble crumbs of it floating by, she was possessed by an irritating urge to compare it with something ... it is like, like, like ... an aircraft carrier, a temple with pillars, the white ridged sole of a Reebok! 'It's hopeless trying to control the flood of metaphor', she exclaimed. Garner wondered if it was because we rage against a landscape without human meaning. She felt guilty about not liking penguins. And she wrote that 'I'm lonely because everyone else is hiding behind a camera ... I am being driven insane by photography.' The camera obliterated social contact. She determined that she 'would go to the icy continent in a state of heroic lenslessness; that I would equip myself with only a notebook and a pen.'[14] On a beach landing, while others looked through their cameras, Garner wanted so badly to pocket a bright orange pebble that signalled to her from among all the big grey ones. Unlike Keneally, she could resist this temptation because she inhabited a different era. She knew she was not allowed to take anything away except what you throw out or throw up, and *that* you have to take. There was a shocking moment when a photographer lost a plastic bag to the wind and the whole party watched, transfixed with horror, as the 'big white bubble of poison' danced away across the water. A man in a zodiac chased and retrieved it.

On their return to South America, the leader, Greg Mortimer, spent a morning in a beech forest outside Ushuaia, absorbing green shadows before sailing south again with a new bunch of tourists that same afternoon. Helen Garner and several of her fellow tourists watched their ship fill up with strangers who were taking their places, so confidently, so soon.

In 1968, when Thomas Keneally took Captain Scott's biscuit, the hut had recently been restored by the New Zealand government. Conservation work was undertaken on all three Ross Island huts in the summers of 1960–61 and 1963–64.[15] The huts had become museums during 30 or 40 years of neglect, the passing of two world wars of time. When people next peered through their windows or forced their doors against interiors of ice, they did so in awe. When Americans from the icebreaker, *Burton Island*, entered McMurdo Sound in 1947, they visited the Cape Evans hut and found a frozen carcass of a dog standing on four legs as if it were alive, and they sampled the biscuits at the Hut Point hut and considered them 'still edible, although rather tasteless'.[16] The huts were undergoing a transformation from shelters to shrines. Shackleton's men had always souvenired Scott's things and Scott's men Shackleton's – and the Ross Sea party everyone else's – but none of them had hesitated to mix their living with that of the original expeditioners. They knew they shared an era. Past and present were seamless. But now everything in the huts seemed sacred. The American Operation Deep Freeze established a tent camp within a few yards of the *Discovery* hut in 1955–56 and Admiral George Dufek issued strict instructions that the hut was to be regarded as a historic shrine. The following summer it had a plaque on it to prove its status, provided by the New Zealand Historic Places Trust. The Antarctic Treaty of 1961 protected these historic monuments. No longer were people allowed to sleep in the huts or move things around. Soon they would be locked. The flow of time through the buildings had been stopped and their heroic moment fixed. Adding to – or subtracting from – the history of the place had become a violation.

Ironically the hut at Hut Point, in particular, had a layered history of casual and impromptu visitation. For the first decade and a half of its life, the *Discovery* hut had been famously makeshift; it was known to be cold and comfortless and a bit of a pigsty. In a thaw it stank. It was a prefabricated building of Australian homestead design, with a sun verandah on three sides. From the time it was unloaded from the *Discovery* in early 1902, it was used as an emergency shelter, a storehouse and a theatre, and the men preferred to live on the ship. Shackleton used it as a staging post and summer hostel during his 1908–09 *Nimrod* expedition (his main hut was established at Cape

Royds), and Scott again used it as a supplementary hut in 1911–12 while establishing his headquarters at Cape Evans. Edward Wilson, 'Birdie' Bowers and Apsley Cherry-Garrard stayed there on returning from their winter journey to collect penguin eggs from Cape Crozier, but they preferred to pitch a dry tent inside the hut and then lit two primus stoves in the tent for warmth. Thus the *Discovery* hut was never the primary hut; it was never snug and homely; it was always a place where you were rearranging the packing cases, raiding someone else's stores, and helping yourself to someone else's biscuits. When, in January 1915, it again came into emergency use by the Ross Sea party supporting Shackleton's *Endurance* expedition, the men entered it by smashing a window, found a stores tally book kept by Birdie Bowers, souvenired a jersey bearing the name of Captain Oates, wrote their names on the wall near the blubber stove, and had a memorable meal there of fried bacon, frozen sardines, biscuits, strawberry jam and tea, 'all from Scott's stores'.[17] Many of the historic stores from which they were snacking were 12 years old. The oldest of Scott's biscuits were 'moth-eaten' even in the absence of moths.[18] Scott took over 16 300 kilograms (36 000 pounds) of biscuits for dog and man in 1901–04 and left so many behind in the *Discovery* hut that tins of them could be stacked as interior walls to make the hut warmer and more cosy. Although Scott ran out of food on the plateau, his stores in the hut sustained other expeditioners for years.

But if the Hut Point hut is a memorial to anyone, it is to the 'burglars' of the Ross Sea party. For they came to need that hut more than anyone before or since. They were forced to make it their home for a while, especially from March to July 1916. They were also the last to stay in it. The tin from which Thomas Keneally extracted a biscuit would have been opened by these men, suffering extreme privation, in 1916. They settled into the *Discovery* hut despite the dirt and soot, read the same books for the third time, played cribbage, and burned blubber against the creeping cold. They became so ingrained in the fabric of this hut that it is hard to tell what is their imprint, what is Scott's. But it was never their favourite refuge and Ernest Joyce recalled that the party 'lived like animals at Hut Point'.[19] The deaths of Aeneas Mackintosh and Victor Hayward in May 1916 occurred at least partly because of their desire to escape the cramped and squalid existence there. Crossing the thin, autumnal sea ice of McMurdo Sound, which stretched between the two huts on their rocky promontories of Ross Island and around the tongue of the Erebus Glacier, was always hazardous. With a blizzard approaching, the two men chose to race the weather to Cape Evans rather than risk being trapped any longer in the *Discovery* hut. The storm swept the ice and the men away. The hut is a memorial to their foolhardiness, and their deaths are memorials to the *Discovery* hut's repellence. Although they longed to escape the hut, it has now been restored to the way they left it. The story of Shackleton's Ross Sea party has been superbly told in a book called *Polar Castaways* by two men, Richard McElrea and David Harrowfield, who were drawn to their subject by devotion to the conservation of the Ross Island huts.

Which past should the huts memorialise? At which moment do you stop the layering of time, and to which era do you strip them back? Professionals are now managing their contents and fabric. Amateur visitors might occasionally have taken a biscuit, but professional visitors have removed unique artefacts for conservation, safekeeping or display, sometimes losing them, sometimes failing to record their existence or new location.[20] The Ross Island huts today offer tableaux of an Edwardian frontier. In their darkened interiors, you will find still lives with rope, harness, biscuit, wood and tin.[21] You can hear footfalls, voices, whispers, shouting in the night. Decades of accumulated ice were hacked out of the huts and blocks of it were left to melt in the summer sun, leaving a residue of artefacts. It is marvellous what the restoration work turned up. In the hut at Hut Point were found letters from Mackintosh and Joyce, a script from the play, *Ticket of Leave*, which was written and performed during the *Discovery* expedition, a plywood snow-shoe made from a biscuit case, ten hand-carved chessmen made possibly from a broom handle, two scones found near the blubber stove, and numerous tins of Huntley & Palmers biscuits manufactured in 1901. The tins still had their colourful labels: 'Huntley & Palmers Biscuits – Captain – Grand Prizes 1878 Paris 1900'. Captain Scott's biscuit was a Captain biscuit.

Since 1987, the Antarctic Heritage Trust has had responsibility for the care of the Ross Sea huts, which include the winter home of Borchgrevink's 1898–99 party at Cape Adare. Conservation plans have recently come to include the reconstruction of ruined outbuildings and the replacement of missing artefacts with replicas. Shackleton's stores, stacked against the outer walls of the Cape Royds hut, are being meticulously removed and archived. Stephen Murray-Smith was appalled at the state of Mawson's Hut in 1985 – filled with ice and trapped in a paralysis of cautious management – and he saw it as a monument to 'bureaucratic mindlessness'. But, as a culture of professional intervention has moved into the other Antarctic huts, the more sedate pace of 'restoration' at Commonwealth Bay may eventually be seen as a blessing.

There is a rich human history of paying pilgrimage to relics in their places of honour *and then removing them*. It may be done from a sense of responsibility, a desire to preserve things that are vulnerable. The object is salvaged from the contingencies of time and place and preserved elsewhere, in some centre of knowledge or empire where more people can see it. There is an attractive democracy in that. But it can also be done from a lack of imagination, an inability to see that preserving by removing so often destroys.

On his geological sledging journey to the Queen Maud Mountains in 1929, Laurence McKinley Gould and his party looked for a cairn which they knew Roald Amundsen had built on his return from the South Pole. After much searching they did find it, and 'with reverent hands' took a few rocks from the side of the cairn 'so that we could see what was in it without in any way disturbing the shape or structure'. What they found was a five gallon tin can of kerosene, a waterproof package containing 20 small boxes of safety matches and a tin can with a tight lid. 'It was the climax, the high spot of the summer for all of us', remembered Gould, when he pried off the lid of the tiny can and took out two little pieces of paper. One was torn from a book and carried the names and addresses of the two men, Oscar Wisting and Helmer Hanssen, who had built the cairn. The other was a page carefully torn from Amundsen's own notebook and it recorded the successful attainment of the South Pole. Gould was deeply moved to be in this spot with that note in his hand: 'I do not think anyone could have appreciated more fully than did the six of us all that lay behind that bit of paper and its simple account ... I think we shall none of us forget the glamour of that moment.' They replaced the rocks 'and left the cairn looking just as it had when Amundsen and his men had built it 18 years before'. But they took the note.[22]

In March 1957, Sir Edmund Hillary and his party from the Commonwealth Trans-Antarctic Expedition visited Cape Crozier, where they hoped to find the humble shelter used in the darkness 46 years earlier by Wilson, Bowers and Cherry-Garrard during the worst journey in the world. With deep respect and Cherry-Garrard's book in their hands, they searched for relics. They found an old pemmican tin and, after renewed searching, they located the legendary rock hut, as powerful a monument to heroic science as exists. An old nine-foot sledge was iced in, as was an old pick axe. They dug the sledge out and found pieces of penguin, a box containing pemmican, cheese, two tins of salt, a larger box with test tubes, bottles of alcohol, alum, several sketching pencils, a storm lantern and several envelopes with the *Terra Nova*'s stamp on them. They removed as many of these objects as possible, and took them back to New Zealand, where they were distributed among various museums. Do we wish they had left them there, or would they then have been lost for all time? By 1977 the rock hut was marked by a plaque. In 1985 there were still fragments of green Willesden canvas, the pelt of an Emperor, the rusty shell of a small lantern, a woollen sock. In

1990, there was little left. 'Unfortunately, every tourist is a collector', remarked Jorge Berguño, the assistant director of the Chilean Antarctic Institute in 2002.[23]

There is a global diaspora of Antarctic relics. I wrote some of this book sitting beside Mawson's desk from the *Aurora*, which is now preserved in the wonderful Australian Antarctic Division Library in Kingston, Hobart. The desk was used by Mawson on board ship in his 1911–14 expedition and by Shackleton's Ross Sea party in 1914–16. When it was auctioned in 1930, its new owner found a sheaf of papers stuffed behind a desk drawer. They turned out to be a diary, letters and messages from the ship during the anxious months after it was blown out to sea and caught in the ice, leaving the Ross Sea party stranded ashore.[24] At the Scott Polar Research Institute in Cambridge, staff and researchers are called to morning and afternoon teas by the bell from Scott's ship on his last expedition, the *Terra Nova*. Amundsen's Antarctic gramophone and the sledging flag he flew at the South Pole are preserved in Norway. The extraordinary *Discovery* is a floating museum at Dundee. Mawson's famous balaclava made a return trip south with the adventurer, Peter Treseder, who completed an unsupported pole trek in 1997, carrying not only the balaclava but also Amundsen's ivory good luck charm and a 160 kilogram sled-load of food and equipment.[25] He did not take any good luck charm of Scott's. In the Tasmanian Museum and Art Gallery there is a pocket medical case, donated in 1946 by Dr R Whishaw, which is believed to have been found on Dr Edward Wilson's body in the final tent on the Great Ice Barrier.[26] In 2002 the *James Caird*, 'arguably the most exalted sailing craft in maritime history', travelled the world to the kind of reverent acclaim it received in 1916 from the Norwegian whalers who dragged it ashore.[27] Normally it is on display at Shackleton's old school, Dulwich College in south London, and once I was fortunate to join the James Caird Society for their annual dinner beside what they rightly call this 'most potent relic'.

The continent itself buried relics and it has also set them free.

Amundsen's base, Framheim, and Richard Byrd's series of Little America townships, dug into the rim of the Ross Ice Shelf, became icebergs. The most devoted archaeology conducted on the ice has been by expeditioners excavating their former selves. Returning to Little America in 1934, Byrd and his men rushed over the rise to exclaim at the sight of the tops of their radio towers and the blackened smoke stacks of their houses still protruding. They dug frantically, coming across 'strange unremembered things that the snows of four winters hadn't covered'. The rooves of their buildings had sagged, the main beams had cracked, a film of ice lay over the walls, and on a table stood a coffee pot, a piece of roast beef with a fork stuck in it, and half a loaf of bread. Calendars from 1929 with the days scratched off hung on the walls. Only four years had passed, but they felt like they had entered a prehistoric tomb. Young Finn Ronne crossed the room to look at the bunk occupied by his father, Martin Ronne, who had been with Byrd's first expedition and had died aged 68 just the year before. Finn found that his father, just before leaving Little America, had written his son's name on the wall, and there it was in the candlelight. 'The old man must have known I'd come down', he exclaimed. While they were standing in the mess hall, the telephone rang. Someone was testing the system from another building buried nearby in the ice. Then one of their party idly flipped a switch. And the lights went on. They were dim but gleaming, a tribute to the battery. On the stove were cooking pans full of frozen food. There was coal in the scuttle. It took them no time to settle in again, although drift made them 'perpetual archae-ologists'. Decades later, perhaps in the 1960s, Little America declared its independence and floated north.[28]

Since their death and burial in the polar tent on the Ross Ice Shelf, Scott, Wilson and Bowers have been moving at just under a kilometre a year. They passed the latitude of the longed-for One Ton Depot years ago, and are heading for the sea.[29]

Many relics and ruins remain *in situ*. On South Georgia, the animals have reclaimed the death camp. Elephant seals now lounge, yawn and scratch in the flensing sheds at Stromness, where Shackleton stumbled down from the mountain and was taken to the manager's hut, which also still stands. At Stonington Island where Harry and Jennie honeymooned, archaeologists found a razor, a jar of hair tonic, boxes labelled 'Vitamins Plus', a toy balloon, a Mormon text, a chess bishop and an eight of clubs. The ice cave inhabited by Scott's northern party in 1912 was found again and again over the years by thrilled expeditioners who located bamboo poles, seal skins, a small blubber stove and a blubber lamp made from a biscuit tin. In the 1980s melting ice destroyed the cave and the wind scattered the artefacts, although some were removed to Scott base for protection where all but two were lost. The site is marked with a plaque.[30] The carpenter's shed at Mawson − the oldest surviving building at the oldest continental base − is a Sistine chapel of *Playboy* centrefolds.[31] Such public posters are disapproved in modern Antarctica, but not if they constitute heritage. The whale bones that litter the beaches of some Antarctic islands − a grisly architecture of skulls, ribs and vertebrae − are 'far more permanent than any human-made structure', observed David Campbell.[32] And what of ventifacts; stones shaped only by the wind? They are classified as artefacts and must not be removed, warns the *Lonely Planet* guide.[33] Bill Green, having souvenired one, later returned it.

In the late 1970s, two ancient Indian projectile points were discovered by Chilean archaeologists in sediments dredged out of a bay on the South Shetland Islands, near the Antarctic Peninsula. The artefacts were flecked quartz of the type used by the Indians of central Chile until about 1500.[34] This exciting find was greeted with profound scepticism, in Chile as elsewhere. This was the period of escalating resource politics, when babies were contrived to be born in Antarctica. Were the projectile points clear evidence of early South American colonisation of the ice, and therefore a startling claim to priority? Or were the stone arrow-heads planted, like the flags?

Whether it was an international conspiracy or a student prank, it *was* a hoax, fortunately exposed by a Chilean, the archaeologist Rubén Stehberg who admits today that his scholarly revelations were regarded at home as unpatriotic.[35] In his first year working for the Chilean Antarctic Institute, Stehberg used archaeological scrutiny to prove that the Chilean projectile points had been mischievously dropped into the sediment samples. In his second year with the institute, his analysis of historic wood that was found at several Antarctic settlement sites proved that the inhabitants had come from North America rather than from South America as had been hoped. In his third year of research, when he again approached the Chilean Antarctic Institute for support, he found that 'they were not interested in us!' Stehberg was saddened by the response: 'They are not archaeologists of the world. They had finished with archaeology in Antarctica. We don't want to prove that the Americans were there!'

But Stehberg was already launched on a different kind of patriotic mission. What about those indigenous South Americans who really did get to Antarctica but have been neglected by history? He argues that they were experienced sailors who accompanied and guided the nineteenth-century voyages in Antarctic seas by British and American sealers and whalers, people whose local knowledge of southern seas was valued but whose names were never recorded. When, in the 1980s, Daniel Torres of the Chilean Antarctic Institute found a skull on the South Shetlands, it was identified as that of a mestizo, a young woman of native South American descent. The institute reluctantly turned to Rubén Stehberg to investigate the historical archaeology of the find. 'So I went south again', he shrugged. He excavated three remaining rock shelters of sealers from the first intensive years of hunting (1819–22) and he found more 'Aboriginal remains' at Desolation Island in the South Shetlands. His forensic curiosity is driven, he admits, by a nationalistic quest. 'How can the British or Americans say they were first, when they were guided by Chilean hunters?'[36]

Will Mawson's Hut – this potent claim to Australian sovereignty – be brought home, as my shipmates hoped? I don't think so. Dismantling it and bringing it back would render it a replica, anyway, so why not replicate it here and leave the original there? But we shouldn't underestimate the strength of feeling about its return to Australia. If it were rubbish, it would have to come home. 'The arc of history's narrative does not end on the Ice', Steve Pyne reminds me. If the hut, like Mawson himself, comes home, then it will complete the voyage, finish the roundtrip. Ian MacNamara (known as 'Macca'), who speaks to Australians right across the country on his Sunday morning radio program 'Australia All Over', put it this way on 15 May 2005: 'That's something John Howard [Australian prime minister] could do. He could do it for all of us. Bring back Mawson's Hut so that we can see it, rather than let it blow out into the Arctic Sea.'[37]

But what does 'return' mean in an age when someone can visit Antarctica 'about 100 times'?

As the years passed, Thomas Keneally began to feel that his biscuit should be returned. When in 1985 he saw the television film, *The Last Place on Earth*, which was inspired by Roland Huntford's infamous book about Scott and Amundsen, he felt that these suffering souls of the south were making a claim on his biscuit. It was about this time that Australians increasingly returned fragments they had earlier taken from Uluru, the massive red monolith at the heart of their continent which was returned to Aboriginal ownership in 1985. They feared that their souvenirs of the sacred Aboriginal rock might bring them bad luck. These gestures of reconciliation became known as 'sorry rocks'. Keneally's souvenir had become a 'sorry biscuit' and was heading south again.

Keneally had found that Antarctica haunted his dreams; it 'produced jolts of insomniac chemicals' into his system. 'It was not landscape, it was not light. It was super-landscape, super-light, and it would not let you sleep.' He had written two novels filled with Antarctic hauntings, about a marooned explorer on the ice, about murder, and about light. Finally, he and his wife boarded the *Kapitan Khlebnikov* from New Zealand in 2003, and soon he was carving the names of his grandchildren into the flat top of an iceberg and sailing into sight of the Transantarctic Mountains. He had been assured he would find someone on his trip associated with the preservation of the huts. But something inexplicable happened. Keneally told several people about returning the biscuit, booked his passage, packed his bag, flew to New Zealand, boarded the ship, and left the biscuit behind.

If he could not return the biscuit in person, this former Catholic priest could at least make a confession. So when in Antarctica again, he found the head of operations of the Antarctic Heritage Trust, Nigel Watson, and revealed his soul. Watson reassured Keneally that American naval personnel had taken much greater plunder from Hut Point, and that one man had returned canned goods and books to him. 'It was agreed', wrote Keneally, 'I would send him the biscuit for assessment.' A professional had absolved him, and the biscuit would now enter the mysterious half-world of conservation analysis.

The Huntley & Palmer biscuit is a wonderful symbol of empire for Scott to take to the end of the Earth. Huntley & Palmers was founded in Reading, England in 1822 and boasted that it was 'The Most Famous Biscuit Company in the World'. Henry Stanley set off in search of Dr Livingstone in Central Africa equipped with Huntley & Palmer biscuits, we presume. The company grew with the British empire, and at the beginning of the heroic era its overseas sales accounted for 75 per cent of the total export of biscuits and cakes from the United Kingdom. The current Huntley & Palmers company website lists Captain Scott as a famous 'client' who was robbed of victory at the pole by 'a delay to the start of their trip'.

The Huntley & Palmer biscuit is also a minor player in the perennial debate about diet on polar journeys. Did the Huntley & Palmer biscuits kill Scott and his party? Cherry-Garrard has told us they were cooked to a secret recipe devised by Dr Wilson and the firm's chemist (no longer a secret – it's on the company website), and we know that biscuits constituted a more significant proportion of Scott's sledging diet on his last expedition. Historians and biochemists have analysed the Huntley & Palmer biscuit and have confirmed that it was indeed different to other hard-tacks of the time. Whereas Amundsen's biscuits were based on wholemeal flour and crude rolled oats with yeast as the leavening, the Huntley and Palmer biscuit contained white flour and sodium bicarbonate and therefore much less Vitamin B. Furthermore, the presence of sodium bicarbonate could have lowered the contents of some of the vitamins on baking, possibly destroying all of the thiamine. Because the biscuits were, for Scott's party, such a crucial source of thiamine, this may have been a critical deficiency. A lack of thiamine can lead to beriberi, a disease which is not unlike scurvy. But, reader be warned, for this is contested ground and the biscuit ultimately returns us to the battlefield over Scott's reputation.[38]

In September 1999, a Huntley & Palmer biscuit came up for auction at Christies in London. I was lucky enough to be there, amidst an excited crowd of polar enthusiasts and heritage experts. It was, we were told, a biscuit taken from the polar tent, removed from the sacred tomb before it was closed and collapsed. It was the last food the polar party had left; one biscuit, cracked, with a corner damaged, and a small bag of rice. The provenance seemed impeccable. The 'potent relics' were removed from the tent by the search party, brought back to England and given to Kathleen Scott. They were kept in a little suitcase in a bank vault and handed down through the family: first to Peter, who in March 1912 at the age of two had told his mother that 'Daddy isn't working now', then to Sir Peter's widow, Lady Philippa Scott, and then to auction.[39] The biscuit sold for almost £4000 that day. The telephone buyer was Sir Ranulph Fiennes, who has since so strenuously and passionately defended Scott's reputation. Fiennes was determined to keep such a relic in Britain, and thanks to his generosity, it is now on loan to the National Maritime Museum in London. In his book about Captain Scott, Fiennes writes in exasperation about the biscuit. He felt that he was tricked by the Antarctic legend-makers. In spite of the alleged provenance – confirmed by the Scott family, printed in the Christies catalogue, and still supported by the National Maritime Museum – Fiennes is sceptical that a biscuit would be left in a tent by starving men.[40] I disagree. It was their solid little symbol of self-respect. Scott wrote on his return journey from the South Pole that they would 'fight it out to the last biscuit'. The last biscuit was his metaphor for heroism in extremity. As these civilised men starved, they retained a commitment to 'seeing the game through with a proper spirit'. And although in their last weeks they decided to run the risk of a full food ration, the last biscuit could not be eaten. Dividing fairly and sustainably was a religion, a commitment to a kind of infinity. The biscuit gave them a future, however spurious, and the culture of that tent was all about faith. While the last Huntley & Palmer remained, hope flickered; or at least the courtesy of hope. Sir Ranulph Fiennes did well to buy it for us all. The last biscuit was like the rocks in the sled and the words in Scott's diary. It was a powerful message that they retained their dignity and humanity, that they might still have made it had luck turned, had a search party found them in time, had the blizzard relented.

## Tuesday, 7 January

**44 degrees 15 minutes south,
145 degrees 10 minutes east**

We are re-enacting the great geological voyage of our
history, doing in nine days what the Australian fragment of
Gondwana did in 45 million years.

Late today we sighted land. The first bit of rock was on
our starboard side, a lone cluster of tiny islands, and then
looming on the port side, a line of hills you thought you
might be imagining, but then began to believe was really
there. I watched that line for a long time before dinner, but
it did not solidify or get closer. But it *is* land.

## Wednesday, 8 January

**43 degrees 1 minute south,
147 degrees 23 minutes east**

We awoke in the Derwent doing about 1 knot, crawling
through still waters to our destiny with the pilot at 8 am.
People were out on deck with their mobile phones! Those
walking the ship are now distracted instead of transfixed.
Our enclosed, floating world is already dispersing.

Tasmania looks beautiful this morning. It is green
and clean and sparkling in the sun. I was beginning to
disbelieve in land. How could there be land in such a
vast, restless globe of water? I have new respect for little
Tasmania, rearing up defiantly out of this great ocean. And
what a formidable southern partner it has, and only the
deep, dark sea between them.

As Fred Middleton's ship, the *Aurora*, approached New Zealand's north island in early February 1917, its captain John King Davis wrote its ending: 'This strange voyage, that had seemed almost to set time at nought and be a voyage into the past, was over.'

'No one on this ship is sane', the doctor told Stephen Murray-Smith as they sighted Tasmania and home in early 1986. It happened to be Doctor Death who was dispensing this further happy advice, the same doctor who had tried to scare us all into the tropics at the start of our journey. He continued to tell Stephen: 'Those with us who have been away their fourteen or sixteen months are certainly not sane; but even you, by the time you've been on this ship in these waters two or three weeks, are not sane either.'

Returning has always been a dangerous moment for polar explorers. In Antarctica they glimpsed, perhaps, a kind of perfection. Last night we discussed insanity over farewell drinks. Three experienced expeditioners explained to me the isolation of the returning winterer. You are finally reunited with family and friends, but you know little of their conversation. Things have changed, people have moved on, a couple has broken up ... you know nothing! So you listen as they talk, as if to another language. And gradually they notice that you are not joining in and they ask you: 'Well, what is it like down there in Antarctica?' And you think, where do I start and what can I say? They know nothing! In any case, you suspect they are asking out of politeness so you don't say very much. People begin to comment on your reticence: 'Oh Antarctica has changed you! You've got quieter.' But you're not quieter; you just don't know what they're talking about. After all, you sailed off the planet for a year. And now, for a little while, you have brought the silence back.

# | ACKNOWLEDGMENTS

I am grateful to the Australian Antarctic Division (AAD) for the opportunity to travel to Antarctica as a humanities fellow in the summer of 2002–03. The director of the AAD, Dr Tony Press, and staff of the AAD – especially Andrew Fleming, Andrew Jackson, Kim Pitt and Andie Smithies – were wonderfully supportive and encouraging. I particularly thank my voyage leaders, Phil Gard and Luke Vanzino, and my fellow-expeditioners for their conversations, questions and kindness.

I would like to acknowledge the scholarly generosity and enthusiasm of a distinguished group of Australian humanities scholars of Antarctica; Brigid Hains, Christy Collis, Bernadette Hince, Marie Kawaja, Stephen Martin, Mark Pharoah, Elizabeth Leane, Alasdair McGregor and Michael Pearson have provided a warm, collegial context for my work. I especially thank Brigid Hains for reading the whole manuscript with such generous insight. I am also grateful to Rob Easther, Andrew Jackson, Grace Karskens, Michael Pearson and Libby Robin for being such perceptive and constructive readers of drafts. Stephen Pyne, an American scholar of Antarctica, has given me wise advice and friendly support. Greg Dening introduced me to the delights of voyaging. Some of the ideas for this book were elicited by the inspiring 'Challenges to Perform' graduate workshops he led with Donna Merwick at the Australian National University. Mike Smith happily joined me in conversation about the history of science and taught me much. Alan Platt was, as ever, greatly encouraging. Michael and Julie Landvogt lived the ideas with me. Peter (Bloo) Campbell, Peter Cook, Mark Forecast, Jenni Mitchell, Brian Murphy, Kim Pitt, John Rich, Geoff Stevens and Eric Woehler each shared my voyage and helped me to write about it. Bill Middleton gave me a copy of his father's diary to take back across the Southern Ocean. I wish to thank the many people who assisted me with references, advice and encouragement, in particular Alessandro Antonello, David Armitage, Bain Attwood, Geoffrey Blainey, Tim Bowden, Darren Boyd, Peter Boyer, Emily Brissenden, Ian Britain, Gary Burns, Jane Carruthers, Vincent Carruthers, Graeme Davison, Kirsty Douglas, Stephen Dovers, Jasmine Foxlee, Lynette Finch, Guy Fitzhardinge, Irina Gan, David Hansen, Janet Hughes, Ken Inglis, Grace Karskens, Marie Kawaja, Brian Kinder, Phillip Law, David Lowenthal, John Magee, Mark McKenna, Tony Marshall, Mandy Martin, John Mulvaney, Kay Nantes, Neville Nicholls, Cassandra Pybus, Peter Read, Libby Robin, Deborah Rose, Tim Rowse, Tim Sherratt, Stefan Sippell, Barry Smith, Rubén Stehberg, Carolyn Strange, Rupert Summerson and Gilbert Wallace. I am also grateful for the research assistance of Greg Bowen, Ana Rubio and Jessie Mitchell. Three inspiring books that I have kept on my desk while writing are Stephen Murray-Smith's *Sitting on Penguins*, Stephen Pyne's *The Ice*, and David Campbell's *The Crystal Desert*.

I wish to thank the staff of the National Library of Australia, the library of the Australian Antarctic Division in Kingston, Tasmania, and the library and archive of the Scott Polar Research Institute in Cambridge, England. Robert Headland and the late William Mills were welcoming to an Australian scholar in Cambridge, and Andie Smithies, Graeme Watt and Meredith Inglis allowed me to establish a regular base camp in their wonderful library at the Australian Antarctic Division.

I am pleased to acknowledge the financial support of the Council for the Centenary of Federation which, in the early stages of my research, enabled the organisation of a conference in 2001 at the National Museum of Australia on *Australians in Antarctica*. Parts of the research for this project were funded by the council as well as by an Australian Research Council Linkage Grant with the Bureau of Meteorology and the National Museum of Australia (LP0347378).

My colleagues and students in the History Program of the Research School of Social Sciences provided a happy and most stimulating environment for my work on Antarctica. I am grateful for their friendship and counsel, and particularly thank Gordon Briscoe, Nicholas Brown, Desley Deacon, Barry Higman, Pat Jalland, Di Langmore, Ann McGrath, Kay Nantes, Tim Rowse, Barry Smith and Karen Smith. My doctoral students – Travis Cutler, Barbara Dawson, Kirsty Douglas, Karen Fox, Christine Hansen, Marie Kawaja, Rani Kerin, Darrell Lewis, Joy McCann, Jessie Mitchell, Lawrence Niewójt, Emily O'Gorman, Maxine Pitts, Tiffany Shellam, Tim Sherratt, Jeremy Taylor, Rebe Taylor and John Thompson – inspired me.

This book took shape in a sustained conversation with the gifted publisher, Ian Templeman. I feel very fortunate indeed to have had the opportunity to work with him and his talented team at Pandanus Books. I am also grateful to Phillipa McGuinness for encouraging this work from its beginnings, and for her friendship, enthusiasm and support as its publisher at UNSW Press. Jessica Perini, Di Quick and Heather Cam have ably seen it through production, and Kathleen McDermott at Harvard University Press has ensured that it will reach an international audience.

My family has made writing this book a pleasure. I thank Libby, Kate and Billy for so enthusiastically sharing my interest in Antarctica and for offering themselves as an early and sympathetic audience for my words. My parents, Ray and Kay Griffiths, have been a constant and wonderful support. Liz, Andrew, David and Margaret Fleming made my visits to Hobart especially enjoyable.

This book is dedicated to another family overseas, and to the kind of international friendship Antarctica fosters. Almost 30 years ago, an extraordinary English family welcomed us into their lives and blessed ours. My first conversation with Charlie Menzies-Wilson was about Antarctica, and I happily continue it in these pages.

# NOTES

## Prologue

1 Stephen Murray-Smith, *Sitting on Penguins: People and Politics in Australian Antarctica*, Hutchinson Australia, Sydney, 1988, pp 229–230.
2 Barry Lopez, 'The gift of good land', *Antarctic Journal*, vol xxvii, no 2, June 1992, pp 1–5.

## The Fire on the Snow

1 Roland Huntford, *Scott and Amundsen*, Hodder & Stoughton, London, 1979.
2 Jean-Baptiste Charcot, *Pourquoi faut-il aller dans l'Antarctique?* (1907), quoted in WA Hoisington Jr, 'In the service of the Third French Republic: Jean-Baptiste Charcot (1867–1936) and the Antarctic', *Proceedings of the American Philosophical Society*, vol 119, no 4, August 1975, p 316.
3 William S Bruce, 'Prefatory note', and RN Rudmose Brown, JH Pirie and RC Mossman, 'Preface', in RN Rudmose Brown, JH Pirie and RC Mossman, *The Voyage of the Scotia*, Mercat Press, Edinburgh, 2002 (first published 1906), pp xiii–xv.
4 Frank Debenham, *The Quiet Land: The Diaries of Frank Debenham*, June Debenham Back (ed), Bluntisham Books, Norfolk, 1992, pp 105, 126.
5 Folder of reviews of Huntford's book, Scott Polar Research Institute (SPRI) Library, Cambridge, including by Gordon Robin in *Polar Record*, 19 (123), 1979, pp 624–626; by Sir Vivian Fuchs in *Geographical Journal*, 146 (2), 1980, pp 272–274; by Wayland Young in *Encounter*, 54 (5), 1980, pp 8–19; and 'Scott and Amundsen: An exchange between Roland Huntford and Wayland Young', *Encounter*, November 1980, pp 85–89. See also Phillip Law, 'Fire on the ice', *Overland*, no 82, 1980, pp 58–63.
6 Alleged by Don Aldridge in *The Rescue of Captain Scott*, Tuckwell Press, East Linton, 1999, p 174.
7 The American-based geographers, Cindi Katz and Andrew Kirby, also dismissed the rock specimens as 'of little importance': Cindi Katz and Andrew Kirby, 'In the nature of things: the environment and everyday life', *Transactions of the Institute of British Geographers*, 16, 1991, note 6, p 269.
8 RJ Tingey, 'Heroic age geology in Victoria Land, Antarctica', *Polar Record*, vol 21, 1983, pp 451–457; AC Seward, 'Antarctic fossil plants', *British Antarctic (Terra Nova) Expedition, 1910. Natural History Report. Geology*, 1 (1), British Museum (Natural History), London, 1914, pp 1–49.
9 Ranulph Fiennes, *Captain Scott*, Hodder & Stoughton, London, 2004, p xiii.
10 Fiennes, *Captain Scott*, p 2.
11 Fiennes, *Captain Scott*, p 24.
12 Barry Lopez, 'Landscape and narrative', *Crossing Open Ground*, Vintage Books, New York, 1989, pp 61–71.
13 JW Gregory, Correspondence, 1899–1904, MS 1329, SPRI.
14 Debenham, *The Quiet Land*; see also Griffith Taylor (also loyal to Scott), *Journeyman Taylor: The Education of a Scientist*, Robert Hale Limited, London,

1958, p 108: 'Scott's total time of journey he estimated at 84 + 53 + 7 days (i.e. 144 days), so that, if he started on 3rd November, he could not be back until March, 27th.'

15 Ursula Le Guin, 'Heroes', in *Dancing at the Edge of the World*, Victor Gollancz Ltd, London, 1989, pp 171–175.

16 Philip Ayers, *Mawson: A Life*, Melbourne University Press, Carlton, 1999, pp 56–57.

17 Sir John Cleland, 'Hypervitaminosis A in the Antarctic in the Australasian Antarctic Expedition of 1911–14: A possible explanation of the illnesses of Mertz and Mawson', *The Medical Journal of Australia*, vol 1, no 26, 28 June 1969, pp 1337–1342; RV Southcott and NJ Chesterfield, 'Vitamin A content of the livers of huskies and some seals from Antarctic and subantarctic regions', *The Medical Journal of Australia*, 6 February 1971, pp 311–313.

18 John C Smith, conversation, 2005.

19 Captain John King Davis, Private journal, 22 January 1913, Davis Papers, Australian Manuscripts Collection, State Library of Victoria, Box 3232/5.

20 Lennard Bickel, *This Accursed Land*, Pan Macmillan, Sydney, 1977, p 185.

21 For this account of Mawson's journey, I have drawn on his *The Home of the Blizzard*, Wakefield Press, Adelaide, 1996 (first published 1915); *Mawson's Antarctic Diaries*, Fred Jacka and Eleanor Jacka (eds), Allen & Unwin, Sydney, 1988; Bickel, *This Accursed Land*; Philip Ayres, *Mawson: A Life*, Melbourne University Press, Carlton, 1999; and Belgrave Ninnis, Correspondence, 1911–12, SPRI.

22 Susan Solomon, *The Coldest March: Scott's Fatal Antarctic Expedition*, Melbourne University Press, Carlton, 2001. The Scott expedition's meteorologist, George Simpson, conducted a similar enquiry in the aftermath of the tragedy and came to the same conclusion as Solomon documents. But Solomon is able to draw also on the records of automated weather stations installed in Antarctica since the 1980s.

23 Solomon, *The Coldest March*, p 288.

24 Solomon, *The Coldest March*, p 327.

## The Breath of Antarctica

1 This account is drawn from Ernest Shackleton, *South: The Story of Shackleton's 1914-17 Expedition*, Heinemann, London, 1919; Frank Worsley, *Endurance: An Epic of Polar Adventure*, P Allan, London, 1931; and Roland Huntford, *Shackleton*, Atheneum, New York, 1986, pp 526–572.

2 Jan DeBlieu, *Wind: How the Flow of Air has Shaped Life, Myth, and the Land*, Houghton Mifflin Company, Boston, 1999, p 3.

3 Greg Dening, *Mr Bligh's Bad Language: Passion, Power and Theatre on the Bounty*, Cambridge University Press, Cambridge, 1992, p 77.

4 Jean-René Vanney, *Histoire des Mers Australes*, Fayard, Paris, 1986, p 22.

5 WLN Tickell, *Albatrosses*, Yale University Press, New Haven, 2000, p 261.

6 Belgrave Ninnis, Correspondence 1911–12, SPRI.

7 Glyndwr Williams, *The Great South Sea: English Voyagers and Encounters 1570-1750*, Yale University Press, New Haven, 1997.

8 Samuel Taylor Coleridge, 'The Rime of the Ancient Mariner', *Poetical Works*, 1797, in Arthur M Eastman (ed), *The Norton Anthology of Poetry*, WW Norton & Co, New York, 1970, p 599.

9   Robert Cushman Murphy, *Logbook for Grace: Whaling Brig Daisy, 1912-1913*, Robert Hale Ltd, London, 1948, p 116.

10  Edward A Wilson, *Aves*, vol II, 1907, quoted in William Jameson, *The Wandering Albatross*, Rupert Hart-Davis, London, 1958, p 23.

11  Jameson, *The Wandering Albatross*; Tickell, *Albatrosses*.

12  JH Parry, *The Discovery of the Sea*, University of California Press, Berkeley, 1981, chapter XII.

13  Parry, *Discovery of the Sea*, pp 86–87.

14  DeBlieu, *Wind*, p 50.

15  Frank Broeze, *Island Nation: A History of Australians and the Sea*, Allen & Unwin, Sydney, 1998.

16  Geoffrey Blainey, *The Tyranny of Distance*, Sun Books, Melbourne, 1966.

17  Ross Gibson, 'Ocean settlement', *Meanjin*, vol 53, no 4, Summer 1994, pp 665–678, at pp 666–667.

18  Grace Karskens, *The Rocks: Life in Early Sydney*, Melbourne University Press, Carlton, 1998, pp 18, 183.

19  Edward Snell, *The Life and Adventures of Edward Snell*, Tom Griffiths and Alan Platt (eds), Angus & Robertson, Sydney, 1988, pp 25–26 (9 October 1849).

20  Snell, *Life and Adventures*, pp 27–28 (13 October 1849).

21  Snell, *Life and Adventures*, p 384 (10 April 1858).

22  Snell, *Life and Adventures*, p 386 (27 April 1858).

23  Blainey, *Tyranny of Distance*, p 177.

24  Matthew Fontaine Maury, *The Physical Geography of the Sea and its Meteorology*, John Leighly (ed), the Belknap Press of Harvard University Press, Cambridge, 8th edition, 1963 (originally published 1861), p 307.

25  Maury, *Physical Geography of the Sea*, p 306.

26  Alan Platt, 'The City of Adelaide', unpublished manuscript on the economic and social history of the tall ships, kindly made available to the author.

27  Les Murray, *The Australian Year*, Angus & Robertson, Sydney, 1985, p 13.

28  Alan Frost, *Botany Bay Mirages*, Melbourne University Press, Carlton, 1994, chapter 7.

29  Quoted in Frost, *Botany Bay Mirages*, p 156.

30  RA Swan, *Of Myths and Mariners*, RA Swan, Sydney, no date, copy in SPRI, Cambridge.

31  Douglas Mawson, 'The unveiling of Antarctica', ANZAAS Presidential Address, in GW Leeper (ed), *Report of the Twenty-Second Meeting of the Australian and New Zealand Association for the Advancement of Science, Melbourne Meeting, January, 1935*, HJ Green, Government Printer, Melbourne, 1935, p 31.

32  Stephen J Pyne, *The Ice: A Journey to Antarctica*, University of Iowa Press, Iowa City, 1986, pp 41–42.

33  David Campbell, *The Crystal Desert: Summers in Antarctica*, Houghton Mifflin Co, Boston, 1992, chapter 8.

34  Jean M Grove, *The Little Ice Age*, Methuen, London, 1988; Campbell, *The Crystal Desert*, p 161.

35  Max Downes, 'First visitors to Heard Island', *ANARE Research Notes No 104*, no date (c 1990s), Australian Antarctic Division, Hobart.

36  John Gardner, 'Stormy weather: A history of research in the Bureau of Meteorology', *Metarch Papers No 11*, December 1997, Bureau of Meteorology, Melbourne, 1997.

37  Maury, *Physical Geography of the Sea*, pp 406–415.

38  Louis Bernacchi, *To the South Polar Regions: Expedition of 1898-1900*,
    Bluntisham Books, Erskine Press, Norfolk, 1991 (first published 1901), p 12.
39  Bernacchi, *To the South Polar Regions*, p 15.
40  Patricia Fara, *Sympathetic Attractions: Magnetic Practices, Beliefs, and Symbolism
    in Eighteenth-Century England*, Princeton University Press, Princeton, 1996.
41  Bernacchi, 'The climate of the South Polar regions', Part II of his *To the South
    Polar Regions*, pp 287–308.
42  Henry Chamberlain Russell, 'Moving anticyclones in the Southern
    Hemisphere', in Ralph Abercromby (ed), *Three Essays on Australian Weather*,
    FW White, Sydney, 1896, pp 1–15.
43  Griffith Taylor, 'Climatic relations between Antarctica and Australia', in
    WLG Joerg (ed), *Problems of Polar Research*, American Geographical Society,
    Special Publication No 7, New York, 1928, pp 285–300.
44  Maury to Robert FitzRoy, 25 February 1859, quoted in WJ Gibbs, *The Origins
    of Australian Meteorology*, Australian Government Publishing Service, Canberra,
    1975, p 9.
45  Robert Marc Friedman, *Appropriating the Weather: Vilhelm Bjerknes and the
    Construction of a Modern Meteorology*, Cornell University Press, New York, 1989.
46  Cecil T Madigan, 'Tabulated and reduced records of the Cape Denison station,
    Adelie Land', in Australasian Antarctic Expedition 1911–14, *Scientific Reports*,
    Series B, vol IV, Meteorology, Alfred J Kent, Government Printer, Sydney,
    1929, p 20.
47  Douglas Mawson, 'Preface to meteorological publications', *Scientific Reports*,
    Series B, vol III, p 5.
48  Morton Henry Moyes, Diary, Mitchell Library, Sydney, ML MSS 388/1,
    CY 3660.
49  Douglas Mawson, *The Home of the Blizzard*, Wakefield Press, Adelaide, 1996
    (first published 1915), p 281.
50  Moyes, Diary, 1912.
51  *Herald*, 22 December 1915.
52  *The Dominion*, 27 August 1913.
53  Ainsworth to Hunt, 12 August 1913, MS, Bureau of Meteorology, kindly
    brought to my attention by Neville Nicholls. See also George Ainsworth's
    chapters in Mawson, *The Home of the Blizzard*, chapters 24–26.
54  Pyne, *The Ice*, p 50.

## The University of the Southern Ocean

1  This narrative is drawn from Robert Cushman Murphy, *Logbook for Grace:
   Whaling Brig Daisy, 1912-1913*, Robert Hale Ltd, London, 1948; Grace E
   Barstow Murphy, *There's Always Adventure: The Story of a Naturalist's Wife*,
   George Allen & Unwin Ltd, London, 1952; and their granddaughter's
   biography, Eleanor Mathews, *Ambassador to the Penguins: A Naturalist's Year
   Aboard a Yankee Whaleship*, David R Godine, Boston, 2003.
2  RI Lewis Smith, 'Early nineteenth century sealers' refuges on Livingston
   Island, South Shetland Islands', *British Antarctic Survey Bulletin*, no 74, February
   1987, p 51.
3  Richard Ellis, *The Empty Ocean*, Island Press/Shearwater Books, Washington,
   2002, p 177.
4  Raymond Rallier du Baty in 1908–09, quoted by Bernadette Hince, 'The teeth

of the wind: An environmental history of subantarctic islands', PhD thesis, Australian National University, Canberra, 2005, p 80.

5   Louis Bernacchi, *To the South Polar Regions: Expedition of 1898-1900*, Hurst & Blackett, London, 1901, pp 43–44.

6   Hince, 'The teeth of the wind', p 78 (Kerguelen), p 162 (Heard).

7   Ellis, *The Empty Ocean*, p 167.

8   Hince, 'The teeth of the wind', p 124.

9   Stephen Martin, *The Whales' Journey*, Allen & Unwin, Sydney, 2001, p 83.

10  Martin, *The Whales' Journey*, p 87.

11  Robert Headland, 'Antarctic odyssey: Historical stages in development of knowledge of the Antarctic', in Aant Elzinga, Torgny Nordin, David Turner and Urban Wråkberg (eds), *Antarctic Challenges: Historical and Current Perspectives on Otto Nordenskjöld's Antarctic Expedition, 1901-1903*, Royal Society of Arts and Sciences, Göteberg, 2004, pp 15–24.

12  Roland Huntford, *Shackleton*, Atheneum, New York, 1986, pp 390–391.

13  Alan Gurney, *Below the Convergence: Voyages Towards Antarctica, 1699-1839*, Pimlico, London, 1998, p 60 for 'pasture of the ocean'.

14  David Campbell, *The Crystal Desert: Summers in Antarctica*, Minerva, London, 1993, pp 97–104.

15  Robert Headland, *The Island of South Georgia*, Cambridge University Press, Cambridge, 1984, p 50.

16  Headland, *South Georgia*, p 50.

17  Rachel Carson, *Silent Spring*, Houghton Mifflin Co, Boston, 1962, pp 103, 155–159.

18  Ernest Shackleton, *South: The Story of Shackleton's Last Expedition 1914-1917*, William Heinemann, London, 1919, p xv.

19  Francis Spufford, *I May Be Some Time*, Faber and Faber, London, 1996, pp 45–46; Huntford, *Shackleton*, pp 597–599, 602; Mathews, *Ambassador to the Penguins*, p 348.

## Great South Lands

1   John Cawte Beaglehole, *The Life of Captain James Cook*, Adam and Charles Black, London, 1974, p 107.

2   Beaglehole, *The Life of Captain James Cook*, pp 107–109.

3   Bernadette Hince, 'Something's missing down there', *Occam's Razor Talk*, ABC Radio National, 24 May 2004; Bill Green, *Water, Ice and Stone: Science and Memory on the Antarctic Lakes*, Harmony Books, New York, 1995, p 57.

4   Glyndwr Williams, *The Great South Sea: English Voyagers and Encounters 1570-1750*, Yale University Press, New Haven, 1997, p xiv.

5   Stephen J Pyne, *The Ice: A Journey to Antarctica*, University of Iowa Press, Iowa City, 1986, p 245.

6   Beaglehole, *The Life of Captain James Cook*, pp 431–436.

7   Beaglehole, *The Life of Captain James Cook*, p 292.

8   Jules S-C Dumont d'Urville, *An Account in Two Volumes of Two Voyages to the South Seas*, vol II, translated and edited by Helen Rosenman, Melbourne University Press, Carlton, 1987, pp 486–487; Charles Wilkes, *Narrative of the United States Exploring Expedition*, vol I, Ingram, Cook and Co, London, 1852, pp 276–277.

9 GS Griffiths, 'Antarctic exploration – The duty of Australia' (1888), quoted in Lynette Cole, 'Proposals for the first Australian Antarctic expedition', *Monash Publications in Geography*, no 39, 1990, p 27.

10 JW Gregory, 'Antarctic exploration', *Popular Science Monthly*, 60 (1901–02), pp 209–217.

11 Stephen J Pyne, 'Heart of whiteness: The exploration of Antarctica', *Environmental Review*, 10, 4, Winter 1986, p 234.

12 Douglas Mawson, 'The unveiling of Antarctica', ANZAAS Presidential Address, in GW Leeper (ed), *Report of the Twenty-Second Meeting of the Australian and New Zealand Association for the Advancement of Science, Melbourne Meeting, January, 1935*, HJ Green, Government Printer, Melbourne, 1935.

13 Laurence McKinley Gould, *COLD: The Record of an Antarctic Sledge Journey*, Brewer, Warren & Putnam, New York, 1931, p 170. Gould had visited the closer Rockefeller Mountains in the autumn but had been disappointed to find volcanic rock. His companions on the journey to Queen Maud Mountains had not been on land for a year.

14 Gould, *COLD*, p 183.

15 Paul A Carter, *Little America: Town at the End of the World*, Columbia University Press, New York, 1979, pp 140–141.

16 Quoted in Douglas Botting, *Humboldt and the Cosmos*, Michael Joseph, London, 1973, p 90.

17 Alan Moorehead, *Darwin and the Beagle*, Hamilton, London, 1969, p 169.

18 W Saville-Kent, *The Naturalist in Australia*, Chapman & Hall Ltd, London, 1897, pp 1–2, 7.

19 Aant Elzinga, Torgny Nordin, David Turner and Urban Wråkberg (eds), *Antarctic Challenges: Historical and Current Perspectives on Otto Nordenskjöld's Antarctic Expedition, 1901-1903*, Royal Society of Arts and Sciences, Göteberg, 2004, pp 179–182.

20 Charles Laseron, *The Face of Australia*, Angus & Robertson, London, 1953, p 87. Kirsty Douglas drew my attention to this quote.

21 Professor GE Nicholls, 'The fauna of WA and its biogeographical relations', Presidential Address, Royal Zoological Society of NSW, no date (c 1932–33), in the papers of Ellis Le Geyt Troughton, Australian Museum Archives, Sydney, Box no 1, Item 2.

22 Quoted in Nicholls, 'The fauna of WA', describing Wegener's view; Pyne, *The Ice*, p 261.

23 Green, *Water, Ice and Stone*, p 55.

24 Quoted in Hal Hellman, *Great Feuds in Science*, John Wiley & Sons Inc, New York, 1998, p 146.

25 Quoted in Hellman, *Great Feuds in Science*, p 153.

26 John McPhee, *Annals of the Former World*, Farrar, Straus and Giroux, New York, 1981, p 131.

27 Richard Fortey, *Life: An Unauthorised Biography*, Harper Collins, London, 1997, pp 217–220.

28 McPhee, *Annals*, pp 126–128.

29 McPhee, *Annals*, p 115.

30 Green uses the ancient scroll metaphor in *Water, Ice and Stone*, p 141.

31 Laurence McKinley Gould, 'Antarctica: The world's greatest laboratory', *American Scholar*, vol 40, no 3, Summer 1971, pp 402–415.

32 Gilbert Dewart, *Antarctic Comrades: An American with the Russians in Antarctica*,

The Ohio State University Press, Columbus, 1989, p 150.

33  Timothy F Flannery, 'Twenty million years of rangelands evolution in Australia and North America', in David Eldridge and David Freudenberger (eds), *People and Rangelands: Building the Future*, Proceedings of the VI International Rangeland Congress, Townsville, 1999, vol 1, pp 2–3.
34  Timothy F Flannery, *The Future Eaters: An Ecological History of the Australasian Lands and People*, Reed Books, Sydney, 1994, chapter 6.
35  I am drawing on Pyne, *The Ice*, chapter 6; Mary E White, *The Greening of Gondwana*, Reed Books, Sydney, 1986; and Eric Rolls, 'The nature of Australia', in Tom Griffiths and Libby Robin (eds), *Ecology and Empire: Environmental History of Settler Societies*, Keele University Press, Edinburgh, 1997, pp 35–45.
36  Jim Bowler and Mike Sandiford, 'Making Australia: Landscape, climate and tectonics', Geological Society of Australia Public Lecture, Melbourne, October 2003.
37  John Calaby, 'Foreword' to M Archer and G Clayton (ed), *Vertebrate Zoogeography and Evolution in Australasia: Animals in Space and Time*, Hesperian Press, Sydney, 1984.
38  This is the word used by White in *The Greening*, p 40.

## Heavenly Bodies

1   Stephen Pumfrey, *Latitude and the Magnetic Earth*, Icon Books, Cambridge, 2002, p 63.
2   Louis Bernacchi, *To the South Polar Regions: Expedition of 1898-1900*, Hurst and Blackett, London, 1901, p 131.
3   Pumfrey, *Latitude and the Magnetic Earth*; and Patricia Fara, *Sympathetic Attractions: Magnetic Practices, Beliefs and Symbolism in Eighteenth-Century England*, Princeton University Press, Princeton, 1996, p 14.
4   William Gilbert, *De Magnete*, quoted in Mary Midgley, *Science as Salvation: A Modern Myth and its Meaning*, Routledge, London, 1992, p 82.
5   Granville Allen Mawer, *South by Northwest: The Magnetic Crusade and the Contest for Antarctica*, Wakefield Press, Adelaide, 2006, pp 2, 109.
6   Harvey J Marchant, Desmond J Lugg and Patrick G Quilty (eds), *Australian Antarctic Science: The First 50 Years of ANARE*, Australian Antarctic Division (AAD), Kingston, 2002, p 206.
7   Douglas Mawson, 'Magnetic observations: The Magnetic Pole and the Aurora', in Ernest Shackleton, *The Heart of the Antarctic*, William Heinemann, London, vol 2, pp 358–361.
8   Philip Ayres, *Mawson: A Life*, Melbourne University Press, Carlton, 1999, pp 69–70.
9   Raymond Priestley, 'The Professor', Paper given in Cambridge c 1921, Scrapbook relating to TWE David, Priestley Papers, SPRI.
10  Marchant et al (eds), *Australian Antarctic Science*, pp 43, 56–58.
11  Stephen J Pyne, *The Ice: A Journey to Antarctica*, University of Iowa Press, Iowa City, 1986, pp 268–278.
12  Bernacchi, *To the South Polar Regions*, pp 138–139.
13  Jorgen Amundsen, 'Puzzle of explorers' tools left at the South Pole', *The Times*, 31 December 2003, p 9.
14  Harry Woolf, *The Transits of Venus: A Study of Eighteenth-Century Science*,

Princeton University Press, Princeton, 1959.

15 JC Beaglehole, *The Life of Captain James Cook*, Adam and Charles Black, London, 1974, p 157.

16 The Australian Solar Physics Committee published a *Memorandum Upon the Proposed Solar Observatory in Australia* in 1909 in which they surprisingly referred to the Sun as masculine. See Tom Frame and Don Faulkner, *Stromlo: An Australian Observatory*, Allen & Unwin, Sydney, 2003, p 19.

17 Beaglehole, *The Life of Captain James Cook*, pp 182–183.

18 Woolf, *The Transits of Venus*, p 195.

19 Griffith Taylor, Field journal, 31 July 1919, SPRI.

20 John Gribbin and Mary Gribbin, *Ice Age*, Allen Lane, Penguin, London, 2001; Brian John, *The Winters of the World: Earth under the Ice Ages*, The Jacaranda Press, Milton, 1979.

21 Johann Reinhold Forster, *Observations Made During a Voyage Around the World*, Nicholas Thomas, Harriet Guest and Michael Dettelbach (ed), University of Hawai'i Press, Honolulu, 1996, p 37.

22 Pyne, *The Ice*, p 278; MJS Rudwick, 'The glacial theory', *History of Science*, 8, 1969, pp 136–157.

23 Quoted in Kirsty Douglas, '"Pictures of time beneath": Science, landscape, heritage and the uses of the deep past in Australia, 1830–2003', PhD thesis, Australian National University, 2004, p 102.

24 John McPhee, *Annals of the Former World*, Farrar, Straus and Giroux, New York, 1981, p 257.

25 MA Smith, 'Palaeoclimates: An archaeology of climate change', in Tim Sherratt, Tom Griffiths and Libby Robin (eds), *A Change in the Weather: Climate and Culture in Australia*, National Museum of Australia Press, Canberra, 2005, pp 176–186.

26 Gribbin and Gribbin, *Ice Age*, Epilogue.

27 Douglas, 'Pictures of time beneath', chapter VII.

28 John McPhee, *Annals of the Former World*, p 32.

29 Keith McConnochie, 'Desert departures', in Xavier Pons, *Departures: How Australia Reinvents Itself*, Melbourne University Press, Carlton, 2002.

30 MA Smith, 'Prehistory and human ecology in Central Australia: an archaeological perspective', in SR Morton and DJ Mulvaney (eds), *Exploring Central Australia: Society, the Environment and the 1894 Horn Expedition*, Surrey Beatty & Sons, Chipping Norton, 1996, pp 61–73, at p 62.

31 MA Smith, 'Biogeography, human ecology and prehistory in the sandridge deserts', *Australian Archaeology*, no 37, 1993, pp 35–50.

32 Bloo Campbell, 'Notes from the Antarctic desert', *BushMag: Journal of the Outback*, 2003 <http://www.bushmag.com.au>, viewed 1 November 2006.

33 Quoted in Ayres, *Mawson*, p 12.

34 Douglas Mawson, *The Home of the Blizzard*, Wakefield Press, Adelaide, 1996 (first published 1915), p 40.

35 Richard Byrd, *Little America*, GP Putnam's Sons, London, 1930, p 353.

36 Pyne, *The Ice*, p 292.

37 Charles Swithinbank, *Foothold on Antarctica: The First International Expedition (1949-1952) Through the Eyes of its Youngest Member*, The Book Guild Ltd, Sussex, 1999, p 212; Gordon de Quetteville Robin, 'The seismic journey 1951–52', Appendix 4 to John Giaever, *The White Desert: The Official Account of the Norwegian-British-Swedish Antarctic Expedition*, Chatto & Windus, London,

1954, pp 281–298.
38 Marchant et al (eds), *Australian Antarctic Science*, p 412.
39 Gordon de Quetteville Robin, 'The ice of the Antarctic', *Scientific American*, vol 207, no 3, September 1962, p 138; 'Obituary: Gordon Robin', *The Darwinian*, Spring 2005, pp 8–9.
40 Gilbert Dewart, *Antarctic Comrades: An American with the Russians in Antarctica*, The Ohio State University Press, Columbus, 1989, p 155; AM Gusev, 'Sledge-tractor expedition into the interior of the Antarctic continent', *Information Bulletin of the Soviet Antarctic Expedition*, no 1, 1958, Elsevier Publishing Company, Amsterdam, 1964; John Béchervaise, *Blizzard and Fire: A Year at Mawson, Antarctica*, Angus & Robertson, Sydney, 1963, p 210; Stephen Martin, *A History of Antarctica*, State Library of NSW Press, Sydney, 1996, p 14.
41 Robert Thomson, *The Coldest Place on Earth*, AH & AW Reed, Wellington, 1969; Linda Clark and Elspeth Wishart, *66° South*, Australian Antarctic Foundation, Launceston, pp 70–74.
42 Phillip Law, 'The IGY in Antarctica', *The Australian Journal of Science*, vol 21, no 9, 1959, pp 285–294, at p 292; William F Budd, 'The Antarctic ice sheet', in Marchant et al (eds), *Australian Antarctic Science*, pp 309–390.
43 VI Morgan, CW Wookey, J Li, TD van Ommen, W Skinner, MF Fitzpatrick, 'Site information and initial results from deep ice drilling on Law Dome, Antarctica', *Journal of Glaciology*, vol 43, no 143, 1997, pp 3–10; Keith Scott, 'Earth's frozen history defrosted with a drill', *Canberra Times*, 21 March 1989, p 16; Budd, 'The Antarctic ice sheet', in Marchant et al (eds), *Australian Antarctic Science*, pp 348–352; Stephen Warren, 'At Vostok station with the 35th Soviet Antarctic expedition', *Aurora*, vol 11, no 3, March 1992, pp 1–7; interview (28 August 2006) with Gary Burns of the AAD, visitor to Vostok, summer 2005–06.
44 Barry Lopez, 'Trouble way down under', *Washington Post*, 27 March 1988, pp C1–2.
45 Pyne, *The Ice*, p 288.

## Planting Flags

1 John King Davis, 13 January 1930, Personal journal kept on board the *Discovery* during 1929–30, JK Davis Papers, Australian Manuscripts Collection, State Library of Victoria (SLV), Box 3236/6a.
2 *Cape Times*, 8 October 1929, newscutting in Davis Papers, SLV, Box 3189/1. See Eugene Rodgers, *Beyond the Barrier: The Story of Byrd's First Expedition to Antarctica*, Airlife Publishing Ltd, Shrewsbury, 1990, p 17.
3 Headline in the *Adelaide Advertiser*, 8 April 1929, quoted in Christy Collis, 'The Proclamation Island moment: Making Antarctica Australian', *Law Text Culture*, vol 8, 2004, pp 39–56, at p 45.
4 Philip Ayres, *Mawson: A Life*, Melbourne University Press, Carlton, 1999, chapter 12.
5 Alan K Henrikson, 'The last place on Earth, the first place in Heaven', in *Imagining Antarctica/Antarktis-Vorstellung und Wirklichkeit*, Nordico, Linz, 1986, p 13; Peter J Beck, 'Securing the dominant "Place in the wan Antarctic sun" for the British empire: The policy of extending British control over Antarctica', *Australian Journal of Politics and History*, vol 29, no 3, 1983, p 459.
6 Minutes of the Australian Antarctic Committee, 12 March 1929, quoted in Ayres, *Mawson*, p 173.

7 Ayres, *Mawson*, p 173.
8 Douglas Mawson, *Mawson's Antarctic Diaries*, Fred Jacka and Eleanor Jacka (eds), Allen & Unwin, Sydney, 1988, 3 January 1930, p 304.
9 Peter J Beck, 'British Antarctic policy in the early twentieth century', *Polar Record*, vol 21, no 134, May 1983, pp 475–483; Christopher Joyner, *Governing the Frozen Commons: The Antarctic Regime and Environmental Protection*, University of South Carolina Press, Columbia, 1998, p 50; Collis, 'The Proclamation Island moment', pp 39–56.
10 Davis, 15 January 1930, Personal journal, SLV, Box 3236/6a.
11 Davis, 18 January 1930, Personal journal, SLV, Box 3236/6a.
12 Reported in the *Cape Times* and quoted in R A Swan, *Australia in the Antarctic*, Melbourne University Press, Carlton, 1961, p 191.
13 *Argus* (Melbourne), 12 April 1930.
14 Stuart Campbell, Diary, Mitchell Library, Sydney, Microfilm CY 4317, 6 February 1931.
15 Stuart Campbell, Diary, 9 February 1931.
16 *Argus* (Melbourne), 14 April 1930.
17 'Sir Douglas Mawson explains polar voyage', *Herald* (Melbourne), 5 August 1929.
18 *Herald* (Melbourne), 18 September 1929.
19 Quoted in A Grenfell Price, *The Winning of Australian Antarctica: Mawson's BANZARE Voyages 1929-31, Based on the Mawson Papers*, Angus & Robertson, Sydney, 1962, pp 56–57.
20 Davis, 26 December 1929, Personal journal, SLV, Box 3236/6a.
21 Fred Jacka and Eleanor Jacka (eds), *Mawson's Antarctic Diaries*, 25 January 1930, p 325.
22 Fred Jacka and Eleanor Jacka (eds), *Mawson's Antarctic Diaries*, 13 February 1931, p 378.
23 RG Simmers, Diary, 18 February 1931, extracted in Price, *The Winning of Australian Antarctica*, p 156.
24 Collis, 'The Proclamation Island moment', pp 39–56.
25 Collis, 'The Proclamation Island moment', p 53.
26 Stuart Campbell, Diary, 12 December 1930, 5 January 1931, 24 December 1930 and 5 January 1931.
27 Bernt Balchen, *Come North With Me: An Autobiography*, EP Dutton & Co, Inc, New York, 1958, p 24.
28 Theodore K Mason, *Two Against the Ice: Amundsen and Ellsworth*, Dodd, Mead & Co, New York, 1982, p 117.
29 Richard E Byrd, *Little America: Aerial Exploration in the Antarctic, The Flight to the South Pole*, GP Putnam's Sons, New York, 1930, p 241.
30 Byrd, *Little America*, pp 47–48.
31 Balchen, *Come North With Me*, p 178.
32 Byrd, *Little America*, p 91.
33 Douglas Craig, *Fireside Politics: Radio and Political Culture in the United States, 1920-40*, The Johns Hopkins University Press, Baltimore, 2000, p xi.
34 Balchen, *Come North With Me*, p 178.
35 Byrd, *Little America*, p 221.
36 Beau Riffenburgh, *The Myth of the Explorer: The Press, Sensationalism, and Geographical Discovery*, Belhaven Press, London, 1993, p 194.
37 Byrd, *Little America*, p 192.

38  Dennis Rawlins, *Peary at the North Pole: Fact or Fiction?*, Robert B Luce Inc, Wellington, 1973, p 272.

39  Rawlins, *Peary at the North Pole*, pp 262–264.

40  Rodgers, *Beyond the Barrier*, Preface.

41  General Billy Mitchell quoted in Balchen, *Come North With Me*, p 60.

42  Balchen claimed that in Chicago in 1927, shortly before his death, Bennett confessed to him that he and Byrd had never reached the North Pole. Balchen had been suspicious about the flight arithmetic of speed and distance. He was later pressured to censor his published autobiography of this revelation, but one or two hints remain. See Balchen, *Come North With Me*, p 300, and Rawlins, *Peary at the North Pole*, pp 271–272.

43  Byrd, *Little America*, p 342.

44  Balchen, *Come North With Me*, p 191.

45  Paul A Carter, *Little America: Town at the End of the World*, Columbia University Press, New York, 1979, pp 204–205.

46  Peter J Beck, 'A cold war: Britain, Argentina and Antarctica', *History Today*, vol 37, 1987, pp 16–23; Klaus Dodds, *Pink Ice: Britain and the South Atlantic Empire*, IB Tauris Publishers, London, 2002, chapter 4; Jason Kendall Moore, 'Bungled publicity: Little America, big America, and the rationale for non-claimancy, 1946–61', *Polar Record*, vol 40, no 212, 2004, p 23.

47  Stuart Campbell to Mary Jocelyn Long, March 1947, in Mary Jocelyn Long – Letters from Stuart Campbell, 1932–1985, Mitchell Library, Sydney, MS 7191.

48  US Navy directive, 26 August 1946, quoted in Carter, *Little America*, p 223.

49  Interview 1987–88, quoted in Tim Bowden, *The Silence Calling: Australians in Antarctica 1947-1997: The ANARE Jubilee History*, Allen & Unwin, Sydney, 1997, pp 15–16; Geoffrey D Munro, '"Waiting on the weather": The ANARE Years 1947–1955', in Ken Green and Eric Woehler (eds), *Heard Island: Southern Ocean Sentinel*, Surrey Beatty & Sons, Sydney, 2006, pp 202–230.

50  Bowden, *The Silence Calling*, pp 19, 9–10.

51  Bowden, *The Silence Calling*, pp 26–27.

52  RG Chittleborough, 'Early days at Heard Island', *Aurora: ANARE Club Journal*, 5, 4 June 1986, pp 19–22.

53  Bowden, *The Silence Calling*, p 29.

54  Bowden, *The Silence Calling*, p 35.

55  Phillip Law, *Australia and the Antarctic*, The John Murtagh Macrossan Memorial Lectures, 1960, University of Queensland Press, St Lucia, 1962, pp 3–4.

56  Alan Attwood, 'Modern hero of Australian Antarctica: Phillip Law', *Australian Geographic*, no 18, January–March 1989, pp 116–117.

57  'Establishment of Mawson station', undated typescript memorandum, AAD Library, 91(*7) 91(091). Marie Kawaja, 'The politics and diplomacy of the Australian Antarctic', PhD in progress, Australian National University, Canberra, manuscript, 2005.

58  Letters of Mawson to Davis, Davis Papers, SLV, Box 3270/9.

59  Fred Elliott, 'Establishment of Mawson Base 1954', typescript of original diary with later comments in parentheses, MS 9442/1, National Library of Australia.

60  Elliott, 'Establishment', Saturday 13 February 1954, and Diary – January 1955–December 1955, MS 9442/2, National Library of Australia; Phillip Law, *Antarctic Odyssey*, Heinemann, Melbourne, 1983, chapter 9.

## Cold Peace

1   GE Fogg, *A History of Antarctic Science,* Cambridge University Press, Cambridge, 1992, p 147.
2   Phillip Law, *Australia and the Antarctic,* The John Murtagh Macrossan Memorial Lectures, 1960, University of Queensland Press, St Lucia, 1962, p 21.
3   Jason Kendall Moore, 'Thirty-seven degrees frigid: US-Chilean relations and the spectre of Polar Arrivistes, 1950–59', *Diplomacy and Statecraft,* vol 14, no 4, December 2003, pp 69–93, and 'Bungled publicity: Little America, big America, and the rationale for non-claimancy, 1946–61', *Polar Record,* vol 40, no 212, 2004, pp 19–30; Klaus Dodds, *Pink Ice: Britain and the South Atlantic Empire,* IB Tauris Publishers, London, 2002, p 84.
4   Stuart Macintyre, *A Concise History of Australia,* Cambridge University Press, Cambridge, 1999, pp 211–213.
5   Irina Gan, '"There was no cold war in Antarctica": Contribution of Dr Phillip Law to Australia's Antarctic endeavours', *Ice Breaker,* no 34, March–May 2006, pp 14–15, and 'The Soviet preparation for the IGY Antarctic program and the Australian response: Politics and science', Conference Paper, September 2006, kindly made available by the author; RG Casey, Diary, 30 July 1955, quoted in Kathleen Ralston, *Phillip Law: The Antarctic Exploration Years, 1954-66,* Australian Government Publishing Service, Canberra, 1998, p 62.
6   Mikhail Mikhailovich Somov, 'The Soviet Antarctic expedition of the Academy of Sciences of the USSR', typescript, 7 June 1957, Soviet expedition files, AAD Library; 'Photo materials of the Soviet Antarctic expedition of 1955–56', AAD Library, 91(08)(*7), 1955–56.
7   Mikhail Mikhailovich Somov, 'Soviet research on the Antarctic continent', *Information Bulletin of the Soviet Antarctic Expedition,* no 1, 1958, Elsevier Publishing Company, Amsterdam, 1964, pp 1–4.
8   Mikhail Mikhailovich Somov, 'Cooperation of scientists in the Antarctic', *Vestnik (Herald) of the Academy of Sciences, USSR,* January 1966, pp 73–80, translated into English, typescript, AAD Library; Gan, 'There was no cold war in Antarctica', pp 14–15.
9   JCG Kevin, Outward teletype messages to Phillip Law, 20 March 1956 and 16 April 1956, National Archives of Australia (NAA: A1838, 1495/1/9/4 PART 3) quoted in Alessandro Antonello, '"Going opposite ways": Phillip Law and Australian Antarctic policy, 1949–1959', Research essay, Australian National University, Canberra, 2006.
10  Nankyoku Tanken Koenkai, 1913 (Account of Japanese Antarctic Expedition 1911–12, probably written by a subordinate of Shirase, translated by Hilary Shibata), typescript, SPRI Library; Ivor Hamre, 'The Japanese South Polar expedition of 1911–1912: A little-known episode in Antarctic exploration', *The Geographical Journal,* vol 82, no 5, November 1933, pp 411–423; David Branagan, *TW Edgeworth David: A Life: Geologist, Adventurer, Soldier and 'Knight in the Old Brown Hat',* National Library of Australia, Canberra, 2005, pp 231–237.
11  Mawson to Davis, 6 April 1956, John King Davis Papers, Australian Manuscripts Collection, SLV, Box 3270/9.
12  John Béchervaise, *Blizzard and Fire: A Year at Mawson, Antarctica,* Angus & Robertson, Sydney, 1963, p 195.
13  John Bunt, *Antarctic Memoirs,* Seaview Press, Adelaide, 2006, p 143.

14  Vivian Fuchs and Edmund Hillary, *The Crossing of Antarctica: The Commonwealth Trans-Antarctic Expedition, 1955-58*, Cassell, London, 1958. This book was given to me by my friend and fellow-historian, Alan Platt, with the inscription: 'This was a present to my father, and I suspect has never been read, a fate which, at a few glances, it seems to deserve.'

15  Joe MacDowall, *On Floating Ice: Two Years on an Antarctic Ice-shelf South of 75°S*, The Pentland Press Ltd, Bishop Auckland, 1999; and Liz Cruwys, 'Review of *On Floating Ice*', in *The Magazine of The Cambridge Society*, no 45, 1999–2000, pp 55–56.

16  H Robert Hall, 'International regime formation and leadership: The origins of the Antarctic Treaty', PhD thesis, University of Tasmania, 1994, and 'Casey and the Antarctic Treaty negotiations', *Australian Antarctic*, no 2, Spring 2001; Dodds, *Pink Ice*, chapter 5; Moore, 'Thirty-seven degrees frigid'.

17  Alan K Henrikson, 'The last place on Earth, the first place in Heaven', in *Imagining Antarctica/Antarktis-Vorstellung und Wirklichkeit*, Nordico, Linz, 1986, p 11.

18  Fogg, *A History of Antarctic Science*, p 177.

19  Gilbert Dewart, *Antarctic Comrades: An American with the Russians in Antarctica*, The Ohio State University Press, Columbus, 1989, pp 9, 81, 102.

20  Dewart, *Antarctic Comrades*, pp 30–31, 38–39.

21  Dewart, *Antarctic Comrades*, pp 47, 66, 67–68.

22  Dewart, *Antarctic Comrades*, pp 101, 113.

23  Dewart, *Antarctic Comrades*, pp 45, 77.

24  Dewart, *Antarctic Comrades*, p 71.

25  Dewart, *Antarctic Comrades*, pp 166, 155–156.

26  Charles Swithinbank, *Vodka on Ice: A Year with the Russians in Antarctica*, The Book Guild, Sussex, 2002.

27  Galen Rowell, *Poles Apart: Parallel Visions of the Arctic and Antarctic*, University of California Press, Berkeley, 1995, p 174.

28  John May, *Greenpeace Book on Antarctica: A New View of the Southern Continent*, Dorling Kindersley, London, 1988, p 131.

29  Alexa Thomson, *Antarctica on a Plate*, Random House Australia, Sydney, 2003, pp 225–233, 266.

30  Irina Gan, 'Russia in Antarctica', *Aurora*, June 2006, p 7.

## Wintering

1  Stephen Martin, *A History of Antarctica*, State Library of NSW Press, Sydney, 1996, p 109.

2  Frederick Cook, *Through the First Antarctic Night, 1898-1899, A Narrative of the Belgica*, Heinemann, London, 1900, pp xx, 290, 282, 296, 370.

3  Cook, *Antarctic Night*, pp 301, 317, 319, 322, 332, page facing 359.

4  Cook, *Antarctic Night*, pp 291, 307, 292.

5  Cook, *Antarctic Night*, p 375.

6  Martin, *A History of Antarctica*, p 111.

7  Dennis Rawlins, *Peary at the North Pole: Fact or Fiction?*, Robert B Luce Inc, Washington, 1973.

8  For a summary of Bernacchi's career, see RA Swan, 'Louis Charles Bernacchi', *Victorian Historical Magazine*, vol 33, no 3, 1963, pp 379–398.

9  Louis Bernacchi, *To the South Polar Regions: Expedition of 1898-1900*, Hurst &

Blackett, London, 1901, p 17.

10 Roland Huntford, *Scott and Amundsen*, Hodder & Stoughton, London, 1979, p 532.

11 Hobart's relationship with southern polar regions has been explored in Lorne K Kriwoken and John W Williamson, 'Hobart, Tasmania: Antarctic and Southern Ocean connections', *Polar Record*, 29 (169), 1993, pp 93–102, and Ann Savours, 'Hobart and the polar regions', in Gillian Winter (ed), *Tasmanian Insights: Essays in Honour of Geoffrey Thomas Stilwell*, State Library of Tasmania, 1992, pp 175–191. See also Gillian Winter, '"Tasmania hails you as her favour'd guest": The British Antarctic expedition in Hobart, 1840–41', *Tasmanian Historical Research Association Papers and Proceedings*, vol 38, no 384, December 1991, pp 138–147.

12 Lady Jane Franklin, letter to her father, quoted in Alan Gurney, *The Race to the White Continent: Voyages to the Antarctic*, WW Norton & Co, New York, 2000, p 228.

13 Bernacchi, *To the South Polar Regions*, p 23.

14 Hugh B Evans, 'The *Southern Cross* expedition, 1898–1900: A personal account', *Polar Record*, vol 17, no 106, 1974, pp 23–30.

15 On Borchgrevink's career, see Hugh B Evans and AGE Jones, 'A forgotten explorer: Carsten Egeberg Borchgrevink', *Polar Record*, vol 17, no 108, 1975, pp 221–235. RK Headland assesses the various claims to be first in 'First on the Antarctic continent?', *Antarctic*, vol 13, no 8, December 1994, pp 346–348.

16 *Mercury*, 6 December 1898, p 2.

17 Janet Crawford, *That First Antarctic Winter: The Story of the Southern Cross Expedition of 1898-1900 as Told in the Diaries of Louis Charles Bernacchi*, South Latitude Research Ltd, in association with Peter J Skellerup, Christchurch, 1998, p 112.

18 Bernacchi, *To the South Polar Regions*, p 110.

19 Crawford, *That First Antarctic Winter*, p 113.

20 Crawford, *That First Antarctic Winter*, pp 121–122.

21 Crawford, *That First Antarctic Winter*, pp 122, 124.

22 Crawford, *That First Antarctic Winter*, p 127.

23 Bernacchi, *To the South Polar Regions*, p 137.

24 For this account of the winter, I have drawn mostly on Bernacchi's published and unpublished accounts (in Bernacchi, *To the South Polar Regions* and Crawford, *That First Antarctic Winter*) as well as Borchgrevink's published record, *First on the Antarctic Continent: Being an Account of the British Antarctic Expedition, 1898-1900*, Australian National University Press, Canberra, 1980 (first published by George Newnes, London, 1901), and Hugh B Evans, 'The *Southern Cross* expedition, 1898–1900: A personal account', *Polar Record*, vol 17, no 106, 1974, pp 23–30.

25 However, one contemporary reported in October 1901: 'Mr Bernacchi's Antarctic experiences have I think not been such as to inspire him with a fervid desire for more, but it seems they were extremely keen on having him.' Audrey Gregory to Edith Chaplin, 15 October 1901, JW Gregory Papers, Scott Polar Research Institute, Cambridge.

26 Bernacchi, *To the South Polar Regions*, pp 64–65.

27 Bernacchi, *To the South Polar Regions*, pp ix–x.

28 Bernacchi, *To the South Polar Regions*, p 139.

29 Bernacchi, *To the South Polar Regions*, p 78.

30  Raymond Priestley, '1913–14 (The aftermath)', Papers written between 1907–14, 1913–14, June 1962, in the John King Davis Papers, Australian Manuscripts Collection, SLV, Box 3241/15.

31  Barry Lopez, 'Trouble way down under', *Washington Post*, 27 March 1988, pp C1-2.

32  I have drawn this abbreviated account of Jeffryes' illness from Philip Ayres, *Mawson: A Life*, Melbourne University Press, Carlton, 1999, pp 89–95. Ayres reproduces a large part of Mawson's prepared speech.

33  Douglas Mawson to Paquita Delprat, 26 December 1913, Nancy Robinson Flannery (ed), *This Everlasting Silence: The Love Letters of Paquita Delprat and Douglas Mawson, 1911-1914*, Melbourne University Press, Carlton, 2000, p 125.

34  Sidney N Jeffryes, Letter to Miss Eckford, July 1914, Mitchell Library, Sydney, CY4164.

35  John Erskine, Mawson station log, 1967, internal report, Australian Antarctic Division (AAD), Commonwealth of Australia.

36  Raymond Priestley, 'The psychology of exploration' (1914), in the Davis Papers, SLV, Box 3241/15.

37  Charles Swithinbank, *Foothold on Antarctica: The First International Expedition (1949-52) Through the Eyes of its Youngest Member*, The Book Guild Ltd, Sussex, 1999, p 226.

38  André Migot, *The Lonely South*, Rupert Hart-Davis, London, 1957, pp 30–37.

39  André Migot, *The Lonely South*, pp 54–55.

40  Stephen Murray-Smith, *Sitting on Penguins: People and Politics in Australian Antarctica*, Hutchinson Australia, Sydney, 1988, p 105.

41  'Casey Station 1985 – OIC Annual Report', internal report, AAD, Commonwealth of Australia.

42  Barry Martin, 'Casey Station, Antarctica – Report of the 1986 Officer-In-Charge', internal report, AAD, Commonwealth of Australia, p 17.

43  Murray-Smith, *Sitting on Penguins*, p 234.

44  Martin, 'Casey Station, Antarctica – Report of the 1986 Officer-In-Charge', p 17.

45  Robert Easther, 'Davis Station 1986 – OIC Annual Report', December 8, 1986, internal report, AAD, Commonwealth of Australia, p 6.

46  Press clippings file for October 1996, AAD Library.

47  Raymond Priestley, *Antarctic Adventure: Scott's Northern Party*, T Fisher Unwin, London, 1914, p 354.

48  Deborah Cameron, 'Deep freeze', *Sydney Morning Herald*, Spectrum, 16 October 1999.

49  Desmond J Lugg, 'Antarctica as a space laboratory', G Hempel (ed), *Antarctic Science: Global Concerns*, Springer-Verlag, Berlin, 1994, pp 229–242, and Desmond J Lugg, 'What did happen to all those specimens, Doc?', in Harvey J Marchant, Desmond J Lugg and Patrick G Quilty (eds), *Australian Antarctic Science: The First 50 Years of ANARE*, AAD, Kingston, 2002, pp 73–104; Desmond J Lugg, 'The adaptation of a small group to life on an isolated Antarctic station', in OG Edholm and EKE Gunderson (eds), *Polar Human Biology*, William Heinemann, Sussex, 1973, pp 401–409.

50  G Palmai, 'Psychological observations on an isolated group in Antarctica', *British Journal of Psychiatry*, vol 109, 1963, pp 364–370.

51  Lugg, 'What did happen to all those specimens, Doc?', in Marchant et al (eds), *Australian Antarctic Science*, p 85.

52 Knowles Kerry, 'Can Antarctic explorers beget (male) Antarctic explorers?', *Aurora*, Spring 1976, p 33.
53 John McLaren, *Free Radicals: Of the Left in Postwar Melbourne*, Australian Scholarly Publishing, Melbourne, 2003, p 1.
54 Stephen Murray-Smith, 'Three islands: A case study in survival', in Imelda Palmer (ed), *Melbourne Studies in Education 1986*, Melbourne University Press, Carlton, 1986, pp 209–224; Murray-Smith, *Sitting on Penguins*, p 108.
55 Greg Dening, *Mr Bligh's Bad Language: Passion, Power and Theatre on the Bounty*, Cambridge University Press, Cambridge, 1992, pp 55–87.
56 Richard E Byrd, *Little America: Aerial Exploration in the Antarctic, The Flight to the South Pole*, GP Putnam's Sons, New York, 1930, pp 197–198.
57 Michael Parfit, 'Reclaiming a lost Antarctic base', *National Geographic*, vol 183, no 3, March 1993, p 125.

## Solitude

1 Laurence McKinley Gould, *COLD: The Record of an Antarctic Sledge Journey*, Brewer, Warren & Putnam, New York, 1931, p 65.
2 Morton Henry Moyes, Diary, Mitchell Library, Sydney, ML MSS 388/1, CY 3660.
3 Gould, *COLD*, p 31; Alexa Thomson, *Antarctica on a Plate*, Random House Australia, Sydney, 2003, p 341.
4 André Migot, *The Lonely South*, Rupert Hart-Davis, London, 1957, p 145.
5 Richard E Byrd, *Little America: Aerial Exploration in the Antarctic, The Flight to the South Pole*, GP Putnam's Sons, New York, 1930, pp 192–193.
6 This account draws on Byrd's book about the ordeal, entitled *Alone*, Putnam, London, 1938, as well as interpretations of the experience and the expedition by Eugene Rodgers, *Beyond the Barrier: The Story of Byrd's First Expedition to Antarctica*, Airlife Publishing Ltd, Shrewsbury, 1990; Stephen J Pyne, *The Ice: A Journey to Antarctica*, University of Iowa Press, Iowa City, 1986, p 191; Paul A Carter, *Little America: Town at the End of the World*, Columbia University Press, New York, 1979; and Robert N Matuozzi, 'Richard Byrd, polar exploration, and the media', *The Virginia Magazine of History and Biography*, vol 110, no 2, 2002, pp 209–236.
7 Richard E Byrd, *Alone*, Putnam, London, 1938, p 7.
8 Byrd, *Alone*, p 120.
9 Byrd, *Alone*, p 120.
10 Byrd, *Alone*, pp 14–15.
11 Gould, *COLD*, p 31.
12 Emmanuel Cohen quoted in Matuozzi, 'Richard Byrd, polar exploration, and the media', p 229.
13 Matuozzi, 'Richard Byrd, polar exploration, and the media', p 235.

## Honeymoon on Ice

1 Paquita Delprat to Douglas Mawson, 21 September 1913, in Nancy Robinson Flannery (ed), *This Everlasting Silence: The Love Letters of Paquita Delprat and Douglas Mawson, 1911-1914*, Melbourne University Press, Carlton, 2000, p 102.
2 Douglas to Paquita, 9 and 10 November 1912, in Flannery (ed), *This Everlasting Silence*, pp 47–48.

3  Paquita to Douglas, 17 August 1913, in Flannery (ed), *This Everlasting Silence*, p 97.
4  Douglas to Paquita, 3 January 1912, 24 June, 15 July 1913, in Flannery (ed), *This Everlasting Silence*, pp 27, 83, 91.
5  Douglas to Paquita, 1 June 1913, in Flannery (ed), *This Everlasting Silence*, p 69.
6  Paquita to Douglas, 17 August 1913, in Flannery (ed), *This Everlasting Silence*, p 97.
7  Paquita to Douglas, 21 April 1913, in Flannery (ed), *This Everlasting Silence*, pp 58–59.
8  Douglas to Paquita from the Antarctic Ocean, 26 December 1913, in Flannery (ed), *This Everlasting Silence*, p 125.
9  Douglas to Paquita, 3 January 1912, in Flannery (ed), *This Everlasting Silence*, p 25.
10  Paquita to Douglas, 9 November 1913, in Flannery (ed), *This Everlasting Silence*, p 112.
11  Quoted in Brigid Hains, *The Ice and the Inland: Mawson, Flynn, and the Myth of the Frontier*, Melbourne University Press, Carlton, 2002, p 21.
12  Douglas to Paquita, 26 December 1913, in Flannery (ed), *This Everlasting Silence*, p 123.
13  Lisa Bloom, *Gender on Ice: American Ideologies of Polar Expeditions*, University of Minnesota Press, Minneapolis, 1993, p 6.
14  Patricia Grimshaw, Marilyn Lake, Ann McGrath and Marian Quartly, *Creating a Nation, 1788-1990*, McPhee Gribble, Melbourne, 1994, p 193.
15  Douglas to Paquita, 1 June 1913, in Flannery (ed), *This Everlasting Silence*, pp 66–68.
16  Hains, *The Ice and the Inland*, pp 12–13.
17  Sara Wheeler, *Terra Incognita: Travels in Antarctica*, Vintage, London, 1997, p 145.
18  Marilyn Lake, 'The politics of respectability: Identifying the masculinist context', *Historical Studies*, vol 22, 1986, pp 116–131.
19  Douglas to Paquita, 5 May 1913, in Flannery (ed), *This Everlasting Silence*, p 63.
20  Hains, *The Ice and the Inland*, p 14.
21  Douglas to Paquita, 3 January 1912, in Flannery (ed), *This Everlasting Silence*, p 27.
22  Douglas to Paquita, 22 July 1913, in Flannery (ed), *This Everlasting Silence*, p 94.
23  Douglas to Paquita, July 1913, in Flannery (ed), *This Everlasting Silence*, pp 86–87.
24  Mawson's letter to Kathleen Scott is quoted in Philip Ayers, *Mawson: A Life*, Melbourne University Press, Carlton, 1999, p 89, and Peter Pan is quoted in Hains, *The Ice and the Inland*, p 13. For more reflections on Mawson's intense friendship with Kathleen Scott, see Brigid Hains' review of Ayres' biography of Mawson in *Historical Records of Australian Science*, vol 13, no 2, 2000, pp 226–228.
25  I have based this narrative on Jennie Darlington's account (as told to Jane McIlvaine) in *My Antarctic Honeymoon: A Year at the Bottom of the World*, Doubleday, New York, 1956.
26  Elizabeth Chipman, *Women on the Ice: A History of Women in the Far South*, Melbourne University Press, Carlton, 1986, p 79.
27  Paul-Emile Victor, *Man and the Conquest of the Poles*, Hamish Hamilton Ltd,

London, 1964 (first published in French in 1962), p 257.

28 Michael Parfit, 'Reclaiming a lost Antarctic base', *National Geographic*, vol 183, no 3, March 1993, p 125.

29 John C Behrendt, *Innocents on the Ice: A Memoir of Antarctic Exploration, 1957*, University Press of Colorado, Niwot, 1998, p 210. See also, pp 107, 155 and 208.

30 Judith Weinraub, 'Memories frozen in time', *Washington Post*, 5 April 1995.

31 Darlington, *My Antarctic Honeymoon*, pp 131–132.

32 Darlington, *My Antarctic Honeymoon*, p 132.

33 John Béchervaise, *Blizzard and Fire: A Year at Mawson, Antarctica*, Angus & Robertson, Sydney, 1963, p 73.

34 Béchervaise, *Blizzard and Fire*, pp 239, 252.

35 Richard E Byrd, *Alone*, Putnam, London, 1938, pp 142–143.

36 Vivian Fuchs, *Of Ice and Men: The Story of the British Antarctic Survey, 1943-73*, Anthony Nelson, Shropshire, 1982, pp 327–328.

37 Richard E Byrd, *Little America: Aerial Exploration in the Antarctic, The Flight to the South Pole*, GP Putnam's Sons, New York, 1930, p 10.

38 Fuchs, *Of Ice and Men*, p 328.

39 Interview published in Charles Neider, *Beyond Cape Horn: Travels in the Antarctic*, Sierra Club Books, San Francisco, 1980, pp 264–265.

40 Esther D Rothblum, Jacqueline S Weinstock and Jessica F Morris (eds), *Women in the Antarctic*, The Haworth Press, New York, 1998, p 213.

41 Chipman, *Women on the Ice*, pp 86–87.

42 See, for example, Gilbert Dewart, *Antarctic Comrades: An American with the Russians in Antarctica*, The Ohio State University Press, Columbus, 1989, pp 179–180.

43 Quoted in Dorothy Braxton, *The Abominable Snow-Women*, AH & AW Reed, Wellington, 1969, p 22. On all-male facilities, see also Walter Sullivan, 'Antarctic, a no-woman's land, to get 6 females', *Polar Times*, December 1969, p 5.

44 Chipman, *Women on the Ice*, p 87.

45 Rothblum et al, *Women in the Antarctic*, p 231.

46 Charles Swithinbank, *Forty Years on Ice: A Lifetime of Exploration and Research in the Polar Regions*, The Book Guild Ltd, Sussex, 1998, p 191.

47 Wheeler, *Terra Incognita*, p 200.

48 Alexa Thomson, *Antarctica on a Plate*, Random House Australia, Sydney, 2003, p 260.

49 Chipman, *Women on the Ice*, p 114.

50 William L Fox, *Terra Antarctica: Looking into the Emptiest Continent*, Trinity University Press, San Antonio, 2005, p 79.

51 Wheeler, *Terra Incognita*, p 35.

52 Wheeler, *Terra Incognita*, p 204.

## Of Huddles and Pebbles

1 Apsley Cherry-Garrard, *The Worst Journey in the World: An Account of Scott's Last Antarctic Expedition, 1910-13*, Penguin Books, Middlesex, 1970 (first published 1922), p 282.

2 Brigid Hains, *The Ice and the Inland: Mawson, Flynn, and the Myth of the Frontier*, Melbourne University Press, Carlton, 2002, pp 19–22.

3    Cherry-Garrard, *Worst Journey*, p 275.
4    Cherry-Garrard, *Worst Journey*, p 284.
5    Cherry-Garrard, *Worst Journey*, p 335.
6    Cherry-Garrard's diary, quoted in *Worst Journey*, p 357.
7    Cherry-Garrard, *Worst Journey*, p 276.
8    The British scientist, Bernard Stonehouse observed a small colony of 300 Emperors at Marguerite Bay from 5 June 1948 to 15 August 1948.
9    Robert Dovers, *Huskies*, G Bell & Sons Ltd, London, 1957, p 218. I am grateful to Professor Stephen Dovers, for the opportunity to talk about his father and Antarctica on 19 April 2006.
10   Robert G Dovers, Diary of 1952, Mitchell Library, Sydney, MS 3812/2/items 1–2, Sunday 27 January 1952.
11   Dovers, *Huskies*, p 7.
12   My account of the 1952 expedition and the behaviour of the Emperors at Pointe Géologie is drawn from the French scientific reports by Jean Prévost, *Ecologie du Manchot Empereur*, Hermann, Paris, 1961, and Jean Rivolier, *Terre Adélie 1952: Eléments de Climatologie Biologique dans l'Archipel de Pointe Géologie*, Expéditions Polaires Françaises, Paris, 1955, as well as the following sources: Jean Rivolier, *Emperor Penguins*, translated by Peter Wiles, Elek Books, London, 1956; Dovers, Diary of 1952 and his book, *Huskies*; Mariana Gosnell, 'The ice nursery', *Orion*, September/October 2005, pp 16–23; David Campbell, *The Crystal Desert: Summers in Antarctica*, Minerva, London, 1993, chapter 4; and Pauline Reilly, *Emperor: The Magnificent Penguin*, Kangaroo Press, Sydney, 1978. I have also benefited from reading a report on Emperors at Auster and Taylor Glacier by the Australian biologist, Graham Robertson, *Report on the Biology Program at Mawson, 1988*, Australian Antarctic Division, Hobart, 1989.
13   Rivolier, *Emperor Penguins*, p 52.
14   Luc Jacquet, *The March of the Penguins*, American edition, Warner Independent Pictures, Burbank, 2005.
15   Donald Kennedy, 'Emperors on the ice', *Science*, vol 309, no 5740, 2 September 2005, p 1494; Marlene Zuk, 'Family values in black and white', *Nature*, vol 439, no 917, 23 February 2006; David Smith, 'How the penguin's life story inspired the US religious right', *The Observer*, 18 September 2005; Jennifer Gold, 'Does March of the Penguins support intelligent design theory?', *Christian Today* <http://www.christiantoday.com>, viewed 6 November 2006; Peter T Chattaway, 'Review: March of the Penguins', *Christianity Today* <http://www.christianitytoday.com/movies/reviews/marchofpenguins.html>, viewed 6 November 2006; 'Movie review: March of the Penguins, and viewer comments', *Christian Spotlight* <http://www.christiananswers.net/spotlight/movies/2005/marchofthepenguins2005.html>, viewed 6 November 2006; Patrick Letellier 'March of the zealots: Penguins and America's cultural wars', *We The People: Voice of the Lesbian, Gay, Bisexual and Transgendered Community in the North Bay* <http://www.gaysonoma.com>, viewed 6 November 2006; Thomas Riggins, 'March of the pinheads: Conservative values and the penguin movie', *PA: Marxist Thought Online* <http://www.politicalaffairs.net>, viewed 6 November 2006.
16   Cherry-Garrard, *Worst Journey*, pp 275, 643.
17   Smith, 'How the penguin's life story inspired the US religious right'.
18   T Micol and P Jouventin, 'Long-term population trends in seven Antarctic seabirds at Pointe Géologie (Terre Adélie): Human impact compared with

environmental change', *Polar Biology*, vol 24, 2001, pp 175–185; EJ Woehler, J Cooper, JP Croxall, WR Fraser, GL Kooyman, GD Miller, DC Nel, DL Patterson, H-U Peter, CA Ribic, K Salwicka, WZ Trivelpiece and H Weimerskirch, *A Statistical Assessment of the Status and Trends of Antarctic and Subantarctic Seabirds*, SCAR, Montana, 2001.

19  Rivolier described them as 'a small and dying race', *Emperor Penguins*, p 13.

20  WG Burn Murdoch, *From Edinburgh to the Antarctic: An Artist's Notes and Sketches during the Dundee Antarctic Expedition, 1892-93*, The Paradigm Press, Bluntisham Books, Norfolk, 1984, pp 229–230.

21  David G Ainley, *The Adélie Penguin: Bellwether of Climate Change*, Columbia University Press, New York, 2002, p 6.

22  Frederick Cook, *Through the First Antarctic Night, 1898-1899, A Narrative of the Belgica*, Heinemann, London, 1900, p 286.

23  Burn Murdoch, *From Edinburgh to the Antarctic*, pp 238–239.

24  Burn Murdoch, *From Edinburgh to the Antarctic*, pp 262–265.

25  Charles Laseron, *South with Mawson: Reminiscences of the Australasian Antarctic Expedition, 1911-14*, Angus & Robertson, Sydney, 1947 (second edition 1957), pp 111–112.

26  Ainley, *The Adélie Penguin*, p 9.

27  Ainley, *The Adélie Penguin*, p 73.

28  Ainley, *The Adélie Penguin*, p 245.

29  Eric Woehler, *Antarctic Seabirds: Their Status and Conservation in the AAT*, RAOU Conservation Statement No 9, Supplement to *Wingspan* (12), December 1993; 'The distribution of seabird biomass in the Australian Antarctic Territory: Implications for conservation', *Environmental Conservation*, vol 17, no 3, Autumn 1990, pp 256–261. I am grateful to Eric Woehler for providing me with several references and for our discussions, both in Antarctica in 2002–03 and on 20 April 2006.

30  EJ Woehler, RL Penny, SM Creet and HR Burton, 'Impacts of human visitors on breeding success and long-term population trends in Adélie Penguins at Casey, Antarctica', *Polar Biology*, vol 14, 1994, pp 269–274; Micol and Jouventin, 'Long-term population trends in seven Antarctic seabirds at Pointe Géologie', pp 175–185; EJ Woehler et al, *A Statistical Assessment of the Status and Trends of Antarctic and Subantarctic Seabirds*.

## The Changeover

1  Wayne Orchiston, 'From the South Seas to the sun: The astronomy of Cook's voyages', in Margarette Lincoln (ed), *Science and Exploration in the Pacific: European Voyages to the Southern Oceans in the Eighteenth Century*, The Boydell Press, London, 1998, pp 55–72, at p 56; Graeme Davison, *The Unforgiving Minute: How Australians Learned to Tell the Time*, Oxford University Press, Melbourne, 1993, pp 7, 11.

2  Michael Dettelbach, '"A kind of Linnaean being"; Forster and eighteenth-century natural history', in Nicholas Thomas, Harriet Guest and Michael Dettelbach (eds), *Observations Made during a Voyage around the World* by Johann Reinhold Forster, University of Hawai'i Press, Honolulu, 1996, p lxvii. See also Simon Schaffer, 'Our trusty friend the watch', *London Review of Books*, 31 October 1996, p 11.

3  Davison, *The Unforgiving Minute*, p 9.

4   Stephen J Pyne, *The Ice: A Journey to Antarctica*, University of Iowa Press, Iowa City, 1986, p 245.
5   Schaffer, 'Our trusty friend the watch', p 11.
6   Davison, *The Unforgiving Minute*, p 11.
7   Davison, *The Unforgiving Minute*, p 12.
8   Davison, *The Unforgiving Minute*, p 13.
9   EP Thompson, 'Time, work-discipline, and industrial capitalism', *Past and Present*, no 38, December 1967, pp 56–97.
10  Stephen Pumfrey, *Latitude and the Magnetic Earth*, Icon Books, Cambridge, 2002, p 64.
11  Derek Howse, *Greenwich Time and the Longitude*, Philip Wilson Publishers and the National Maritime Museum, London, 1997, p 130.
12  Howse, *Greenwich Time*, pp 133–143.
13  Pumfrey, *Latitude*, p 37.
14  John King Davis, *High Latitude*, Melbourne University Press, Carlton, 1962, p 59.
15  Frederick Cook, *Through the First Antarctic Night, 1898-1899, A Narrative of the Belgica*, Heinemann, London, 1900, p 355.
16  *Sydney Morning Herald*, 3 March 1936, p 11.
17  John Erskine, Mawson station log, 24 February 1967, 1 March 1967, AAD Library.
18  Laurence McKinley Gould, *COLD: The Record of an Antarctic Sledge Journey*, Brewer, Warren & Putnam, New York, 1931, p 240.
19  Bill Green, *Water, Ice and Stone: Science and Memory on the Antarctic Lakes*, Harmony Books, New York, 1995, p 50.
20  Green, *Water, Ice and Stone*, p 97.
21  Green, *Water, Ice and Stone*, p 88.
22  Bill Green and Craig Potton, *Improbable Eden: The Dry Valleys of Antarctica*, Craig Potton Publishing, Nelson, 2003, p 18.
23  Barry Lopez, 'Trouble way down under: Exploiters threaten the eerie beauty of Antarctica', *Washington Post*, 27 March 1988, pp C1–C2.
24  Jennie Darlington (as told to Jane McIlvaine), *My Antarctic Honeymoon: A Year at the Bottom of the World*, Doubleday, New York, 1956, p 9.
25  Helen Garner, 'Regions of thick-ribbed ice', in *The Feel of Steel*, Pan Macmillan Australia, Sydney 2001, pp 13–34, at pp 22–23.
26  Galen Rowell, *Poles Apart: Parallel Visions of the Arctic and Antarctic*, University of California Press, Berkeley, 1995, p 43.
27  John Béchervaise, *Blizzard and Fire: A Year at Mawson, Antarctica*, Angus & Robertson, Sydney, 1963, pp 19, 47, 62–63.
28  'Slope of time' is Barry Lopez's beautiful phrase: *Arctic Dreams: Imagination and Desire in a Northern Landscape*, Picador, London, 1987, p 172.
29  Green, *Water, Ice and Stone*, p 179.
30  David Thomson, *Scott's Men*, Allen Lane, London, 1977, p 292.
31  Thomson, *Scott's Men*, p 294.
32  Thomson, *Scott's Men*, p 288, quoting Lord Curzon at a gathering of the Royal Geographical Society at which Amundsen was present in November 1912.
33  Sara Wheeler, *Cherry: A Life of Apsley Cherry-Garrard*, Jonathan Cape, London, 2001, p 152.
34  Thomson, *Scott's Men*, p 295.
35  Michael Parfit, *South Light: A Journey to the Last Continent*, Macmillan, New York, 1985, p 35.

36 Barry Lopez, *Arctic Dreams*, pp 176, 202.
37 Green, *Water, Ice and Stone*, p 46.
38 Pyne, *The Ice*, p 150.
39 Barry Lopez, *About This Life: Journeys on the Threshold of Memory*, The Harvill Press, London, 1998, p 68.
40 David Campbell, *The Crystal Desert: Summers in Antarctica*, Houghton Mifflin Company, Boston, 1992, p 51.
41 Erskine, Mawson station log, internal report, Australian Antarctic Division (AAD), Commonwealth of Australia, 17–19 February 1968.
42 Béchervaise, *Blizzard and Fire*, pp 238–239.
43 Robert Easther, 'Davis Station 1986 – OIC Annual Report', internal report, AAD, Commonwealth of Australia, 8 December 1986, p 26.
44 Peter Cook, 'WA footprints in Antarctica', manuscript article submitted to the *West Australian*, 2003, kindly made available by the author.
45 John Rich, 'Annual Report Casey 2002', internal report, AAD, Commonwealth of Australia; Ivor Harris, *Station Leader's Annual Report, Casey Station 2003*, internal report, AAD, Commonwealth of Australia.
46 Stephen Murray-Smith, *Sitting on Penguins: People and Politics in Australian Antarctica*, Hutchinson Australia, Sydney, 1988, p 113.
47 Stephen Murray-Smith, *Indirections: A Literary Biography*, Foundation for Australian Literary Studies, James Cook University of North Queensland, Townsville, 1981, p 23; *John McLaren, Free Radicals: Of the Left in Postwar Melbourne*, Australian Scholarly Publishing, Melbourne, 2003, p 49.
48 Murray-Smith, *Indirections*, p 14.
49 Murray-Smith, *Indirections*, p 27.
50 McLaren, *Free Radicals*, p 193.
51 McLaren, *Free Radicals*, p 330.
52 Murray-Smith, *Sitting on Penguins*, p vii.
53 Murray-Smith, *Sitting on Penguins*, p 195.
54 Murray-Smith, *Indirections*, p 60.
55 McLaren, *Free Radicals*, pp 48, 350.
56 Stephen Murray-Smith with Jack Jones and MA Marginson, 'South West Island, and other investigations in the Kent Group', *Victorian Naturalist*, 87/12, 1970, pp 344–371, quoted in McLaren, *Free Radicals*, p 312.
57 McLaren, *Free Radicals*, p 254.
58 Response to a military interviewer, 1944, quoted in McLaren, *Free Radicals*, pp 48, 350.
59 Murray-Smith, *Indirections*, p 39.
60 Murray-Smith, *Sitting on Penguins*, p 116.
61 Editorial, 'Australia on thin ice', *Canberra Times*, 5 January 1986; Editorial, 'Thin ice', *West Australian*, 31 December 1985; 'Australia is falling behind in Antarctic', *The Examiner*, 31 December 1985; Jane Ford, 'Lack of research could freeze us out of Antarctic Treaty', *Australian*, 13 January 1986.
62 Joseph MacDowall, *On Floating Ice: Two Years on an Antarctic Ice-shelf South of 75° S*, The Pentland Press, Edinburgh, 1999, pp 286–296.
63 Ford, 'Lack of research could freeze us out of Antarctic Treaty'.
64 Jeffery Rubin, 'White Australia', *Time*, 30 May 1988.
65 Barry Jones, 'Getting Antarctica on the domestic and global agenda', Second Annual Phillip Law Lecture, Hobart, 2003. Dr Jones' lecture quotes his 1984 memo.

66  Keith Scott, 'A presence first, and science comes later', *Canberra Times*, 22 March 1989.

67  Allison is quoted in Scott, 'A presence first'.

68  Graham Robertson, 'Report on the Biology Program at Mawson', 1988, internal report, AAD, Commonwealth of Australia, 1989, p 37.

69  Quoted in Bernadette Hince, 'The teeth of the wind: An environmental history of subantarctic islands', PhD thesis, Australian National University, Canberra, 2005, p 109.

70  Jean-Paul Kauffmann, *The Arch of Kerguelen: Voyage to the Islands of Desolation*, translated by Patricia Clancy, Four Walls Eight Windows, New York, 1993, p 163.

71  Kauffmann, *The Arch of Kerguelen*, p 178.

72  McLaren, *Free Radicals*, p 230.

73  Stephen Murray-Smith, *Behind the Mask: Technical Education Yesterday and Today*, The 1987 Beanland Lecture, Footscray Institute of Technology, Melbourne, 1987, pp 11, 18.

74  Murray-Smith, *Sitting on Penguins*, p 125.

75  Stephen Murray-Smith, 'Memories of Melbourne University', in Hume Dow, *Memories of Melbourne University: Undergraduate Life in the Years Since 1917*, Hutchinson of Australia, Richmond, 1983, pp 119–136.

## Green Crusaders

1  Phillip Law, *Antarctica – 1984*, Sir John Morris Memorial Lecture (1964), The Adult Education Board of Tasmania, Hobart, 1964, p 9.

2  Bernadette Hince, 'The teeth of the wind: An environmental history of subantarctic islands', PhD thesis, Australian National University, 2005, p 190.

3  Philip Ayres, *Mawson: A Life*, Melbourne University Press, Carlton, p 245.

4  MJ Peterson, 'Antarctica: The last great land rush on Earth', *International Organization*, 34, 3, Summer 1980, pp 377–403.

5  Peterson, 'Antarctica: The last great land rush on Earth', p 385.

6  Stephen Murray-Smith, *Sitting on Penguins: People and Politics in Australian Antarctica*, Hutchinson Australia, Sydney, 1988, p 178.

7  Stephen Martin, *A History of Antarctica*, State Library of NSW Press, Sydney, 1996, pp 231–234.

8  WN Bonner, 'Conservation and the Antarctic', in RM Laws (ed), *Antarctic Ecology*, vol 2, Academic Press, London, 1984, pp 830–836.

9  Robert C Murphy, 'Slaughter threatens end of whales', *Polar Times*, March 1940, p 24.

10  Stephen Martin, *The Whales' Journey*, Allen & Unwin, Sydney, 2001, pp 80–93.

11  Robert C Murphy, 'The urgency of protecting life on and around the great southerly continent', *Natural History*, vol LXXVI, no 6, June–July 1967, pp 18–31.

12  Justin McCurry, 'Big sushi', *The Monthly*, August 2006, pp 42–48; 'Japan barbarian: Puplick', *Canberra Times*, 21 March 1989.

13  Kelly Rigg in John May, *Greenpeace Book on Antarctica: A New View of the Southern Continent*, Dorling Kindersley, London, 1988, p 170.

14  Maj de Poorter, 9 February 1987, 'Voyage to Antarctica', in May, *Greenpeace Book on Antarctica*, p 163.

15  Gudrun Gaudian, 29 March 1987, in May, *Greenpeace Book*, p 164.

16 Justin Farrelly, 14 June 1987, in May, *Greenpeace Book*, p 167.

17 Kevin Conaglen, 23 June 1987, in May, *Greenpeace Book on Antarctica*, p. 167.

18 May, *Greenpeace Book* on Antarctica, p 169.

19 John Béchervaise, *Blizzard and Fire: A Year at Mawson, Antarctica*, Angus & Robertson, Sydney, 1963, p 23.

20 R Hancock, 'Biology of a continental Antarctic ecosystem, 1985–86', AAD Library.

21 Dr Ian Snape, quoted in AAD <http://www.aad.gov.au>, viewed 6 November 2006.

22 Hancock, 'Biology of a continental Antarctic ecosystem, 1985–86'.

23 Architecture of McMurdo station, Antarctica <http://www.glasssteelandstone.com/ZZ/McMurdo.html>, viewed 7 November 2006.

24 Charles Swithinbank, *An Alien in Antarctica: Reflections upon Forty Years of Exploration and Research on the Frozen Continent*, The McDonald & Woodward Publishing Company, Blacksburg, 1997, p 21.

25 *Polar Times*, June 1970, p 6.

26 Quoted in Swithinbank, *An Alien in Antarctica*, p 24.

27 Michael Parfit, *South Light: A Journey to Antarctica*, Bloomsbury Publishing, London, 1988, p 102.

28 William L Fox, *Terra Antarctica: Looking into the Emptiest Continent*, Trinity University Press, San Antonio, 2005, p 82.

29 Murphy, 'The urgency of protecting life'.

30 Nicholas Johnson, *Big Dead Place: Inside the Strange and Menacing World of Antarctica*, Feral House, Los Angeles, 2005, p 45; Sara Wheeler, *Terra Incognita: Travels in Antarctica*, Vintage, London, 1996, p 18.

31 Fox, *Terra Antarctica*, p 80.

32 Alexa Thomson, *Antarctica on a Plate*, Random House Australia, Sydney, 2003, pp 233, 266.

33 'Islands to Ice' exhibition, Tasmanian Museum and Art Gallery, 2006; Hince, 'The teeth of the wind', p 143.

34 Richard Woolcott, *The Hot Seat: Reflections on Diplomacy from Stalin's Death to the Bali Bombings*, HarperCollins Publishers, Sydney, 2003, pp 210–211; Peter J Beck, 'The United Nations and Antarctica', *Polar Record*, vol 22, no 137, May 1984, pp 137–144, and 'Antarctica enters the 1990s', *Applied Geography*, vol 10, 1990, pp 247–263 (Mahathir quote on p 251); Barry Lopez, 'Trouble way down under', *Washington Post*, 27 March 1988, pp C1–2; S Chaturvedi, 'Antarctica and the United Nations', *Indian Quarterly*, vol 42, no 1, January–March 1986, pp 1–26; Patrick Walters, 'Antarctica, under the world's microscope', *Sydney Morning Herald*, 11 January 1986.

35 Tim Bowden, *The Silence Calling: Australians in Antarctica 1947-1997: The ANARE Jubilee History*, Allen & Unwin, Sydney, 1997, pp 410–415; Anthony Bergin, 'Australia and the politics of CRAMRA', Paper presented to the Annual Meeting of the Australasian Political Studies Association, Hobart, September 1990, pp 12–13; Lorraine Elliott, *Protecting the Antarctic Environment: Australia and the Minerals Convention*, The Australian Foreign Policy Publications Programme, Australian National University, Canberra, 1993, p 29.

36 Andrew Jackson, 'Modern politics of the Antarctic', Paper presented to the *Australians in Antarctica* conference, organised by the History Program, Research School of Social Sciences, Australian National University and the National Museum of Australia, Canberra, 5 October 2001. Copy kindly made available by the author.

37 Richard Laws, 'Unacceptable threats to Antarctic science', *New Scientist*, 30 March 1991, p 4; Phillip Law, 'The Antarctic wilderness – A wild idea!', The Australian Institute of International Affairs, 16th National Conference, *Antarctica's Future: Continuity or Change?*, 18–19 November 1989, Hobart.
38 Jackson, 'Modern politics of the Antarctic'.
39 Roff Smith, *Life on the Ice*, Allen & Unwin, Sydney, 2002, p 12.
40 Douglas Mawson to Bob Dovers, 12 August 1956, in Robert G Dovers, Letters from Sir Douglas Mawson 1955–58, Mitchell Library, Sydney, MS 3812/2, item 3.
41 Andrew Darby, 'Pragmatism at the Pole', *Age*, 11 December 1999, and 'All eyes on the last wilderness', *Sydney Morning Herald*, 17 July 2006; information from Andrew Jackson, AAD.
42 Ken McCracken, Neal Young and Ian Bird, 'The stimulus of new technologies', in Harvey J Marchant, Desmond J Lugg and Patrick G Quilty (eds), *Australian Antarctic Science: The First 50 Years of ANARE*, AAD, Kingston, 2002 p 27.
43 Patrick Quilty was quoted in Bowden, *The Silence Calling*, p 470.
44 Tim Flannery, *The Weather Makers: The History and Future Impact of Climate Change*, Text Publishing, Melbourne, 2005, pp 95–96.
45 Jim Hansen, 'The threat to the planet', *New York Review*, 13 July 2006, pp 12–16.
46 JR McNeill, *Something New Under the Sun: An Environmental History of the Twentieth Century*, Allen Lane, The Penguin Press, London, 2000, p 4.

## Feeding Body and Soul

1 Jeff Rubin, *Antarctica*, Lonely Planet Publications, Melbourne, 2nd ed, 2000, p 103.
2 Apsley Cherry-Garrard, *The Worst Journey in the World: An Account of Scott's Last Antarctic Expedition, 1910-13*, Penguin Books, Middlesex, 1970 (first published 1922), p 275.
3 Ernest Shackleton, *The Heart of the Antarctic*, William Heinemann, London, 1910, p 230.
4 Raymond Priestley, *Antarctic Adventure: Scott's Northern Party*, T Fisher Unwin, London, 1914, p 119.
5 Shackleton, *The Heart of the Antarctic*, p 204.
6 Richard E Byrd, *Little America: Aerial Exploration in the Antarctic, The Flight to the South Pole*, GP Putnam's Sons, New York, 1930, p 35.
7 Shackleton, *The Heart of the Antarctic*, p 221.
8 Cherry-Garrard, *The Worst Journey in the World*, p 306.
9 Shackleton, *The Heart of the Antarctic*, p 233.
10 Shackleton, *The Heart of the Antarctic*, pp 233–234.
11 Byrd, *Little America*, p 35; Laurence McKinley Gould, *COLD: The Record of an Antarctic Sledge Journey*, Brewer, Warren & Putnam, New York, 1931, p 76.
12 Priestley, *Antarctic Adventure*, pp 149–150.
13 Frank Wild, 31 January 1909, quoted in Josef and Katharina Hoflehner, *Frozen History: The Legacy of Scott and Shackleton*, Josef Hoflehner, Wels, 2003, p 118.
14 Priestley, *Antarctic Adventure*, pp 256–257.
15 Cherry-Garrard, *The Worst Journey in the World*, p 306.
16 Shackleton, *The Heart of the Antarctic*, p 222.
17 Cherry-Garrard, *The Worst Journey in the World*, pp 386–367.

18 Frederick Cook, *Through the First Antarctic Night, 1898-1899, A Narrative of the Belgica*, Heinemann, London, 1900, p 334.
19 Cook, *Antarctic Night*, p 298.
20 Galen Rowell, *Poles Apart: Parallel Visions of the Arctic and Antarctic*, University of California Press, Berkeley, 1995, p 167.
21 C Bertram, 'Antarctica sixty years ago: A reappraisal of the British Graham Land Expedition of 1934-37', *Polar Record*, vol 32, no 181, 1996, pp 98-183.
22 Richard McElrea and David Harrowfield, *Polar Castaways: The Ross Sea Party (1914-17) of Sir Ernest Shackleton*, Canterbury University Press, Christchurch, 2004, p 201.
23 The debate can be followed in Roland Huntford, *Scott and Amundsen*, Hodder & Stoughton, London, 1979; Susan Solomon, *The Coldest March: Scott's Fatal Antarctic Expedition*, Melbourne University Press, Carlton, 2001; Ranulph Fiennes, *Captain Scott*, Hodder & Stoughton, London, 2004, and Robert E Feeney, *Polar Journeys: The Role of Food and Nutrition in Early Exploration*, University of Alaska Press and American Chemical Society, Fairbanks, 1997.
24 Phillip Law, *Antarctic Odyssey*, William Heinemann Australia, Melbourne, 1983, p 166.
25 Sara Wheeler, *Terra Incognita: Travels in Antarctica*, Vintage, London, 1997, p 113.
26 Philip Ayers, *Mawson: A Life*, Melbourne University Press, Carlton, 1999, p 22.
27 Station log quoted in Tim Bowden, *The Silence Calling: Australians in Antarctica 1947-1997: The ANARE Jubilee History*, Allen & Unwin, Sydney, 1997, p 56.
28 Phillip G Law, 'Nutrition in the Antarctic', *The Medical Journal of Australia*, vol 1, no 20, 18 May 1957, pp 676-679.
29 Cook, *Antarctic Night*, pp 302-303, 306.
30 Robert Cushman Murphy, *Logbook for Grace: Whaling Brig Daisy, 1912-1913*, Robert Hale Ltd, London, 1948, p 158.
31 Finn Ronne, *Antarctic Conquest: The Story of the Ronne Expedition, 1946-48*, Putnam's Sons, New York, 1949, p 40.
32 Australian Antarctic Division (AAD) <http://www.aad.gov.au>, viewed 7 November 2006, click on 'Going South' and 'FAQs on Food'.
33 Ian Sutherland, 'Casey 1997: Station Leader's Annual Report', internal report, AAD, Commonwealth of Australia, p 25.
34 Douglas Mawson, *The Home of the Blizzard*, Wakefield Press, Adelaide, 1996 (first published 1915), p 183; Lennard Bickel, *This Accursed Land*, Pan Macmillan, Sydney, 1977, pp 115, 197.
35 Martin J Riddle and Paul M Goldsworthy, 'Environmental science and the environmental ethos of ANARE', in Harvey J Marchant, Desmond J Lugg and Patrick G Quilty (eds), *Australian Antarctic Science: The First 50 Years of ANARE*, AAD, Kingston, 2002, p 563.
36 Alexa Thomson, *Antarctica on a Plate*, Random House Australia, Sydney, 2003, p 225.
37 Dean Fosdick, 'Green fingers at work in Antarctica', *Canberra Times*, 29 January 2003.
38 Linda Clark and Elspeth Wishart, *66° South: Tales from an Antarctic Station*, Queen Victoria Museum and Art Gallery, Launceston, 1993, p 73.
39 Interview with former AAD plumber, 16 August 2006.
40 Eric N Webb, 'Magnetic Polar Journey, 1912', Mitchell Library, Sydney, typescript, ML MSS 6812, CY4413.

41 Michael Parfit, 'Reclaiming a lost Antarctic base', *National Geographic*, vol 183, no 3, March 1993, p 119.
42 Elizabeth David, *Harvest of the Cold Months: The Social History of Ice and Ices*, Viking, New York, 1994; Ingrid D Rowland, 'The empress of ice cream', *New York Review of Books*, vol 43, no 6, April 1996.
43 Cook, *Through the First Antarctic Night*, pp 181–182.
44 Bowden, *The Silence Calling*, p 288.
45 Law, *Antarctic Odyssey*, p 224.
46 Thomson, *Antarctica on a Plate*, p 242.
47 Bernadette Hince, *The Antarctic Dictionary: A Complete Guide to Antarctic English*, CSIRO Publishing and Museum of Victoria, Melbourne, 2000, and 'The teeth of the wind: An environmental history of subantarctic islands', PhD thesis, Australian National University, Canberra, 2005.
48 Ranulph Fiennes, *Mind over Matter: The Epic Crossing of the Antarctic Continent*, Sinclair-Stevenson, London, 1993, p 171.
49 Shackleton, *The Heart of the Antarctic*, p 219.
50 Carl Skottsberg, quoted in Stephen Martin, *A History of Antarctica*, State Library of NSW Press, Sydney, 1996, p 125.
51 Mawson, *The Home of the Blizzard*, pp 175–176.
52 Alberto Manguel, *A History of Reading*, Flamingo, London, 1997, p 171.
53 Gould, *COLD*, pp 240–241.
54 Roland Huntford, *Shackleton*, Atheneum, New York, 1986, p 244.
55 Shackleton, *The Heart of the Antarctic*, p 209.
56 Raymond Priestley, 'The Professor', Paper given in Cambridge (c 1921), Priestley Papers, Scrapbook relating to TWE David, MS 507, SPRI.
57 Captain John King Davis, Copy of *Aurora* log, 14 August–2 November 1911, 9 September 1911, Davis Papers, Australian Manuscripts Collection, SLV, Box 3232/5A.
58 Davis, Copy of *Aurora* log, 14 August–2 November 1911, 10 September 1911, Davis Papers, SLV, Box 3232/5A.
59 For stimulating work on Antarctic reading and writing, especially Antarctic fiction, see Elizabeth Leane, 'Romancing the Pole: A survey of nineteenth-century Antarctic utopias', *Futures Exchange: ACH (Australian Cultural History)*, vol 23, 2004, pp 147–171, and 'Locating the thing: The Antarctic as alien space in John W Campbell's "Who Goes There?"', *Science Fiction Studies*, vol 32, 2005, pp 225–239; Elizabeth Leane and Stephanie Pfennigwerth, 'Antarctica in the Australian imagination', *Polar Record*, vol 38, no 207, 2002, pp 309–312; and Bill Manhire (ed), *The Wide White Page: Writers Imagine Antarctica*, Victoria University Press, Wellington, 2004.
60 John Béchervaise, 'Men and motives in Antarctica', *Quadrant*, vol 1, no 4, Spring 1957, p 40.
61 Mordecai Richler, 'Amateurs of the cold regions', *Times Literary Supplement*, 26 July 1996, pp 5–6.
62 Gould, *COLD*, p 64.
63 Sara Wheeler, *Cherry: A Life of Apsley Cherry-Garrard*, Jonathan Cape, London, 2001, p 171.
64 Gould, *COLD*, pp 64–65; Byrd, *Little America*, pp 210, 241–242.
65 Copy of letter to Captain Davis from A Stevens, 20 January 1917, Davis Papers, SLV, Box 3289/1.
66 Brigid Hains, *The Ice and the Inland: Mawson, Flynn, and the Myth of the Frontier*,

Melbourne University Press, Carlton, 2002, pp 11–15, 24–26.

67  *Argus* (Melbourne), 31 August 1929, *Herald* (Melbourne), 13 September 1929, 'Antarctic Expedition Press Cuttings', Davis Papers, SLV, Box 3240/8.

68  'Dickens novels in Antarctic', *Argus* (Melbourne), 2 August 1929, 'Antarctic Expedition Press Cuttings', Davis Papers, SLV, Box 3240/8.

69  Bill Green, *Water, Ice and Stone: Science and Memory on the Antarctic Lakes*, Harmony Books, New York, 1995, pp 14–15.

70  Peter Medawar, 'Two conceptions of science', in his *The Strange Case of the Spotted Mice and Other Classic Essays on Science*, Oxford University Press, Oxford, 1996, pp 63, 71.

71  Green, *Water, Ice and Stone*, p 30. See Philip Ball, *$H_2O$: A Biography of Water*, Weidenfeld & Nicolson, London, 1999, pp 156–158.

72  Robert Falcon Scott, *The Voyage of the Discovery*, Smith Elder, London, 1905, vol II, p 293.

73  Green, *Water, Ice and Stone*, p 88. See also Bill Green and Craig Potton, *Improbable Eden: The Dry Valleys of Antarctica*, Craig Potton Publishing, Nelson, 2003.

74  Green, *Water, Ice and Stone*, p 105.

75  Green, *Water, Ice and Stone*, p 22.

76  Stephen J Pyne, *The Ice: A Journey to Antarctica*, University of Iowa Press, Iowa City, 1986, p 137.

77  Green, *Water, Ice and Stone*, p 259.

78  Green, *Water, Ice and Stone*, pp 227–228.

79  Green, *Water, Ice and Stone*, p 256.

80  David L Harrowfield, *Vanda Station: History of an Antarctic Outpost, 1968-1995*, New Zealand Antarctic Society Inc, Wellington, 2003, p 37.

81  Green, *Water, Ice and Stone*, p 199.

82  Green, *Water, Ice and Stone*, p 169.

83  Green, *Water, Ice and Stone*, pp 161, 215.

## Captain Scott's Biscuit

1  Thomas Keneally, 'Captain Scott's biscuit', *Granta*, no 83, 2003, pp 129–144.

2  Jennifer Byrne, 'Antarctica: Continent of superlatives', *Age*, Saturday Extra, 7 April 2001, pp 1, 6. Byrne claims that the Cape Adare visitors' book recorded 500 tourists in a single day.

3  Greg Mortimer, 'Antarctic Tourism; Past, Present and Future', Third Annual Phillip Law Lecture, Hobart, 19 June 2004.

4  Raymond Priestley, 'Obituary of Sir Edgeworth David', *Nature*, 6 October 1934, pp 523–524, in the Priestley Papers, Scrapbook relating to TWE David, MS 507, SPRI, Cambridge.

5  Barry Lopez, 'Standing on the South Pole', *Washington Post*, 27 March 1988, pp C1–2.

6  Dorothy Braxton, *The Abominable Snow-Women*, AH & AW Reed, Wellington, 1969, p 28.

7  Rosamunde J Reich, 'The development of Antarctic tourism', *Polar Record*, vol 20, no 126, September 1980, pp 203–214; Mortimer, 'Antarctic Tourism'; William L Fox, *Terra Antarctica: Looking into the Emptiest Continent*, Trinity University Press, San Antonio, 2005, p 176; International Association of Antarctic Tour Operators (IAATO), *IAATO Overview of Antarctic Tourism*,

2004-05 Antarctic Season (Revised January 2006), Information paper to the 28th Antarctic Treaty Consultative Meeting, Stockholm, 2005 <http://www.iaato. org>, viewed 14 November 2006. I am grateful for the advice of Mike Pearson.

8   Andrew Darby, 'Cool tourism hotspot raises scientists' temperatures', Age, 2 September 2006, p 3.

9   Roger Mear and Robert Swan, In the Footsteps of Scott, Jonathan Cape, London, 1987; Lord Shackleton, George Bishop, Vivian Fuchs, John Hemming, John Hunt and Peter Scott, 'Obstacles in the steps of Scott', Letter to the Editor, Times, 29 March 1986; Stephen Murray-Smith, Sitting on Penguins: People and Politics in Australian Antarctica, Hutchinson Australia, Sydney, 1988, p 104.

10  Mear and Swan, In the Footsteps of Scott, p 239.

11  Stephen Moss, 'Antarctica, the coolest tourist destination on the planet', Sunday Age, 23 February 2003, p 5.

12  Peter J Beck, 'Regulating one of the last tourism frontiers: Antarctica', Applied Geography, vol 10, October 1990, pp 343–356, at p 350.

13  Andrew Dodds, 'Something rich and strange', Eureka Street, vol 9, no 2, March 1999, pp 23–26.

14  Helen Garner, 'Regions of thick-ribbed ice', in The Feel of Steel, Pan Macmillan Australia, Sydney 2001, pp 13–34.

15  David Harrowfield, 'Historical archaeology in Antarctica', New Zealand Antarctic Record, vol 1, no 3, 1978, pp 45–50.

16  LB Quartermain, Two Huts in the Antarctic, Government Printer, Wellington, 1963, p 63.

17  These were two distinct trips in January 1915: Richard McElrea and David Harrowfield, Polar Castaways: The Ross Sea Party (1914-17) of Sir Ernest Shackleton, Canterbury University Press, Christchurch, 2004, pp 50, 56.

18  McElrea and Harrowfield, Polar Castaways, p 201.

19  McElrea and Harrowfield, Polar Castaways, p 202.

20  Neville Ritchie and Alexy Simmons report on this problem in 'Management of historic sites in New Zealand's Ross Dependency, Antarctica', Archaeology in New Zealand, vol 31, no 1, March 1998, pp 12–27.

21  Josef and Katharina Hoflehner (photos) and David Harrowfield (text), Frozen History: The Legacy of Scott and Shackleton, Josef Hoflehner, Wels, 2003, p 20.

22  Laurence McKinley Gould, COLD: The Record of an Antarctic Sledge Journey, Brewer, Warren & Putnam, New York, 1931, pp 218–221. Each of the men also took a little bit of rock from the cairn, and they left a note of their own.

23  Tierramérica, 28 November 2002, p 1.

24  Alan Doble, 'Mawson's desk', Aurora, vol 25, no 2, December 2005, pp 2–3.

25  Australian Geographic, no 53, January–March 1999, p 67.

26  Ann Savours, 'Hobart and the polar regions', in Gillian Winter (ed), Tasmanian Insights: Essays in Honour of Geoffrey Thomas Stilwell, State Library of Tasmania, Hobart, 1992, p 185.

27  Steve Meacham, 'Incredible voyager', Sydney Morning Herald, 15 July 2002.

28  Richard E Byrd, Antarctic Discovery: The Story of the Second Byrd Antarctic Expedition, Putnam, London, 1936, pp 73–77.

29  Aisling Irwin, 'Scott reaches base camp … 86 years after his death', Sydney Morning Herald, 12 November 1998.

30  David L Harrowfield, Icy Heritage: The Historic Sites of the Ross Sea Region, Antarctica, Antarctic Heritage Trust, Christchurch, 1995, pp 29–31.

31  Roff Smith, Life on the Ice, Allen & Unwin, Sydney, 2002, p 10.

32  David G Campbell, *The Crystal Desert: Summers in Antarctica*, Houghton Mifflin Co, Boston, 1992, p 13.

33  Jeff Rubin, *Antarctica*, Lonely Planet Publications, Melbourne, 2nd ed, 2000, p 314.

34  Campbell, *Crystal Desert*, p 147.

35  Rubén Stehberg, *Arqueología Histórica Antártica: Aborígenes Sudamericanos en los Mares Subantárticos en el Siglo XIX*, Centro de Investigaciones Diego Barros Arana, Santiago de Chile, 2003 (translated with assistance from Ana Rubio); discussion with author, 29 November 2004.

36  For Stehberg's recent archaeological work with Michael Pearson in the South Shetlands, see Michael Pearson and Rubén Stehberg, 'Nineteenth century sealing sites on Rugged Island, South Shetland Islands', *Polar Record*, 42 (4), 2006, pp 1–13.

37  For an example of the arguments of 'bring-it-backers', see Peter Meredith, 'Mawson's Hut: Take it or leave it?', *Australian Geographic*, no 18, April–June 1990, p 29, and for a perceptive essay on the meaning of Mawson's Hut, see Christy Collis, 'Mawson's Hut: Emptying post-colonial Antarctica', *Journal of Australian Studies*, no 63, 1999, pp 22–29. An exchange between Michael Pearson and Bill Burch about the management of the hut can be found in *Aurora*, vol 8, nos 3 and 4, 1989.

38  AF Rogers, 'The death of Chief Petty Officer Evans', *The Practitioner*, no 212, 1974, pp 570–580; Robert E Feeney, *Polar Journeys: The Role of Food and Nutrition in Early Exploration*, American Chemical Society and the University of Alaska Press, Fairbanks, 1997, chapter 14; and Roland Huntford, *Scott and Amundsen*, Hodder & Stoughton, London, 1979, pp 510, 581.

39  Peter Scott, *The Eye of the Wind*, Hodder & Stoughton, London, 1961, p 9; Elspeth Huxley, *Peter Scott: Painter and Naturalist*, Faber and Faber, London, 1993, p 13.

40  Ranulph Fiennes, *Captain Scott*, Hodder & Stoughton, London, 2003, pp 390–391; information from the National Maritime Museum, August, 2006.

# INDEX